SOCIAL GEOGRAPHIES

This timely exploration of social geography reveals how inequality and justice unfold across interconnected scales around the globe. Through embodied, relational encounters, the book unpacks how everyday spaces shape who we are and what we might become. Grounded in care and hope, it invites readers to imagine fairer futures and remain committed to meaningful, long-term change.

The text offers a dynamic, case-driven and sometimes image-driven exploration of pressing social issues in our increasingly unequal, post-pandemic world. Designed for student engagement, the book provides real-world case studies that span the globe, critical concepts and thought-provoking questions to help readers analyse injustice, build conceptual 'toolkits', and imagine routes toward meaningful social change. Rather than offering easy answers, this book equips readers with flexible, practical frameworks – encouraging selective, critical thinking to better understand and communicate complex social problems. Ideal for classroom use and independent study, it empowers students to connect theory with action and to think like geographers committed to justice. The book also includes Engagement Questions, encouraging readers to reflect on the subject of the chapter and the wider topics within the field of social geography.

This book is essential reading for students of social and political geography, and allied fields of urban, health and well-being and political geographies.

Geoff DeVerteuil is Professor of Social Geography at Cardiff University, in the School of Geography and Planning.

Andrew (Andy) Power is Professor of Geography at the School of Geography and Environmental Science at University of Southampton.

'Fresh, insightful and highly accessible, this book brings Social Geography to life through personal stories and vibrant visuals. It encourages deep reflection on human connections, shared encounters, and the profound search for hope amidst the challenges of our complex world. Essential for any student seeking to truly understand their social landscape.'

Julie MacLeavy, *Professor of Economic Geography, University of Bristol*

'An accessible and engaging guide to social geographies today. With examples and case studies from across the globe, this book brings key topics to life – from housing and inequality to care and citizenship. A brilliant resource for students looking to connect theory with the pressing issues shaping our world.'

David Conradson, *University of Canterbury*

SOCIAL GEOGRAPHIES

Relationalities, Encounters and Hope

Geoffrey DeVerteuil and Andrew Power

Routledge
Taylor & Francis Group
LONDON AND NEW YORK

Designed cover image: Photograph by Geoffrey DeVerteuil. Buenos Aires, 2015.

First published 2026
by Routledge
4 Park Square, Milton Park, Abingdon, Oxon OX14 4RN

and by Routledge
605 Third Avenue, New York, NY 10158

Routledge is an imprint of the Taylor & Francis Group, an informa business

British Library Cataloguing-in-Publication Data
A catalogue record for this book is available from the British Library

ISBN: 9781032540450 (hbk)
ISBN: 9781032540443 (pbk)
ISBN: 9781003414889 (ebk)

DOI: 10.4324/9781003414889

Typeset in Helvetica
by Newgen Publishing UK

CONTENTS

FIGURES

Cover photograph: ELEFANTE BLANCO (White Elephant) This image was taken by Geoff in September 2015, and illustrates the built environment of Ciudad Occulta, an informal settlement in Buenos Aires. More particularly, it shows a modest church in the foreground and the massive, abandoned *Elefante Blanco* hospital building in the back ('White Elephant'), which has since been demolished. This juxtaposition shows informality in the shadows of state failure, and brings up wider issues of inequality, hope, the social and state welfare. All images are copywritten to Geoff DeVerteuil unless otherwise noted.

TABLE

BOXES

PREFACE

This book was written in the wake of the global pandemic, on the branch of history (and geography) that it had created. For all the attempts to quickly forget this watershed event, it certainly made the 'straight path' (in Dante's words) harder to find, but more important to find as well. Beyond making sense of social geography, this textbook also tries to make some sense of the post-pandemic (social) world, a world increasingly messy and inarticulate that offers no easy answers.

ACKNOWLEDGEMENTS

Geoff and Andrew would like to acknowledge a wide variety of support in the writing of this textbook. First are the many students who took their Social Geography classes at the University of Southampton and Cardiff University (Geoff taught at Southampton for a time before moving to Cardiff), and through which they were able to refine and finesse many of the ideas that permeate the chapters.

They would also like to thank those who influenced their thinking further back in time. Geoff would like to underline the impact that David Ley's *A Social Geography of the City* had when it was first read it in 1989 at McGill University as an undergraduate, even before he took Peter Jackson's course the following year (see Chapter 1). Therein he was exposed to the intricate relationships between city and society. Since that time, and whenever he visits a new city, he is always interested in what its social geography looks like – where are the rich areas, the poor areas, and most importantly, the areas in between or those in transition?

For Andrew, Gill Valentine's book (2001) *Social Geographies*, was influential in his early years of teaching. In particular, it helped him (and the students he taught) see the importance of everyday places that people use and occupy in their lives, including the body (as a site we embody!), the home, the street, institutions, and the city. Across these different spaces, the theme of social justice loomed large, and which informed a core strand of his research and this book.

Both authors would also like to thank their partners. Geoff would like to thank Stephanie for her patience with this process. In the final year of writing the book, he endured several crises, including developing a permanent eye condition (non-arteritic anterior ischemic optic neuropathy). Her support was unwavering.

Somewhat ominously, Andrew was also ill-fated with his health during the writing of this book, after experiencing a fractured hip from a road cycling accident. He would like to thank Jenny for being the carer during the long recovery process and for supporting the work-life juggle at a diminished capacity.

Finally, both authors would like to thank the reviewers for sharpening the book's many (perhaps too many) messages and for challenging them in constructive ways.

PART I

FOUNDATIONS OF THE TEXTBOOK AND A RELATIONAL SOCIAL GEOGRAPHY

The title – *Social Geography: Relationalities, Encounters and Hope* – maps on to the three parts of the book. Part I focuses on developing a *relational* social geography, across a variety of terms, relating it to intersectionality, inequality, citizenship, rights, embodiment, care, social inclusion and social exclusion, social identifiers, and so forth. Part II centres on the various *encounters* (e.g. home, neighbourhoods, institutions, public spaces) that are framed by these relationships, across core social identifiers (gender, race, age, class, (dis)ability). Finally, Part III is about hope and change, about how various approaches to social justice (quiet, loud) can have important impacts on society and cities.

In more detail, Part I of this book explores the broad foundations and structures (including social, political, and economic) that underpin the field of social geography. First, we apply a critical geographical lens to consider the *everyday* geographies of difference between people within the social fabric of society, using a variety of core concepts in social geography. We explore three key themes around social geography – the social, inequality

DOI: 10.4324/9781003414889-1

and social justice – and tease out how they are experienced, lived and felt by people across a very differentiated society.

We take a detailed look at the various dominant and relational factors that determine how these three key themes are shaped. We first consider the role of the state in shaping the laws, policies and resources which can determine how differences between people are defined. These laws and policies and what they enact take place under the banner of citizenship, which shapes our responsibilities and entitlements. We then consider how the welfare state and local care systems (including charities and community groups) help people experience and practice citizenship in everyday places. These 'external' actors, and the institutional and material structures they work through, build on the connections people have with family, neighbours and friends. Taken together, they serve as the initial foundations of the various spaces of encounter that people experience in their daily, 'ordinary' lives. Collectively, they shape the contexts in which we live and mediate our experiences of the social and social justice on the ground.

1 INTRODUCTION

1.1 Social Geographies in Our Contemporary Times: Themes of the Social, Inequality and Social Justice

According to Smith et al. (2010: 1), social geography is defined as 'the study of social relations and the spatial structures that underpin those relations'. To Del Casino (2009), the social remains an incredibly important organising framework across the world, despite calls for a complete individualisation where everyone takes care of their own and where everyone theoretically has the same opportunities. Social identities and groupings – organised through race, gender, (dis)ability, class, nation states, neighbourhoods or social networks – are to this present day still used to distinguish both physical and social differences.

Our own approach is first and foremost informed by *three key themes* of social geography developed by Smith et al. (2010). The first is the truism that social geography has always been committed to 'the idea of *the social*', meaning the need to document the structures and processes that connect people, societies and space, and how these intersect with issues of personal identity. The second theme is the longstanding preoccupation with the hard edge of *inequality* and the uneven experience of state-provided services involving a multidimensional view, and where class is but one axis of inclusion (Smith et al., 2010). Third and finally, social geography has always been a moral enterprise, characterised by an appreciation for the principles and practicalities of *social justice*, whether as critique or increasingly drawing upon the normative turn in social research, of what ought to be and should become.

How did we, as authors of this textbook, become social geographers? For Geoffrey (Geoff), his personal story began at McGill University in September 1990, when the then up-and-coming social geographer, Peter Jackson, came from the UK to teach a course on that very topic, centred on cities and diversity. For Geoff, discovering social geography went hand-in-hand with discovering the city at the same time, since one of the assignments was to explore an inner-city area of Montréal. In his case, it was Pointe-St-Charles, which by 1990 had suffered two decades of deindustrialisation (DeVerteuil, 2004). The

DOI: 10.4324/9781003414889-2

fieldwork involved immersing himself into an unfamiliar neighbourhood, and included ethnography, talking to people and taking pictures. This took him out of his comfort zone, having grown up in a leafy suburb 10 miles to the west. The wide-ranging nature of his primary research would influence his subsequent trajectory, which became even more varied. It is this very diversity of interests which has helped enormously with the writing of this textbook. After a break from social geography with a planning degree and diploma in Geographic Information Systems, he returned to the fold, studying with the consummate social geographers Michael Dear and Jennifer Wolch, whose focus on social issues exposed him to a different side of social geography – of inequality, state-provided services and homelessness in the context of Los Angeles.

For Andrew (Andy), his introduction to social geography came via a module of the same name from a UK geographer (working in Ireland), Rob Kitchin, whose early work focused on the geographies of disability. Rob had developed close connections with a local disability organisation, embracing their mantra 'nothing about us without us' which informed Andy's later commitment to research *with* rather than *on* disabled people and the use of inclusive research methods. Following his undergraduate degree, he worked as a disability assistant for an Irish national disability organisation and spent time learning about their experiences of exclusion and disempowerment, and their politics of activism. At this time, there was a call for 'rights' among the disabled people's movement, and a rejection of the idea of 'care', based on claims that this was disempowering. This sparked an interest to undertake a PhD on the experiences of family care and disability, exploring the tensions between the rights of disabled people and caregivers. Andy returned to his alma mater to undertake a PhD under Rob Kitchin's supervision and once immersed in this field, began to teach social geography and was exposed to wider debates about social justice – working with Geoff and others. Following the PhD, he undertook a postdoctoral fellowship with Christine Milligan at Lancaster, UK, a leading scholar on geographies of care who helped him refine his thinking on how everyday geographies are shaped by encounters with care and the state.

While their personal trajectories differed over time, certain common experiences also marked their emergence as social geographers. A recent example was the 2020–2022 pandemic, which acted as both an avoidable tragedy but also as a way to underline the importance of the first theme, that of the social.

The pandemic is now for many a distant memory, yet we are living in the branch of history (and geography) that it created. At the very least, it provided a common point of reference for the entire planet – one of those rare, truly universal events, albeit globally and generationally uneven. In some nations (US, UK, Latin America) the death rates were higher than in nations which only opened up once the vaccination programme was complete (East Asia, Australia, New Zealand). But most nations implemented some kind

FIGURE 1.1 Covid memorial wall, London, 2021 © G DeVerteuil

of lockdown, and in this way the pandemic underscored the absence of the social in our everyday lives, particularly the 'in-person' version that came with the realisation that our lives could not be entirely online. Humans are essentially social beings, despite what Margaret Thatcher (UK Prime Minister in the 1980s) famously said that there was no such thing as 'society', only individuals and families. In its enforced absence, the pandemic made the importance of society even clearer. Further, the pandemic raised the notion that key *social identifiers* – class, race, gender, age, sexuality, (dis)ability, and beyond – influenced the degree to which people were able to adapt to lockdowns, as well as emerge from them in 2021 and especially 2022. For some people, the lack of social interaction during the pandemic led to a rise in *anti*-social behaviour after the pandemic, resulting in an increase in the inability to see beyond selfish motives.

Five core social identifiers mentioned above will be given working definitions early in Chapter 2, and then will form the backbone of several subsequent relationalities and encounters in Chapters 6 (age, youth), 7 (race), 8 (class), 9 and 10 (disability) and 12 (gender). While it is better that these populations do encounter each other rather than staying in their bubble, there is still no guarantee of understanding. Crucially, public spaces have become more fragmented with social media controlled in such a way that different groups end up remaining only among themselves. This freezes opportunities for social interaction and worse still, has given rise to right-wing populist, right-wing extremist, and conspiracy-theoretical groups in recent years. As such, the idea of the social has been challenged but certainly not thrown out.

For the second theme, there has been a broad trend of rising income inequality across countries over the past four decades. Contemporary global inequalities are close to the peak levels observed in the early 20th century, at the end of the pre-First World War era (variously described as the Belle Époque or the Gilded Age) that saw sharp increases in global inequality (WID, 2023). The pandemic also left its legacy in further concentrating wealth and income to a smallish clique of people who were able to work from home via technology and hidden subsidies – the so-called 'remote' or 'hoarding' class compared to a 'syndemic' comprising people experiencing long-term, entrenched inequality and the desperation of the pandemic. For young people, educational gaps increased in student achievement (not to mention socio-emotional and motivational development). This coincided with a decline in face-to-face state-provided services as part of the welfare state, a trend that we will pick up on in Chapter 4 in greater detail, as well as focusing on notions of care (Chapter 5), which placed new burdens on women in particular.

In order to avoid making this textbook all about (and only about) the deprived, the downtrodden, and the outsiders, we also wanted to use the second theme to highlight groups which have increased their privileges and entitlements as societies became more unequal. A key group here is the 1%, that is the 1% highest income earners (or in terms of wealth), and how they threaten social cohesion in a variety of ways (Dorling, 2014). At least in the UK, they do it in terms of hoarding more and more of the income and wealth in a society, to the detriment of the 99%. They pay less tax than they used to (but also rarely use state services like education and health) yet place considerable burdens on housing (e.g. invariably owning multiple dwellings and inadvertently driving gentrification, explored in Chapter 11), have undue influence on government policies (e.g. tax policies, privatisation), make it more difficult for the 99% to exercise social mobility to move up the social ladder and maintain unfair advantages and privileges in terms of life chances. These life chances relate to scarce resources such as private schooling and university places for young people (Chapter 5), health, as well as top jobs in the labour market, especially the private sector. As we shall see in Chapter 13, this hoarding can lead to social corrosiveness, strife and protests; in other words, the 1% extract more than they create. Crucially, Danny Dorling argues that the 1% is 'more than just economics – it is the culture that divides and makes social mobility so painful, both for those dropping down the income and wealth scales and for those going up' (p.24). This is a culture of creeping elitism, with a disproportionate influence on politics (given that many top politicians are originally from, and work on behalf of, the 1%) that denigrates the poor, celebrates unearned riches, calls for lower taxes (already far lower than they were in the 1970s) and cuts state services, with predictably negative effects upon the most impoverished who own no wealth.

Finally and crucially, social injustice has risen substantially around the globe, including housing displacement, homelessness and poverty (as signposted in the second theme), which further implicates social geography as a moral enterprise. But what is social justice for the purposes of this textbook? At its core, justice refers to the standards used in assessing what is fair; measuring social justice is assessing what is fair, good or moral across society, especially the distribution of benefits and burdens between different population groupings (see Chapter 2). Definitions of a just society are necessarily wide-ranging. These musings on social justice are not simply academic; they are also reflected in the real world. For instance, the post-1945 period has featured a trend towards codifying social justice in terms of individual 'rights' (ideals against which behaviour may be measured, as explored in Chapter 3), the most prominent being the 1948 United Nations Universal Declaration of Human Rights. The preamble recognises the equal and inalienable rights of all members of the human race, and that this recognition forms the basis for freedom, peace and critically, justice in the world. Accordingly, Article One states unequivocally that 'all human beings are born free and equal in dignity and rights'. The Declaration proved to be the foundation for the Convention on the Prevention and Punishment of the Crime of Genocide (also 1948) and, in 2007, the Declaration on the Rights of Indigenous Peoples. Subsequently, many democratic nation states have legally incorporated explicit notions of social justice into their own human rights charters.

Beyond this starting point, however, there is much deliberation over exactly what constitutes fairness, both in descriptive terms (how to measure what is fair) and normative terms (what ought to be fair?), and how to move beyond the individual to incorporate larger groupings. Even if the requirements for justice are clear, the solutions are often not, and must be contextualised and understood as always both personal and political (see also Wilton and Schormans, 2020). Social (and spatial) arrangements may be seen to be unjust if, for example, the privileged class take more than their 'fair share' of resources, or if state policies are enacted that worsen the situation of the worst-off. For many thinkers, the starting point to social justice is the determination of basic needs, such as food, shelter, hygiene and so forth. Much more complicated is moving beyond these basic needs to identify other cross cutting requirements that ought to be met in order to ensure a more equal society. This exercise necessarily reveals a tension between what is considered just for all humans (such as the aforementioned UN Declaration), and what is considered just for certain groups at particular times and places.

Social geographers have engaged with issues of social justice intermittently, usually in terms of critiquing the unequal spatial arrangements of society, such as exclusion and segregation. Throughout, social geography distinguishes itself from other subsets of human geography in its search for socio-spatial justice, as well as its insistence on accessibility rather than on notions of mobility. Further to this point, socio-spatial justice generally refers to the fact that space matters in terms of both generating/differentiating the inequalities facing society and also in the solutions to that same inequality. Here, social geography is political, in so far as it is concerned with how power relations operate behind social relations and identities, and how different groups struggle for recognition and claim rights for addressing social injustice. It also focuses on social movements and political struggles going on in everyday life.

Going forwards, we signpost that social justice is composed of the following *three strands*: (1) distribution, (2) recognition and (3) procedure. Distribution is very much focused on who gets what, and where. This implicates the welfare state first and foremost (Chapter 4), but also other social realms such as the community, the voluntary sector and even the private sector. Recognition means recognising different needs and identities among different populations, leading to different understandings of social justice, and definitely not treating everyone the same. Recognition extends beyond merely making visible; it also includes a more positive representation. Procedure involves how social justice is obtained, through institutional and social pathways, such as state services and entitlements. It must be noted that these three aspects do not always work together, and the emphasis placed on each is not always the same. While more will be written on these relations in Chapters 2, 4 and 13–15, suffice it to say that the focus in politics has shifted from distribution to recognition and procedure over the past 20 years, as many governments increasingly shrink away from considering tax raises and fear a rise in national debt. Yet for each chapter in this book, we will equally ponder each aspect as it relates to the topic at hand, pulling out the various implications of social justice as an abiding concern.

For the next section, we show how these three themes of social geography are crosscut by the fact that social geographies are always relational, and that they can produce encounters across different people and places.

1.2 Crosscutting Threads: Relationality and Encounter

Now that we have established the importance of the social, inequality and social justice as structuring themes, we also need to understand how these key themes are crosscut with issues of *relationality* and *encounter*. These two *crosscutting threads* necessarily involve grounding the themes to particular places that are both ordinary and common, but also and always connected to broader contexts. Doreen Massey (1993a, 1993b) outlined what are now well-rehearsed arguments around the relational aspects of place and movement:

> Different social groups have distinct relationships to mobility: some are more in charge of it than others; some initiate flows and movement, others don't; some are more on the receiving end of it than others; some are effectively imprisoned by it.
>
> (1993a: 61)

This 'power-geometry' opens up considerations of place-spanning and place-bound agents. In other words, some places are quite bounded while others are related to various 'elsewheres' though migration, remittances, travel and investment. This relates to Massey's 'progressive sense of place'. She argues that people can feel attached to places and value their uniqueness or sense of rootedness, without being closed off or backward-looking. Instead, she suggests that places can also be open and connected to the wider world. This relationality means that 'interdependence and identity, difference and connectedness, uneven development and the character of place, are in each pairing two sides of the same coin … in constant tension with each other' (Massey, 2011: 4).

Relationalities can exist in many forms, in terms of creating relationships that range from the optional to the obligatory, from the extraordinary to the ordinary and every day, and in various directions – mutually beneficial to parasitic. More on this is explored in Chapter 4 with regards to the welfare state and those who must depend on it. A relational lens can help provide a more nuanced and lived sense of geography, where places are not necessarily fixed (e.g. as either 'inclusive' or 'exclusive'), but are continually mediated by the people who use and occupy them in particular moments. Relationalities can have a temporal quality, as either being longstanding in nature (e.g. a person will chronic mental ill-health who requires long-term care) or as more fleeting moments in particular places, which can be nonetheless influential (e.g. a one-off hate crime incident).

This relates to an important allied concept within social geography, that of *spaces of encounter* (which understandably shifted during the pandemic), and will be threaded through the book. While the two concepts are related, it is worth stating that not all relationalities involve encounters, in some cases they are hidden or take place at a wider or institutional scale, yet can remain influential in framing how encounters unfold. In her landmark paper on spaces of encounter, Gill Valentine (2008: 323) underlines 'the role of shared space in providing the opportunity for encounter between "strangers"'. She wanted to unpack the notion that the more different people are in contact with each other, the more they will see past their differences, whether racial, gender, age, and so on. In this sense, the city becomes a 'site of connection' where different people interact on a level playing field. Yet Valentine also noted that spaces of encounter remain uneven and tinged with suspicion, competition, resentment and mis-understandings. Spaces of encounter do not automatically produce understanding and respect, and chance encounters are not always pleasant ones!

An example of this is public space (defined as spaces that are theoretically accessible to all, including green spaces, sidewalks, streets, squares) shared by a very wide range of populations, diverse in terms of class, race, gender, sexuality and (dis)ability, and beyond. Public space can produce very mixed feelings, from togetherness and belonging to alienation and fear, depending on the viewpoint, and circumstances. For every person who feels comfortable in a park, there is another who feels fear of crime or being accosted by the police for being at the wrong place at the wrong time. Some (more powerful) people try to limit any encounters with those they consider 'others' or outsiders, including avoiding shared spaces altogether. Spaces of encounter are indeed complex and contradictory, places where social difference is apprehended, negotiated and sometimes ignored in favour of social sameness. Boundaries can be broken but also erected. An example that we explore in Chapter 11 is during gentrification in urban neighbourhoods. New public spaces like cafés, parks or art galleries may become shared zones where people from different backgrounds interact, temporarily softening social divisions. At the same time, rising rents and property prices can displace long-time residents, leading to social exclusion and a loss of community.

How does relationality and encounter crosscut the three themes of social geography? While the two threads will only become apparent with a full reading of the textbook, we can signal at this point that face-to-face encounter is a key component of the social, and one that has arguably been denuded with the digital mediation of social life, and the increasingly scripted or guarded nature of encounters in commercial and institutional settings (e.g. customer service, local welfare offices). Encounters and relationality are always marked by various inequalities, as signposted by Massey and Valentine. In more polarised societies, people may feel less comfortable engaging with strangers. In fact, the inequality that pervades society invariably structures encounters and relations. For social geographers, this inequality across places and people is a key aspect of the sub-discipline, and will be revisited in multiple ways across the textbook. This reinforces the theme of social justice, in that some groups will want to counter structural inequalities that structure the social. For the remainder of the book, we will thread in relationality and encounter to differing effect, with more emphasis on the former in Part I of the book, and more emphasis on the latter in Part II of the book, more on which is found in the section below.

1.3 The Structure of the Book

The structure of the book is divided into an introduction and three core parts (comprising 15 chapters). This corresponds to the structure of a term-based module, with initial foundational sessions on key concepts and social structures, followed by case-study based sessions. Part I explores the broader foundations and structures (social, political and economic) that underpin the field of social geography. It comprises four chapters which outline how societal relationships and relationality are established across various facets – inequality, the political, the social, the welfare state, care and the voluntary sector. These will be covered at various scales, from the body to the household to the nation state (see Section 1.4 for more on these various scales).

Chapter 2 asks you to consider how your identity is shaped by broader structures in society, such as the geographies of production and social reproduction, in ways that may be inequitable and are often out of your control. Chapter 3 considers the role of the state – as a series of actors and institutions with statutory duties – in shaping our citizenship in our everyday lives. Chapter 4 provides a more detailed account of the geographies of state-centred support that provides a safety net including the welfare state. Chapter 5 shifts our lens to examine the informal care provided by neighbours and the wider community, as well as the more formal channels of the voluntary sector.

To help you to apply these insights, Part II, entitled 'Social Geographies of Spaces of Encounter', explores the various encounters that social groups deemed to be at the margins of society experience based on their intersectional characteristics and societal positioning. Building on the relational perspective from Part I, we consider how certain lives are bound up in relationships which may feel like a space of confrontation, mistrust and control. In some cases, these encounters may be well-meaning, yet nonetheless harmful due to wider constraints. Equally, we remain open to consider spaces of more positive encounter, or at least how these positionings can be blurred in more ambivalent ways. Here, we pay particular attention to the significance of the local and the everyday; on the relationships within and between social groups and the spaces that people occupy and use.

This relational approach to encounters offers a novel way for you to engage with a range of social groups, such as young people (Chapter 6), race (Chapter 7), impoverished and marginalised groups (Chapter 8), disabled people and family carers (Chapter 9), people with mental health and addiction issues (Chapter 10), those experiencing housing precarity (Chapter 11) and gendered groups (Chapter 12). Some are on more clearly defined social groups, others are about how wider social problems (like access to housing) unequally affect a broad swathe of people. Through navigating the various spaces of encounter, the experiences of people in each chapter are shaped by relations with wider communities, their peers, the state, the voluntary sector and so on.

Part III, entitled 'Geographies of Hope for Social Justice' provides an important counterpoint to the difficult experiences of people covered in Part II by inviting you to consider the geographies of hope and possibility, building on the three strands of social justice (distribution, recognition and procedure). We highlight the spaces of care and hope which can sustain and enrich these encounters, as well as seek to transform them. Chapters 13 through 15 will give notions about how to shift and nudge relationships and encounters in more socially just directions. Here, we examine the role of rights, welfare and an ethic of care in supporting and repairing people's lives. We also seek to disentangle how people's feelings of being ill-served can breathe life into forms of resilience and activism, which ultimately seeks to drive urban justice and the right to the occupy public space. Equally, for those in care and support roles (e.g. for disabled people in Chapter 9) or allies to those seeking justice (in Chapters 13 and 14), we look at the emotional geographies of their care or advocacy work, which can motivate and sustain their provision. Chapter 13 focuses on the geographies of resilience and slow/quiet activism. Here, we emphasise how groups have sustained their lives in spite of adversity and have exercised resilience or sought to remake their immediate conditions around them. Chapter 14 examines the geographies of resistance and direct action, and takes a closer look at protests and unrest around the world, and critically appraises how these forms of agency have not always been in the name of social justice. Chapter 15 is our conclusion which offers final reflections and a more hopeful way forward for putting the geographies of social justice into practice.

1.4 Moving Forward: Pedagogic Approach

The pedagogic assumptions of this textbook are many, and are worth listing here so that you are well-aware of the various underpinnings, both ethical and material. First and foremost, we want to articulate a 'buffet' (all-you-can-read) approach to the material, showing off as many perspectives as possible and letting you, as the student, choose the one(s) that best suit your disposition (previous or emerging), identity and circumstances. Such a buffet approach allows for at least two sides of the debate to emerge, enabling a more critical perspective that eschews a 'one-size-fits-all' path. While we are undeniably selective in what we present in this book, and that our actual selection has no doubt favoured material that is more critical, emancipatory and even radical, we are still committed to providing a range of experiences and viewpoints. We provide conceptual tools to make sense of the social world, but not to fully explain it, nor to foreclose alternative explanations. One way to ensure that these explanations remain open is via the Engagement Questions, which are found in select chapters.

We can further illustrate this buffet approach with the troublesome issue of what are deemed *social problems*, themselves linked to the aforementioned idea of the anti-social. These 'problems' animate a substantial portion of the book (especially Part II), from mental health to homelessness, addiction, violence and marginality. Back in 1989, an entire book called *Social Problems and the City* was published with chapters by luminaries in the field of UK social geography: Chris Philo, David M Smith, Jon Mohan, Sarah Curtis, Susan J Smith and Michael Keith. To them, social problems consisted of unmet social needs (e.g. deprivation, ill-health) and that these were particularly concentrated *in* cities (but not always *of* cities – we will make the argument throughout the book that many social problems are exactly that, insofar as they are issued from fundamental debilitations generated by society as a whole). This concentration of social problems has always been a source of fascination for social geographers (and other social scientists), mapped, and studied as early as 1920s Chicago with such works as *The Hobo: The Sociology of the Homeless Man* (1923), *The Gang* (1927), *The Ghetto* (1928), *A Delinquent Boy's Own Story* (1930) and *The Taxi-Hall Dance* (1932). As part of the Chicago School of Sociology, these works chronicled, using close-in ethnographic studies, the underbelly of Chicago in somewhat lurid terms. Nowadays, this kind of street-based research must strike a balance between acknowledging misery (and privilege), but not piling on even more exploitation and stigma. Viewed through the prism of social problems, we acknowledge that the social world is a messy place. It is not immaculate. There are not always easy answers, and there are certainly no answers at the end of this textbook! Rather, we provide several student Engagement Questions in several chapters to encourage debate, rather than rote learning.

Giving a variety of perspectives means that we cannot and will not validate all of them – that is up to you as a thinking person. It is not up to professors and universities to mandate (or validate) everything that you feel, only to give you the tools to make critical choices. In that respect, we do provide some paths to understanding, as well as *trigger warnings* when the material becomes rather challenging. While we are at it, trigger warnings apply especially to the substantive chapters in Part II: Chapters 6, 7, 8, 9, 10 and 11. The topics covered by those chapters are very much focused on social problems and the grounded experiences of marginalised groups desperately trying to survive at the periphery of society, which necessarily involves signposting behaviour and places that are perhaps well outside of your lived experience.

While the focus on the margins is a preoccupation of social geography and comprises an important proportion of the book, we acknowledge that this presents but a partial picture of social geography. Consequently, we think it is important to temper this focus with an appreciation that everyone may experience the effects of social injustice, either directly or indirectly, at some point in their lives, e.g. middle-income people in the 'squeezed middle' or men experiencing different masculinities, whose story is not always one of privilege or power. The oppressed can become oppressors, and vice-versa, sometimes with indecent haste. Second, we seek to carve out a space for hope, drawing on the many and varied ways that groups have advanced causes for social justice or found motivation to sustain their support or way of life in the belief that they can make a difference.

One way to illustrate our overall pedagogic approach (and a justification for using material that may trigger distress) is through the visuals used in this textbook, almost all of which were taken by Geoff. In effect, beyond a focus on the themes of the social, inequality and welfare, as well as a moral enterprise, social geography is further known for its methodological eclecticism (Smith et al., 2010), and this can include the enthusiastic use of visuals. As Wyly (2010) put it, images begin conversations about social conditions. And while visualising social geography does not give the full story and could even be accused of being at best impressionistic, it certainly addresses a neglected side of the story, of images as central to the project of documenting the themes of social, inequality and social justice. The added value of visuals lies in capturing a grounded, visceral materiality in non-verbal ways. Rose (2016: 308) argued persuasively that photographs are seen as valuable in urban social research because they can convey something of the feel of urban places, spaces and landscapes, specifically of course those qualities that are in some way visible: they can suggest the layout, colour, texture, form, volume, size and pattern of the built environment, for example, and can picture people too. Photographs can thus capture something of the sensory richness and human inhabitation of environments (though not all, of course: they cannot convey sound and can only suggest touch). Not everything is embodied or material, and as such not everything can be visualised or made tangible – photography is never more than partial and always situated. So while we focus on the social as something tangible in the city, we cannot visualise the intangibles – the mistrust, the tension, the dissidence that also speak to the social, inequality and social justice (DeVerteuil, 2022).

The visual approach in the book also very much adopts the 'both sides now' approach mentioned in the pedagogic approach above. Note that like the Engagement Questions, the images are unevenly distributed across the book, mostly found in chapters that feature social processes that are material and visible – 4 (the welfare state), 8 (marginality and class) and 11 (housing). And so we try to expose social fault lines but also with an eye to seeing the emancipatory and dissident side of society – the promise of the social amid demise and inequality. More to the point, 'visual images communicate their messages very powerfully, either because they can carry lots of information, or because images can be more affectively powerful than academic writing' (Rose, 2016: 331). This bearing witness is very much about documenting and evoking the visual, while recognising that the visual always exceeds what text and numbers can give us (Rose, 2014). We must avoid what Brickell (2024) calls 'helicopter research', preferring instead (when possible) a more long-term connection and promises to various places and people. These issues will be discussed in particular in the photo essays in Chapters 8 (marginality and poverty) and 13 (counter-places).

All this discussion is set within an era of visual glut(tony), with its associated distractions, disposability and silences that undercut the supposed democratisation of photography where images circulate a lot more than they used

FIGURE 1.2 Mumbai airport

Note: This image was taken by Geoff in February 2019, and shows two worlds in close geographic proximity but socially, worlds apart – on the left is the edge of the Jarimari informal settlement and its focus on everyday survival – and on the right is the take-off strip for the international airport, connected to the world but used by very few Mumbaikers and certainly not those dwelling in Jarimari, even if they might work at the airport. This chasm between two worlds is at the heart of social geography, particularly concerning deep-seated urban inequalities and a relational sense of the world that is both local and global. But it also shows how the informal settlement continues to exist in the shadows of global power, a source of livelihood and perhaps even hope.

to. By contrast, this textbook engages *against* this increasingly 'Instagram' approach to visualising society and cities. The 'Instagram city' has no sense of bearing witness, with so many shots of the same thing and so few shots of what really matters in cities, and what really matters to social geography and social geographers. Seeking out the 'untagged city' has its own issues, of course. Here we try to avoid problematic street photography that focuses on the underbelly of cities; much of the photography in this book is accusatory, moving beyond just exposing inequality and unfairness to pointing fingers at the powerful actors sustaining inequality, but also finding dissidence and even joy in the everyday. This also necessarily involves delving deeply into the urban margins. In so doing, we wish to condemn larger power structures, rather than pile on even more stigma upon subordinated places and citizens via photographic depictions. We can never ignore these margins either, lest we give in to the Instagram version of cities and society. These issues will be put into practice particularly in Chapter 8 around race, poverty and marginality, where we will revisit questions of privilege, accusation and ties to specific places.

1.5 The Geographies of this Book

The geographical range of this book is, not surprisingly, focused on the Global North, to the contexts that we know best (e.g. Australia, Canada, Ireland, New Zealand, UK, USA). But part of this book's remit is to also build a more *global* social geography, which becomes apparent in Chapters 8 and 11, where we move beyond our comfort zone to embrace contexts in the Global South and the Global East. The Global South (e.g. Brazil, Cambodia, Peru, South Africa) encompasses those nations that are marked by higher birth and death rates, lower incomes per capita, as well as greater degrees of informality (e.g. lack of direct government involvement in day-to-day living), inequality and rates of urbanisation. The Global East (e.g. China, Japan, Korea, Taiwan, Vietnam, etc.) is more similar to the Global North, but marked by very high levels of state intervention in the built environment, as well as high levels of social homogenisation and coherency when compared to the Global North. We also understand that some places combine all three, such as Dubai – whose scale of development rivals the Global East, with birth and death rates approaching those of the Global North, but also features a large immigrant population from the Global South (especially the Indian Subcontinent).

Moreover, the scales of the book run the gamut, from the mind and body themselves (this is especially the case in the chapters on race, youth, mental health and gender) to the neighbourhood (Chapter 5 on care and the voluntary sector, Chapter 8 on Poverty and Marginality, Chapter 11 on housing) to the city and eventually the nation (Chapter 4 on the welfare state, for instance). As geographers, you are likely quite used to dealing with various materialities – buildings, neighbourhoods, cities and even regional/national and international scales. But it is worth also considering the body itself, of how it co-constitutes the neighbourhood, the city, the nation – gendered bodies, racialised bodies, disabled bodies, sexualised bodies. This connects to the well-established concept of *embodiment*, that is 'the notion that disability [or other forms of social identity] as a lived experience is, by its very nature, experienced through the body' (Wilton and Evans, 2009: 205). More on this will be found in the chapter on disability (Chapter 9). The visuals of the book are also a form of embodiment and embodied knowledge-making, since someone (usually Geoff) had to physically go somewhere and decide to take certain pictures. In this respect, taking pictures is a space of encounter, between the image-taker and the scene.

However, we do not cover the international scales of geopolitics, war and so on, since this is more the domain of political geography. We do acknowledge the importance of these, but also the fact that scale itself is a socially constructed device that nonetheless has important social impacts on what happens where. We also go beyond the national scale to encompass global concerns around the *environment* itself, that the social cannot escape the larger issues of environmental degradation, climate change and calls for environmental justice. These issues tie into class, race, gender, mental health, but also activism that we will explore at the end of the book, including changing one's behaviour, but not engaging in loud protests.

Finally, the intellectual scale of this book is firmly within the geographical tradition, yet any work on social geography will invariably owe a debt to the sociological world. As such, there are nods in the textbook to some key sociologists that have underpinned some of the work of social geographers: Goffman (Chapter 2); Durkheim, Bourdieu and Foucault (Chapter 10), Klinenberg (Chapter 5), Omi and Winant (Chapter 7) and the Chicago School of Urban Sociology (across the book, but especially in Chapters 7 and 8). We acknowledge this debt, but always present sociological theory through the eyes of geography.

Further Reading

Anderson, N. (1923) *The Hobo: The Sociology of the Homeless Man.* Chicago: University of Chicago Press.

Cressey, P. (1932) *The Taxi-Hall Dance.* Chicago: University of Chicago Press.

World Inequality Database (2023) Home – WID – World Inequality Database.

Harvey, D. (1973) *Social Justice and the City*. Baltimore: Johns Hopkins University.

Katz, C. (2004) *Growing Up Global: Economic Restructuring and Children's Everyday Lives.* Minneapolis, MN: University of Minnesota Press.

Kearns, G. and Reid-Henry, S. (2009) Vital geographies: Life, luck and the human conditions. *Annals of the Association of American Geographers*, 99(3), 554–574.

Lees, L. and Hubbard, P. (2021). "So, Don't You Want Us Here No More?" slow violence, frustrated hope, and racialized struggle on London's Council Estates. *Housing, Theory and Society*, 39(3), 341–358.

Lefebvre, H. (1991) *The Production of Space*. Oxford: Blackwell.

Migdal, J. (2001) *State in Society: Studying How States and Societies Transform and Constitute Each One Another*. Cambridge UK: Cambridge University Press.

Mitchell, D. (2003) *The Right to the City: Social Justice and the Fight for Public Space*. London: Routledge.

Rawls, J. (1971) *A Theory of Justice*. Cambridge MA: Harvard University Press.

Smith, D.M. (1994) *Geography and Social Justice*. Oxford: Blackwell.

Thrasher, F. (1927) *The Gang.* Chicago: University of Chicago Press.

Trudeau, D. (2012) Constructing citizenship in the shadow state. *Geoforum*, 43(3), 442–452.

Wilkinson, E. and Ortega-Alcázar, I. (2019) The right to be weary? Endurance and exhaustion in austere times. *Transactions of the Institute of British Geographers*, 44, 155–167.

Wirth, L. (1928) *The Ghetto.* Chicago: University of Chicago Press.

References

Brickell, K., Lawreniuk, S. and McCarthy, L. (2024) Qualitative longitudinal methodologies for crisis times: Against crisis exceptionalism and 'helicopter' research. *Area*, 56, e12924.

Del Casino, V. (2009) *Social Geography: A Critical Introduction*. London: Wiley.

DeVerteuil, G. (2004) The changing landscapes of Southwest Montréal: A visual account. *Canadian Geographer*, 48(1), 76–82.

DeVerteuil, G. (2022) Slow scholarship, the slow city, and counter-visual practices. *Environment and Planning F*, 1(2–4), 226–249.

Dorling, D. (2014) *Inequality and the 1%*. London: Verso Books.

Massey, D. (1993a) Power-geometry and a progressive sense of place. In J. Bird, B. Curtis, T. Putnam, G. Robertson and L. Tickner (Eds.), *Mapping the Futures: Local Cultures, Global Change* (pp. 59–69). New York, NY: Routledge.

Massey, D. (1993b). Politics and space/time. In M. Ketih and S. Pile (Eds.), *Place: The Politics of Identity* (pp. 141–161). London: Routledge.

Massey, D. (2011) A counterhegemonic relationality of place. In E. McCann and K. Ward (Eds.), *Mobile Urbanism* (pp. 1–14). Minneapolis: University of Minnesota Press.

Rose, G. (2014) *Visual Methodologies: An Introduction to the Interpretation of Visual Materials*. 3rd edition. London: Sage.

Rose, G. (2016) *Visual Methodologies: An Introduction to Interpreting Visual Materials*. 4th edition. London: Sage.

Smith, D.M. (2000) Social justice revisited. *Environment and Planning A*, 32, 1149–1162.

Shaw, C. (1930) *A Delinquent Boy's Own Story*. Chicago: University of Chicago Press.

Soja, E. (2010) *Seeking Spatial Justice*. Minneapolis MN: University of Minnesota Press.

Smith, S.J., Pain, R., Marston, S. and Jones, J.P. (Eds.) (2010) *The Sage Handbook of Social Geographies.* London: Sage Publications.

United Nations (1948) *The Universal Declaration of Human Rights.* New York: Office of the High Commissioner.

Valentine, G. (2008) Living with difference: Reflections on geographies of encounter. *Progress in Human Geography*, 32, 323–337.

Wilton, R. and Evans, J. (2009) *Disability and Chronic Illness*. London: Elsevier.

Wilton, R. and Fudge Schormans, A. (2019) 'I think they're treating me like a kid': Intellectual disability, masculinity and place in Toronto, Canada. *Gender, Place & Culture*, 27(3), 429–451.

Wyly, E. (2010) Things pictures don't tell us: In search of Baltimore. *City*, 14, 497–528.

2 EVERYDAY SPACES OF DIFFERENCE, IDENTITY AND INEQUALITY

2.1 Geographies of Difference, Production and Social Reproduction

We all have unique identifiers which can shape the perceived differences between ourselves and people around us. An identifier refers to a feature or quality belonging typically to a person, place or object and serving to identify them. Examples include our embodied or compositional characteristics such as our age, gender and ethnicity. Embodiment is understood as a lived experience that is, by its very nature, experienced through the body. Beyond that, the perceived differences between people can be shaped by contextual characteristics, such as our socio-economic status and our geographic background. These are overlaid and affect each other in complex ways.

We accept that some identifiers, such as gender and race, are social constructs. This is not to demote them as inconsequential, as David Harvey (1996: 127), the most cited human geographer in academic circles, contends, 'even though we are dealing with a social construction we are not dealing with something purely subjective or ideal, that is outside of the material world in which we have our existence'. As we explore in Chapters 7 and 12, the social construction of gender and race can manifest in powerful ways through shaping patterns of habitual behaviour, sanctioning and institutionalised practices and rules. What is more, differences can depend on broad external factors including climatic conditions, societal conditions (health care, education systems, prevalence of crime, community relationships) and cultural customs and conventions, among other factors.

Although we will treat the five core identifiers in more detail in subsequent chapters, we can provide some working definitions at the outset here. Gender goes beyond just the binaries of biological differences between men and women's bodies ('sex') to capture how men, women and non-conforming genders present themselves, and are in turn presented through society (Chapter 12). Chapter 12 also covers sexualities which refer to the diverse ways people experience and express sexual attraction, desire, identity and behaviour. This includes variations in who individuals are attracted to (or not), how they define their sexual identity, and the social, and cultural

DOI: 10.4324/9781003414889-3

meanings attached to those experiences. Race is a difficult term to quickly define, given its multiple social meanings but we take inspiration from earlier work on racial formation theory by sociologists Michael Omi and Howard Winant (1994), in which racial categories are created, transformed and given meaning in society. This happens through what they call 'racial projects' – efforts by individuals, institutions or governments to organise and interpret race, often in ways that support or challenge existing power structures (Chapter 7). Disability is not a homogenous group and can include differences in functioning at the level of the body (including physical, sensory, neurodiversity and cognitive differences as well as chronic ill-health) or through participation restrictions that people with disabilities may experience. This acknowledgement of participation recognises that the functioning of a person is determined by their social interactions with society and experiences of ordinary citizenship (Chapter 9). Class involves how people are slotted by their income and wealth across society, which creates an economic-based (spatial) identity, and status, including working class or upper class (Chapters 2 and 8). While more of a contextual characteristic, class can become internalised and become a pseudo-embodied characteristic (but this is neither fixed or absolute), and constitutes an important measure of status within society. In contrast, age is absolute but relational, in that all people have an age but also that a person's age slots them into various categories, from children/youth/middle-age and elderly, but also generational categories which each have their own particular concerns and pressures (Chapters 6 and 11).

As social geographers, we are particularly attuned to how our characteristics are shaped by place. The human world is wildly uneven, and this unevenness has profound impacts on life, and life chances across the globe. The place of our upbringing is undoubtedly a large factor in shaping our experiences, norms and values – and how we are perceived by others. As this textbook explores, place can shape our spatial horizons and mobility, through different zones of encounter, different relationalities and especially different relationships. For instance, Melinda Gates, an American philanthropist, former multimedia product developer and manager at Microsoft, and the ex-wife of its co-founder Bill Gates, who have both compared their own upbringings to a girl in the Sahel, a semi-arid, grassland region located in north-central Africa. Here, the environmental and socio-political characteristics of a place like the Sahel versus Texas and Seattle respectively are also highlighted. In the Sahel, which has experienced conflict and environmental crises, place can have a more overt and forceful role in shaping our lives. We will further connect to this point in a few pages – the sense that the more unequal a society is, the more important place becomes in terms of birth, upbringing, and life chances.

Our characteristics also vary in terms of their durability. Arguably, none are permanent, apart from our birth date, assuming our calendars do not change. Our gender, skin colour, sexuality and especially disability can change over the life course, depending on specific circumstances, preferences, treatments and life experiences. Just as our relationships with each other are not fixed, our characteristics evolve continually in our everyday encounters. Our interests can also wax and wane as we progress through life, and we can choose how to represent ourselves to a certain extent. Nonetheless, some characteristics are more durable than others – and some can appear to stick. For example, our economic status (e.g. class) is shaped not solely by our own education, upbringing, hard work (not easily changeable!) and local amenities, but can be mediated by external factors, such as the availability of work and travel. This means that our personal characteristics and opportunities can be shaped in profound ways by macro factors involving the geographies of production.

The geographies of production focus on where things are produced (goods and services) which can affect where and how we work (or not). It shapes wider geographies in terms of the availability of housing, the extent of prosperity, poverty, inequality and the physical and mental health of the population. To help understand how these factors inter-relate in shaping individual characteristics and places, it is helpful to trace in more detail the ways in which our everyday lives are affected by the geographies of production. David Harvey was acutely aware of this connection, as he previously worked during the 1970s at John Hopkins University, Baltimore – a city that was experiencing wholesale de-industrialisation and disinvestment at that time. Consequently, Baltimore became a casualty of footloose capital that was uprooted and invested overseas, partly exacerbated by globalisation, offshoring and deregulation. This led to the relegation of certain communities (especially working-class areas surrounding the Downtown) who felt that they were left to the 'scrap heap', and which Harvey argued exacerbates issues of social justice due to extensive job losses, rising crime rates, housing, which have since persisted over decades and generations. According to Harvey, the spatial fix of capital is 'capable of holding for a time, though not necessarily resolving, tensions between capital and labour, and economic development and collective consumption' (cited in While et al., 2004: 551).

The above example provides a potent account of how the geographies of production shapes individual livelihoods and identities/characteristics in certain communities. These macro factors can help provide opportunities for young people, provide more security against ill-health, and better community services – assuming the forces of production do not exploit local workers, as it has been shown to do in certain regions and at different times. Indeed, postcolonial scholars have sought to expose the longstanding impacts of the inequitable legacies of slavery and disempowerment, particularly in the Global South. The collapse of the Rana Plaza factory building in Bangladesh in 2013, for example, was the worst industrial incident in recent times to hit the garment industry, and brought worldwide attention to deathtrap workplaces within the industry.

While the role of production is clear, it is also important to consider how production itself is shaped by the local geography, political governance and population of a region. Work opportunities can emerge in connection to the local talent pool, age, education and native language of the local population. Production can also be shaped by the location of nearby resources such as (accessible) land, proximity to an urban location and local minerals, and resources. Governments further determine the extent of investment, regulation, tariffs and access to networked and technological infrastructures, which may shape economic investment and relationally sustain productivity in a region (and mediate against the ill-effects of private sector exploitation). Graham and Marvin (2001) in their book, *Splintering Urbanism*, provide a detailed account of how globalisation and the city have both been shaped through uneven and competing access to these resources, and highlight the impacts caused by fragmenting the experience of the city.

At the local level, production also has a role in shaping our spatial mobility. But the local level shows most obviously the differences between individuals based on our individual characteristics. Spatial mobility refers to the relative ease or difficulty individuals have in moving in and through space, and shapes the various social encounters that they experience. A broader comparison can be made between the mobility of people and of the capital which underpins the availability of work for people. According to Gough (1992), people's dependence on capital and competition is intrinsically geographical. What he means is that the scale of capital's mobility is more footloose, whereas by contrast, territorial ties vary enormously between groups in the population. What kind of mobility do you think you have in comparison to each of the following six people in the Engagement Questions below?

Box 2.1 Engagement Questions

How much can each of these people move freely and what constraints on their mobility can you think of?

A migrant farmer from the Global South working overseas in the Global North

A young adult woman who is pregnant and due to go on parental leave from her employer

A young adult man who is unemployed and on unemployment benefits

A young adult woman who has recently graduated from your University with your degree

An older man who has worked in a professional role, is a homeowner and nearing retirement

A middle-aged disabled man who uses a wheelchair

The point of the above question is to consider the relative different spatial horizons between people. As noted, each individual's spatial mobility affects the social encounters that they are likely to experience, in terms of their relationships with friends, family, neighbours, the wider community and the state. In some cases, despite the uprooting of capital, the territorial ties that people have with a place can remain strong. One example of this is in Newcastle and Durham in north-west England. Despite the closure of the coal mining industry in the 1980s, which precipitated a period of de-industrial decline, a Miners Gala event still takes place each year to celebrate the music and culture of the mining communities. Here, Walter Firey's notion of 'sentiment and symbolism' offers a strong counter point to an assumption that people make decisions based purely on economic rational logic to move location after capital has uprooted (Knox and Pinch, 2010: 159). For this reason, geographers are also interested in social reproduction, which we turn to below.

Social reproduction is concerned with understanding how society 'reproduces' itself, from one generation to another and within generations. It offers a lens through which to analyse the continued evolution and changing patterns of society over time shaping our characteristics. It can organise ever-changing things and people into categories, positions and patterns of behaviour that exceed their individual existence (Weiss, 2021). This includes our customs, values and opportunities across generations. It can also be understood as the passing on of certain characteristics across generations. The geographies of social reproduction thus seeks to consider questions about the role of space and place in shaping these changing patterns and determining what and how things get passed on. As we will signpost in Chapter 5, social reproduction tends to be the domain of women and unpaid labour.

To understand how the geographies of production and social reproduction work in tandem to reinforce differences between people in society, it is useful to go back and look at how they have changed over time. Pre-industrial production involved a period where cottage industries and agrarian work dominated. This meant people's subsistence was generated through their own labour within the household, thus social reproduction and production were deeply intertwined and in close proximity with each other. By tracing income disparity in wealth over this time, wealth gaps between and within countries were relatively modest. Industrial capitalism instituted a clear divide between 'productive work' and the household. The need to work routinely and dependably at predetermined times was completely new for many nations. The second industrial revolution lasted from the mid-19th century until the early 20th century and mainly took place in Britain, continental Europe, North America and Japan. It widened gaps in income both between countries and within them. The household no longer supported itself through what it directly produced.

People depended on wages to buy commodities, which were then transformed through household labour. By the 1930s in America, and 1940s in Europe, income disparity declined. These dates are significant, as America experienced its Great Depression in the 1930s and the European continent was affected by the disruptions of the Second World War (until the Marshall Plan, 1948 which generated a resurgence of European industrialisation and brought extensive US investment into the region). By the 1980s, income disparities grew again, and have accelerated greatly since the 2000s, depending on the nation.

As signposted in Chapter 1, due to recent technological changes over the last 15 years and the global pandemic in 2020, there has been a marked change once again in how many people work. People are increasingly working from home (or perhaps feel that they now 'live in work') and this has blurred the boundary between these two domains. A modern-day equivalent is our craft industries, which has been re-invigorated through online consumer practices (e.g. Etsy). Again, the imprint of the geographies of production (and now consumption) is evident in our everyday lives. By contrast, many still work in sectors that require in-person contact, for example care work, nursing and service/trade roles. This has meant that these changes have also increased inequalities, for example through fuel or commuter costs for people working in such professions. Moreover, production is still by and large shaped by wages. For those seeking work in online or hybrid flexible jobs, inequality can comprise digital and occupational divides based on access to broadband and computing technology and between professional/technical roles.

These prompt us to think once again of the difference between our compositional characteristics – our age, gender and ethnicity – and our contextual characteristics – our place in the world and our local geographies of production and social reproduction. Moreover, it allows us to consider how these are overlaid and affect each other. Collectively, these compositional and contextual differences interact and shape our identity, which we turn to next.

2.2 Identity and Intersectionality

A key concept in social justice is the recognition of various identities, or as we put it at the beginning of the chapter, social identifiers. While we may consider our own personal identity as something that we cultivate ourselves across our life course, through our lifestyle choices, style of clothes, food choices, and so on, our identity is also shaped largely by external factors, in relation to other people, and eventually to larger society. This understanding of our sense of self and identity as relational is a fundamental underpinning of this book, and relates to the first theme of social geography – that of the social. Increasingly, understandings of the social are processed at the individual level through identity, rather than at the collective level. As Massey (2004: 15) describes, 'we do not have our beings and then go out and interact'. Rather, she argues that:

> our beings, our identities, are constituted in and through those engagements, those practices of interaction. Identities are forged in and through relations (which include non-relations, absences, and hiatuses).

In her quote, she leaves it open for us to consider how people have different relational networks, and in some cases, people have small, absent or disempowering networks. For example, in Chapter 9, we explore the experiences of adults with learning disabilities in England, where they reflect on their own limited social networks. These gaps and absences can be as relevant in shaping a person's identity as those relationships that they deem close and empowering. This framing highlights the ways in which identities are shaped by accreting, unfolding and emerging relations within the immediate locale over time, and in relation to individual characteristics (Power et al., 2021). A relational model of self, the development and exercising of individuality and agency is a complex competence, requiring ongoing interpersonal, social and institutional scaffolding. Identities are thus understood as multiple and always in a state of becoming.

Extending beyond individuals, we can also understand the public in a more plural sense, of multiple publics that comprise a multiplicity of voices, identities and experiences that criss-cross within and across communities. This understanding moves beyond a monolithic view of the 'public' to acknowledge the public sphere as dynamic, contested and comprising geographically diverse groups whose interests and experiences are shaped by spatial processes. Geographers also analyse how power relations, social movements and everyday practices intersect with space to create and shape these different relational publics. This framing also recognises that some publics run counter to or seek shelter from the mainstream, as explored in Chapter 13.

Building on this thread of relationality that crosscuts the textbook, and of great relevance to the notion of identity, is the concept of intersectionality (Hopkins, 2019). Intersectionality has become a cornerstone in social geography to understand how oppression and injustice intersect in various ways and contexts. While we explore this concept in more depth in the next chapter, for now we highlight its basic principle that the various social identifiers – gender, race, (dis)ability, sexuality, class, age, religion, and so forth – can mutually constitute a person's identity in various ways. For instance, racism and sexism can interact in ways that can make the experiences of Black women much more oppressive than those of White women, as they belong to not just one but two oppressed groups. But intersectionality is more than just adding one oppression on to the other. It also involves how various social identifiers can interlock in various ways, creating entrenched social injustice.

So, rather than seeing each respective group as a separate set of single identity holders, based on for example, 'class', 'race', and 'gender', it is also important to think of the ways these overlap and coalesce. Intersectionality is thus a metaphor for understanding the ways that multiple forms of disadvantage sometimes compound themselves and create obstacles that often are not understood among conventional ways of thinking about identity. It was first coined by Kimberlé Crenshaw in 1989, when she was a law graduate of Cornell University seeking to contribute to an emerging critical race theory. Crenshaw sought to challenge the optimistic belief that racism would cease to exist with the passage of the Civil Rights Act in the US. She wanted to unpack the way structures of law and society could be intrinsically oppressive. Crenshaw argues in her paper (1989) that by treating Black women as purely women or purely Black, in terms of the form of oppression they faced, as the courts did in 1976, it repeatedly ignored specific challenges that face Black women as a group. As an example, in May 1976, in the case *DeGraffenreid v. General Motors* (described in Coasten, 2019), Judge Harris Wangelin ruled against the plaintiffs, writing in part that 'Black women' could not be considered a separate, protected class within the law, or else it would risk opening a 'Pandora's box' of minorities who would demand to be heard in the law.

Crenshaw argued that the lived experiences – and experiences of discrimination – of a Black woman will be different from those of a White woman, or a Black man, for example. But Crenshaw was not seeking to build a racial hierarchy with Black women at the top; rather, it was to help our understanding of how groups' and individuals' social and political identities result in unique combinations of discrimination and privilege. The concept intersectionality gained widespread attention during the 2017 Women's March on Washington in the wake of Donald Trump's first presidential election, an event whose organisers noted that women's 'intersecting identities' meant that they were 'impacted by a multitude of social justice and human rights issues' (Coaston, 2019).

Space becomes critical to understanding intersectionality, as places can reveal some identity intersections over others. Rather than adding or multiplying categories of social difference, an intersectional approach involves understanding how different social identities are co-constituted in everyday life (Worth, 2015). Given their theoretical heft, pre-given analytical categories – such as gender, race or class – tend to dominate explanatory narratives and can sometimes be used as shorthand, fixed and all-encompassing descriptors. Rather, it is worth trying to discern the ongoing negotiations that may surround them. Gill Valentine (2007: 19) enriches this insight by exploring how intersectionality – the relational complexity of lived experience – is inherently spatial, as identities are highly contingent and situated accomplishments:

> In other words space and identities are co-implicated. Where [we are] located is constitutive of [our] identity, not incidental to it, so that [we] understand ourselves differently when at school than when in at home, or when in the office environment .... The identity of particular spaces – the home, the school, the workplace, or a community space ... – are in turn produced and stabilised through the repetition of the intersectional identities of the dominant groups that occupy them such that particular groups claim the right to these spaces.

Intersectionality is not without controversy though, particularly among the conservative right in the US. Coaston (2019) argues that the political right accepts the idea that people experience discrimination differently depending on their overlapping identities. However, their concerns are with the perceived privileging of groups through recognising their intersectionality. In essence, it is feared that addressing power imbalances inevitably means people at the top of a perceived identity hierarchy would consequently 'lose their spot' (Coaston, 2019: n.p.). This has been used as a method to discredit some social movements that seek to address social disadvantage, such as Black Lives Matter, to which we return in Chapter 14.

The issue of privilege is always a prickly one, since the two authors and most if not all of our readers are in some ways privileged, at least privileged enough to work at or attend university! The privileges that Bill and Melinda Gates acknowledged earlier in the chapter are frequently left unacknowledged, giving people licence to consider their own achievements solely through their own merit within an appreciation of the opportunities they have gained via their societal positioning or the lack of obstacles they have faced through their life course. This can in turn lead to prejudice against minority groups for receiving 'unfair' cultural support or legal protection and so on. We will pick up on some of these issues of privilege in Chapter 8, part of a 'both sides now' approach to topics such as marginality and power. Moreover, the drive towards intersectional understandings should not always be assumed across various places. In a nation like France, the goal is to ensure that everyone is 'French', rather than claiming some ethnic or racial identity, and that as such this makes everyone ostensibly 'equal', even when it is clearly not the case. This integrationist model values social coherency above individual identity and difference. As a result, the social landscape in France may arguably be richer in terms of resources for social coherency than in the US or UK, but this same social landscape erases difference in favour of a universal White French identity, to the point where crucial social identifiers such as race are not even collected in the decennial census of the population.

To varying extents, our identities can also become politicised in a way that informs specific government actions (e.g. creating laws which shape access to rights) and rhetoric (i.e. in the way that a government creates a narrative about who does or does not belong) about a particular group. Thus, our political identity can determine to an extent how people are perceived at a societal level, and how they may be 'assigned their essential character prior to social interaction' (Massey, 1993: 111). What Massey refers to here is that sometimes socio-cultural norms and institutional boundaries can create expectations over how a person ought to be identified. However, this is not static but an ongoing negotiation between people and the wider public. Moreover, it is crucial that we challenge a priori identities themselves and the relations through which those identities have been established. Consequently, identity politics is a term used to describe how population groups are conceptualised and how these conceptions can shape cultural norms and expectations and determine entitlements and levels of access to public services on the one hand and how these conceptions are countered and challenged by groups seeking more valued recognition. We thus understand identity as an inherently relational concept, operating at the embodied, personal and a social-political level.

Erving Goffman, a sociologist who has made a major contribution to geographic thinking on the concept of identity, in his *Notes on a Spoiled Identity* (1963) referred to our 'real identity' and 'virtual social identity' to denote the difference between our own perceived personal identity and our public externally shaped 'virtual' social identity. He argued that when there is a large gap between these identities, it can be a source of stigma. For Goffman, individuals that may share certain attributes or do certain activities do not necessarily have equal levels of stigma. Sometimes intersectionality plays a part. Goffman provides a comparison of a middle-class boy who is unlikely to feel any fear of reprisal for being seen going to the library; a criminal, however, Goffman writes, may feel stigmatised by onlookers:

> I can remember before now on more than one occasion, for instance, going into a public library near where I was living, and looking over my shoulder a couple of times before I actually went in just to make sure no one who knew me was standing about and seeing me do it. (cited in Goffman, 2009: 3)

Building on this insight, Goffman argues that stigma should be seen as a language of relationships, not attributes, thus inadvertently providing an early acknowledgement of the relational geographies which underpin our book. He does however acknowledge that there are important attributes that are discredited almost everywhere in our society and therefore transcend space. We will revisit stigma in Chapter 8 when dealing with marginality and urban space.

A broader societal understanding of identity is pivotal to understanding the ways that particular social groups, based on different characteristics, are governed. This informs our subsequent chapters in Part II which explore a variety of different groups and the various relational networks they may encounter, for example young people, homeless people, disabled people. In some cases, people may be drawn into relational encounters with government agencies, such as social services or long-term health care. This serves as a transition to considering difference at a more societal level, moving from individual meetings to how those meetings are determined in advance by societal inequalities.

2.3 From Difference to Inequality: Distributional and Recognitional Understandings

Returning to the earlier point that differences between people are fundamental to being human, and to some extent expected when considering different socio-spatial compositional contexts, it does not necessarily follow that there should be inequality between different people. This section considers how differences translate into inequalities between people, transitioning from the theme of the social to the theme of inequality. Further, inequality must be seen as relative – that is, it expresses the relationship among different social groups. As signposted in Chapter 1, inequality refers to the state of not being equal, especially in status, rights, and opportunities. David M. Smith (1984; 1994) carefully defined inequality as a special type of differentiation of characteristics among members of a designated group. It is a concept very much at the heart of social justice theories. We focus in this chapter on the management of class inequality, of contrasting wealth and (especially) need, but we recognise other forms of social inequality which we follow up on in subsequent chapters. Inequality is most commonly understood in a distributional sense; some groups are positioned at the top, relatively speaking (such as the 1%), others are positioned nearer the bottom. This section unpacks this understanding of inequality further, but it also poses an alternative recognitional framing, drawing on the work of Iris Marion Young by suggesting that it can be a form of oppression in its own right (irrespective of wealth distribution).

Given our previous discussions of class inequality as well as intersectionality, we focus on economic and social inequality, in that a majority of the world lives in capitalist societies, whereby key elements of social well-being – including nutrition, shelter, health, education, employment opportunities, clean environments, leisure, security, social stability and so forth – are strongly determined by purchasing power, itself linked back to the uneven geographies of production. Some populations are denied access to these basic elements of social well-being, and accordingly suffer from social exclusion, denial of basic human rights and a lower quality of life. Economic and social inequalities are most obviously related to class but can also interact with a variety of other key social identifiers that relate to power, opportunity, and life chances, such as (but not limited to) race, gender, sexual orientation, (dis)ability, religion and age. Here, the alternative recognitional framing proves useful. We present both these framings below.

Distributional Inequality

As signposted in Chapter 1 on the three strands of social justice, distribution refers to the different resources and opportunities that are extended to meet people's needs and seek to address material inequality. Distributive justice is the term to describe the concern over how to allocate resources fairly among members of a society. To acknowledge the (often retrospective) adjustments that are made to address inequality, the term 'redistribution' is more commonly used. For Nancy Fraser (1998), redistribution does not solely mean government taxation and redistribution of monetary benefits (narrowly defined as 'welfare', see Chapter 4). She argues that it can include access to income and property ownership, access to paid work, education, health care and leisure time, but also more starkly in exposure to environmental toxicity, hence in life expectancy and rates of morbidity and mortality

(1998: 166). Distributional concerns have often been put aside, as growth was understood by the analogy that 'a rising tide lifts all boats'. This has been an argument that has been discredited and proven incorrect, yet has occasionally resurfaced to justify reduced investment in public services.

For measuring inequality in (re)distribution, various methods have been used. These is no single metric to capture all the intricacies of economic (income or wealth) and social inequality; by income, we mean money earned through formal work, while wealth includes various assets, including dwellings, cars, investments, pensions and the like. As you may appreciate, wealth inequality is more difficult to measure and is likely to be even more unequal. While most adults earn income, a large segment do not own property of any kind, or have pensions.

Arithmetic inequality can refer to a situation in which populations receive unequal amounts of attributes (e.g. income, education, health care) within an overall distribution. One common method that geographers have used is the Gini coefficient, which provides a measure of inequality of income within a country. The Gini index provides a numerical 'score' based on how equal a society is between 0 and 1 (or 0% and 100%): 0 equates to complete equality and 1 equates to complete inequality. In a perfectly equal country, 50% of the cumulative percentage of national income is owned by 50% of the cumulative percentage of the population. Likewise 80% of the cumulative income is owned by 80% of the cumulative population and so on. The 45-degree line provides a visual illustration of this equal distribution of cumulative income. In a more unequal society, the ownership of income becomes more skewed or sags, so that more income is owned by a smaller percentage of the population. Line B in the figure, known as the Lorenz curve, shows the degree of 'skew' in income ownership. By tracing line B, 80% of the cumulative income is now owned by 90% of the cumulative population. This means that the richest 10% own a larger percentage of national income (20% to be exact). The final 'index' between 0 and 1, referred to as the Gini coefficient, is calculated by dividing A by A+B.

The coefficient has proven particularly effective in spatial expressions of inequality, which we return to in Section 2.4. For now, we present an alternative framing of inequality based on recognition.

Recognitional Inequality

As noted in the opening to this section, inequality can also relate to social matters beyond distributional concerns to include notions of recognition. In this respect, Young (1990) defined social justice within the context of social processes but she did not limit the source of oppression to the economic system, and thus not limited to (unequal) distributions of income or wealth (e.g. various assets such as housing, pensions, cars, investments, etc.). In much the way that Marxists might argue that social justice cannot be achieved without transforming the mode of production and redistribution of wealth, Young argues that social justice remedies that are solely based on (re)distribution will not produce social justice if there are additional sources of oppression embedded in social processes. No amount of distribution will produce equitable outcomes if oppression and domination are embedded structurally because the inequalities will be reproduced through distribution. Moreover, distributing things like wealth and income does not address alternative sources of oppression that lie within the realms of recognition and procedure, the second and third strands of social justice outlined in Chapter 1.

In 1990, Young developed five faces of oppression to frame the intersection of inequality and social injustice: exploitation, marginalization, powerlessness, cultural imperialism and violence. As a set of societal relations, the first three connect primarily to the way in which people participate in the labour market (although relate to broader inequality in society as a whole), and their ability for self-determination and self-development, and are related primarily to the geographies of production already outlined in this chapter, but also to the chapters on welfare (Chapter 4) and poverty (Chapter 8). The latter two extend beyond the labour market to overshadow the everyday social lives of people.

Exploitation can be seen in unequal labour relationships, such as women's unpaid labour within the household which benefits men. It can also be seen when firms use people's labours to produce profit while not compensating them fairly, or how certain groups, such as ethnic and racial minorities, work in jobs that are less likely to allow for self-determination and self-development.

Marginalization is the act of relegating or confining a group of people to a lower social standing or outer limit or edge of the labour market or society. Marginalised people, such as those who receive welfare benefits, are left out of the labour system entirely. In her view, because marginalised people are rendered invisible (e.g., through racism or ableism) in the labour force themselves, they are also unable to practice full citizenship rights in the economic system. Young viewed this face of oppression as especially insidious; more on this in Chapter 8.

Powerlessness is the lack of autonomy, authority or voice, particularly in the workplace but also society. Those who are powerless typically have low-status jobs in which they have little leeway and must work for others such as professionals compared with labourers. It extends to the lack of authority, status and sense of self which would permit a person to be listened to with respect and little opportunity for advancement. For instance, this can manifest in feeling powerless to the effects of government initiatives that demonise poverty.

Cultural imperialism involves taking the culture of the ruling class and establishing it as the norm, thus controlling how the people in that society interpret and communicate. It emerges when a dominant group imposes their ideas and stereotypes. People deemed 'other' are stigmatised through the use of damaging cultural representations and myths – often presented in the media. In reality, this is often subtly manifested, for example, being treated with suspicion while shopping, or stared at and made to feel 'out of place' in public life.

This form of oppression was particularly acute during the 1960s, after a theory called the 'culture of poverty' became popularised by the anthropologist Oscar Lewis in his best-selling ethnographic realist books on family life among the urban poor. Proponents of this theory argued that the poor were not simply lacking resources, but also have a unique (and flawed) value system that perpetuates their position at the margins. According to Oscar Lewis, the culture of poverty tends to sustain itself over time, particularly because it shapes the outlook, goals and personalities of the children raised within it. Fortunately, the concept sparked a backlash, pushing scholars to abandon cultural justifications and negative descriptions of poverty, fearing such analysis may be read as 'blaming-the-victim' (Bourgois, 1995). In the late 1990s, the culture of poverty witnessed a resurgence in the social sciences, although most scholars now reject the notion of a monolithic and unchanging culture of poverty and attribute destructive attitudes and behaviour not to inherent moral character, but to sustained racism and isolation.

Finally, violence is the remaining face of oppression. It is the intention to damage, humiliate or destroy a person or group. This has been expanded to consider forms of structural/systemic violence (DeVerteuil, 2015) including inadequate health care, wages and precarious work contracts, police brutality and class discrimination. Poverty is also a key factor in the level of behavioural violence (e.g. ongoing stares and condescension, sometimes interpersonal). The violence of poverty also becomes internalised which can result in addictive behaviour and suicide. According to Young (1990: 61), 'What makes violence a face of oppression is less the particular acts themselves… than the social contexts surrounding them, which makes them possible and even acceptable'.

Overall then, Young's work has helped shift the dial away from purely distributional justice towards embracing recognitional justice.

Nonetheless, Fraser (1995) offers a helpful critique of Young's work, that social movements have abandoned claims for the redistribution of resources in favour of the recognition of cultural difference (e.g. identity politics). She argues that Young too narrowly focuses on representation (of those who are oppressed). Group identity (which we explore in more depth in the next chapter), Fraser argues, has problematically replaced class interest as the chief focus of political mobilisation. The central plank of Fraser's critique is that by making demands for a 'recognition of difference', many marginalised groups fail to challenge the underlying causes of oppression. For Fraser, redistribution does not solely mean 'welfare'. She is also interested in redistribution to address other forms of material inequality, outlined at the start of this section. Her conclusion is to devise an overarching conception of justice that is defensible and can accommodate both recognition and redistribution.

This more balanced view helps us to understand how recognition is imbricated within redistribution and is reflected in how we consider the inequality of outcomes. According to the UN, inequality of outcomes occurs when individuals do not possess the same level of material wealth or overall living conditions (UN, 2015). Actions that seek to address unequal outcomes refer to efforts to rectify these imbalances. This may be in the form of redistribution, such as a government sickness benefit, for people with specific health needs or a life-limiting condition, for example. Or it may be in the form of recognition of diverse characteristics, for example, through positive action, which refers to efforts to create a more level playing field and enable individuals to compete fairly, while not providing preferential treatment such as encouraging the use of inclusive language in a job advert to widen the diversity of candidates. Positive action thus relates more closely to recognition. Under this action, the selection process for the job is the same for every candidate and the successful candidate is appointed on their ability for the job, irrespective of race or gender etc. By contrast, positive discrimination is a more contested (and lesser used) form of redistribution, which refers to discrimination that favours someone by treating them differently in a positive way (e.g. selecting someone solely based on a characteristic).

2.4 From Inequality to Inequity: Spatial Expressions

Inequality frequently overlaps with the term inequity. Here we move beyond arithmetic inequality and build on the recognition strand of social justice, introduced at length in Chapter 1 and already threaded through this chapter. According to the World Health Organization (WHO) definition, inequities are unjust and avoidable inequalities (e.g. due to cultural exclusion, poor governance, corruption) (WHO definition; see also: www.bmj.com/content/356/bmj.j556/rr-8). This mirrors Starfield's (2011: 1) understanding, which she articulates as 'the presence of systematic and potentially remediable differences among population groups defined socially, economically, or geographically'. There are two key phrases in this definition. Systematic refers to the ways in which expectations over a person's different status become entrenched in the institutions of society, including education, health and social care, state services and benefits. The second key phrase is potentially remediable. Here the emphasis is on how differences among population groups can be avoidable. Do we just accept, for example, that women should not work after becoming mothers, as was often assumed in the early to mid-20th century? This forces us to consider whether differences are imposed, by low expectations, cultural beliefs, ignorance or discrimination, and thereby whether they comprise forms of inequity? This may be the result of circumstances beyond one's control (ethnicity, family background, gender, and so on) as well as talent and effort.

Influenced by David M. Smith (1987; 1994), most social geographers of inequality focus on simultaneously unequal and unfair distributions. More specifically, social geographers have always been cognizant that economic inequality is spatially expressed and spatially variable; in other words, some places are more unequal than others, which leads to important inequality at the intra-regional and intra-national scale (Kodras, 1997; Sassen, 2001). The first systematic mappings of the areal differentiation of inequality began with the Chicago School of Sociology and its concentric zone model. They used a variety of social indicators and measures of social deprivation (disorganisation); when an unequal pattern, most of the places understood as 'problem areas' were confined to the zone in transition around the Central Business District. It was in this zone in transition where a striking pooling of crime, vice, mental ill-health or low income could be observed. In recent times, economic inequality remains the single strongest determinant of spatial variations in international rates of imprisonment, as it does for mental health and social mobility (Dorling, 2014).

Kodras (1997) went beyond mapping of (uneven) patterns to see how place and inequality constitute each other through both dominant market forces and the secondary, potentially offsetting impacts of the state. In other words, the market inherently generates inequality through differential trading agreements and wage structures between the Global North, East and South, and the state can then decide whether to implement certain measures – such as minimum wage laws for low-earners, and higher taxation for high-earners – that will minimise overall inequality, but not eliminate it entirely. For instance, it has been noted that most oil-producing nation states, such as Venezuela and Nigeria, tend to have high levels of inequality because oil wealth rarely trickles down to most citizens, especially if a government is corrupt and beholden to transnational petroleum companies. Inequality is dependent on local and national conditions, creating a patchwork of affluence and poverty, with each different place representing a distinct distillation.

A more contemporary example of how inequality is spatially expressed relates to opportunities in life, or life chances. We contend that the more unequal a society is, the more geography becomes important. If you live in a nation where there is little or no inequality, then it does not matter where you live or were born or educated, because opportunities will not vary that much across geographical space, and people will get ahead more on their natural talents and hard work than any sort of advantages bestowed at birth (e.g. a meritocracy). However, in a highly unequal society, place can make a huge difference – between being born, growing up and educated in a well-off area with great amenities, schools and neighbourhoods with the 'right kind of people' and networks, and between being born, growing up and educated in a very deprived area with poorly resourced schools, high unemployment, social isolation and few obvious links to what is known as social mobility. Social mobility is the ability of the lesser well-off to move up the ladder, usually based on a combination of better educational and work opportunities. Social mobility has of late flagged, especially when compared to the 1950s–1970s in the Global North, when millions of formerly working-class people were able to obtain free (social) housing, free university, free medical care and an expansive economy for the first time in history. To sum up, place seems to have an important role in cumulating, reinforcing and perpetuating inequalities over time. Place can consolidate and sediment inequality: disadvantage can become intergenerational through spatial and social exclusion. Conversely, advantage can be sustained through the simple chance of being born in the right place (Smith, 2000; Kearns and Reid-Henry, 2009).

We can begin to illustrate the spatial expression of inequality using the measure of life expectancy at birth, which captures something perhaps less clear than the Gini coefficient, with the assumption being that highly unequal nations will have lower life expectancies (or at least, drag down the average).

We can see that the lighter blues (which indicate shorter life expectancy) are mostly in the Global South, while the darker blues indicating longer life expectancy are mostly in the Global North and Global East. The longest life expectancy is South Korea and Japan, at around 84 years, while the shortest tend to be in sub-Saharan Africa (in the 50s), which is lower than the UK's life expectancy in the immediate post-war period. How unfair is that? Just

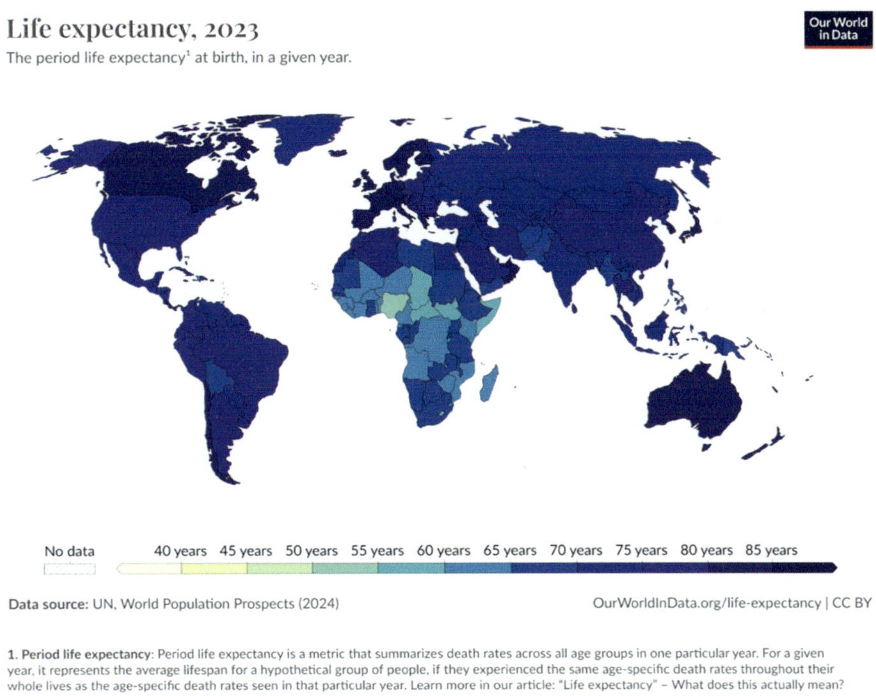

FIGURE 2.1 Worldwide Life Expectancy, 2023

Source: Our World in Data (2023) Saloni Dattani, Lucas Rodés-Guirao, Hannah Ritchie, Esteban Ortiz-Ospina, and Max Roser (2023) – "Life Expectancy" Published online at OurWorldinData.org. Retrieved from: 'https://ourworldindata.org/life-expectancy' [Online Resource]

because you're born in Japan you will likely live more than 30 years longer than someone in Nigeria, on average at birth.

The Gini coefficient mentioned in Section 2.3 has served as a useful tool for geographers to compare income inequality across countries. Which countries do you think are the most 'equal' and 'unequal'? In the table below, we show Gini scores from the World Inequality Database, for 2023:

As noted, these trends resonate across the Global North and South. But it is across the latter sphere where we see the most unequal countries in the world. These have tended to be low- and medium-income nations (but often rich in natural resources) in the Global South. These include the United Arab Emirates, South Africa and Brazil, rather than very poor nations in Africa or Asia that have limited amount of income inequality given the relative lack of well-off inhabitants. Latin America has many of the most unequal societies in the world, with stunning wealth mixed in with desperate poverty. The images below capture some of this inequality as it is expressed spatially, and vertically – many of the informal settlements are found on steep hillsides, rather than on flat land. In both Lima and Rio de Janeiro, we can see the juxtapositions between rich and poor, the formal and the informal and the established with the precarious.

Why are there such deep-seated (class) inequalities? These began at colonisation and first contact with Western powers (e.g. Spain and Portugal, respectively) which led to the subjugation of the indigenous peoples as well as, in the case of Brazil, the importation of African slaves. These original sins were baked into the foundations of many Latin American nations, and ensured a perpetual mal-distribution of resources, particularly land, even though the ancestors of the colonisers are now a minority in both cities. These issues will be explored further in Chapter 3 (citizenship) and Chapter 11 (housing).

Inequality captures the fact that many societies in the Global North mirror those in the Global South – they move at two speeds. The elites move quickly, picking up opportunities, languages, accents, education,

TABLE 2.1 Range of global Gini scores 2023

	GINI SCORE 2023, based on pre-tax national income	RELATIVE RANKING
Norway	0.33	low
Japan	0.36	low
France	0.46	low
Canada	0.50	medium
China	0.57	medium
United States	0.59	medium
Brazil	0.69	high
South Africa	0.75	high

Source: World Inequality Database. World – WID – World Inequality Database

FIGURE 2.2 Lima, 2022 © G DeVerteuil

FIGURE 2.3 Rio de Janeiro, 2006 © G DeVerteuil

networks, while the rest increasingly languish at a much slower speed, lack these same opportunities and become resentful against the larger system, politicians, immigrants, and so forth. For instance, French society has always been quite hierarchical, despite claims of 'liberty, equality, and fraternity'. The elites reserve the best areas, the best schools, the best opportunities for themselves, even in the absence of a private school system or universities. Many of the frustrations that come out of this monopolisation of power and influence will be further explored in Chapter 4 (the welfare state, which aims apparently to counter this frustration), Chapter 8 around poverty and Chapter 11 around housing.

2.5 Conclusions

The chapter builds on this premise and sets out a framework for understanding how our identities are each shaped by our internal, embodied characteristics and desires, and externally through societal norms and expectations. These are also hardwired into the institutional contexts surrounding the geographies of production and social reproduction, which can set guard rails around people's opportunities and spatial horizons. As we saw, these inequalities extend out to the global scale with spatial expressions evident across the spectrum of social life. Sometimes these types of encounters are hidden and embedded in the institutions of society, including education, health and social care. An important principle to hold onto at this stage in the book though is that these inequitable barriers are potentially remediable and frequently are faced and challenged, which we return to in Part III. To help build on the themes explored in this chapter, we next turn to the political role of the state and explore how differences in how we are and *ought to be* are recognised and controlled.

Further Reading

Cheshire, J. (2012) Lives on the Line: Mapping life expectancy along the London Tube network. *Environment and Planning A*, 44, 1525–1528.

Elden, S. (2001). Politics, philosophy, geography: Henri Lefebvre in recent Anglo-American scholarship. *Antipode*, 33(5), 809–825.

Goh, S. (2017) Rapid response to immorality of inaction on inequality. *BMJ,* 356, j556. www.bmj.com/content/356/bmj.j556/rr-8

References

Bourgois, P. (1995) *In Search of Respect: Selling Crack in El Barrio*. Cambridge: Cambridge University Press.

Coaston, R. (2019) *The Intersectionality Wars*, Vox. www.vox.com/the-highlight/2019/5/20/18542843/intersectionality-conservatism-law-race-gender-discrimination

Crenshaw, K. (1989) *Demarginalizing the Intersection of Race and Sex: A Black Feminist Critique of Antidiscrimination Doctrine, Feminist Theory and Antiracist Politics*, University of Chicago Legal Forum: Vol. 8, (1). http://chicagounbound.uchicago.edu/uclf/vol1989/iss1/8

DeVerteuil, G. (2015) Conceptualizing violence for health and medical geography. *Social Science and Medicine*, 133, 216–222.

Fraser, N. (1995) Recognition or redistribution? A critical reading of Iris Young's justice and the politics of difference*. *Journal of Political Philosophy*, 3, 166–180

Fraser, N. (1998) *Social Justice in the Age of Identity Politics: Redistribution, Recognition, Participation*. London: Sage Publications.

Goffman, E. (1963) *Stigma: Notes on the Management of Spoiled Identity*. New York: Penguin.

Goffman, E. (2009) *Stigma: Notes on the Management of Spoiled Identity*. 2nd edition. New York: Touchstone.

Gough, J. (1992) Workers' competition, class relations, and space. *Environment and Planning D: Society and Space*, 10(3), 265–286.

Graham, S. and Marvin, S. (2001) *Splintering Urbanism: Networked Infrastructures, Technological Mobilities and the Urban Condition*. London: Routledge.

Harvey, D. (1996) *Justice, Nature and the Geography of Difference*. Cambridge MA: Blackwell.

Hopkins, P. (2019) Social geography I: Intersectionality. *Progress in Human Geography*, 43(5), 937–947.

Kearns, G. and Reid-Henry, S. (2009) Vital geographies: Life, luck and the human conditions. *Annals of the Association of American Geographers*, 99(3), 554–574.

Knox, P. and Pinch, S. (2010) *Urban Social Geography*. London: Taylor and Francis.

Kodras, J. (1997) The changing map of American poverty in an era of economic restructuring and political realignment. *Economic Geography*, 73(1), 67–93.

Massey, D. (1993). Power-geometry and a progressive sense of place. In J. Bird, B. Curtis, T. Putnam, G. Robertson, and L. Tickner (Eds.), *Mapping the Futures: Local Cultures, Global Change* (pp. 59–69). New York, NY: Routledge.

Massey, D. (2004) Geographies of responsibility. *Geography Annals*, 86B(1), 5–18.

Our World in Data (2023) *Worldwide Life Expectancy, 2023.* https://ourworldindata.org/grapher/life-expectancy-unwpp

Omi, M. and Winant, H. (1994) *Racial Formation in the United States*. London: Routledge.

Power, A., Coverdale, A., Croydon, A., Hall, E., Kaley, A., Macpherson, H. and Nind, M. (2021) Personalisation policy in the lives of people with learning disabilities: A call to focus on how people build their lives relationally, *Critical Social Policy,* 1. doi: 10.1177/02610183211004534

Sassen, S. (2001) *The Global City*. Princeton, NJ: Princeton University Press.

Smith, D.M. (1984) Alternative perspectives on 'urban inequality'. *Geoforum*, 15(1), 75–82.

Smith, D.M. (1987) *Update: Apartheid in South Africa*. Cambridge UK: Cambridge University Press.

Smith, D.M. (1994) *Geography and Social Justice*. Oxford: Blackwell.

Starfield, B. (2011) The hidden inequity in health care. *International Journal for Equity in Health*, 10, 15. www.equityhealthj.com/content/10/1/15

United Nations (2015) *Concepts of Inequality*, Department of Economic and Social Affairs, UN. www.un.org/en/development/desa/policy/wess/wess_dev_issues/dsp_policy_01.pdf

Valentine, G. (2007) Theorizing and researching intersectionality: A challenge for feminist geography. *The Professional Geographer*, 59(1), 10–21.

Weiss, H. (2021) Social reproduction. In F. Stein (Ed.), *The Open Encyclopedia of Anthropology*. Facsimile of the first edition in The Cambridge Encyclopedia of Anthropology. Cambridge, UK. http://doi.org/10.29164/21socialrepro.

While, A., Jonas, A. and Gibbs, D. (2004) The environment and the entrepreneurial city: Searching for the 'urban sustainability fix' in Manchester and Leeds, *International Journal of Urban and Regional Research*, 28(3), 549–569. doi: 10.1111/j.0309-1317.2004.00535.

Worth, N. (2015) Youth, relationality, and space: Conceptual resources for youth studies from critical human geography. In J. Wyn and H. Cahill (Eds.), *Handbook of Children and Youth Studies*. Singapore: Springer. https://doi.org/10.1007/978-981-4451-15-4_8

Young, I.M. (1990) *Justice and the Politics of Difference*. Princeton NJ: Princeton University Press.

3 THE STATE, CITIZENSHIP AND RIGHTS

3.1 Introduction

In tracing the lineage of geographic scholarship which underpins this chapter, we draw on a second, but related, discrete strand of our discipline: political geography. While there are obvious divergences from social geography (in terms of its preoccupation with boundaries and subdivisions of political units), one notable point emerges – both bodies of social and political geographic scholarship cite and engage with each other, in terms of how political decisions affect everyday life and vice-versa. We thus begin with the conceit that just as everything is social (and spatial), everything is also political (Painter and Jeffrey, 2009). For students who may be inclined to feel disengaged in matters they deem 'political', there are two counterpoints to consider. First, some people who may feel this way towards 'politics' nonetheless make daily conscious decisions which can inadvertently help determine political orientations, e.g. shopping sustainably, caring for one's family and friends, raising debate among ones' peers and family, helping one's communities. We might call these activities 'everyday politics', which we return to in part three of this book. Second, some of this feeling may be more about feeling disenfranchised or disgruntled about politics, rather than political apathy per se. People who sometimes feel disempowered or 'weary' (Wilkinson and Ortega-Alcázar, 2019) about the shape of the world, and their circumstances, may have a justified reason for feeling alienated and discontent.

For those who seek to remain unmindful of either of these forms of engagement, we contend that an active choice to be 'apolitical' is usually an endorsement of the status quo. In this case, we encourage people aligning to this outlook to approach this book with an openness and curiosity, and hopefully by the end of the book, you will at least feel more informed and have gained an appreciation of the geographies of social justice. In this chapter at least, the focus is on how social justice is manifested through political means, making demands on the state.

The thread of relationality is very much embedded in this chapter, in particular between the individual and the state as a linchpin of a functioning society. An individual's relationship with the state is underpinned by the

DOI: 10.4324/9781003414889-4

concept of *citizenship*, which as we explore below determines the relationship between individuals, the state and the wider society, through the balance of rights and responsibilities. One of the most powerful tools people have used to better recognise and respond to social issues has been to seek or call on the state for rights – and to use these rights to hold the state, private sector and individuals to account. As this chapter shall explore, various types of rights have emerged which have each sought to achieve different political, social or economic aims. Yet, McCann (2006) reminds us that rights are not all inherently positive, as they can be equally used to support more powerful private interests. So, it is not enough to advocate for rights; it is also important to ask whose needs are being met.

Rights are contingent on or balanced with people's responsibilities to the state and wider society, for example to pay one's taxes, vote in elections or to not break the law. Yet, as we shall discover, these responsibilities may deviate into moral obligations of how certain people *ought* to live their lives. As such, they can be defined by societal norms, which can at times protect the status quo, and thereby inadvertently replicate social inequalities. Consequently, social movements which have sought to advocate for new or amended rights typically are met with challenges which seek to stymie change.

The chapter begins with an overview of the role of the state and governance. We outline a historical focus on the early adoption of rights, from more individual rights such as civil liberties to more collective rights such as social rights. While this history spans several centuries and draws on examples from different countries, a key takeaway is that each right emerged in response to a specific historical call for reform and to give a sense of how hard fought each right has been. The different *blend* of rights that each country enshrines provides us the basis for determining a geography of rights. As noted, this blend is typically balanced with a set of responsibilities, which together define people's citizenship. While this blend naturally differs by country, we propose three conceptual framings for understanding citizenship throughout the chapter. First, *social citizenship* is a term to characterise a form of citizenship marked by the favouring of collective social rights and the state's role in ameliorating inequalities. Second, *neoliberal citizenship* characterises a form of citizenship arguably most prevalent today, where the axis has shifted towards individual responsibilities and management of the poor over collective rights. Finally, *lived citizenship* serves as a concept for making sense of the ways that citizenship is lived and felt in everyday life. While the first two concepts provide a structural, political-economic framing to define specific forms of citizenship, the latter provides the basis for understanding the ongoing social and spatial relationship with the state and how it shapes people's lives in a more embodied, daily sense. The chapter ends with an exploration of the spaces of exception that can persist in spite of rights, where groups still occupy positions at the margins of citizenship, and where the politics of recognition are crucial for bringing these spaces to light.

3.2 The State and Governance

At the heart of determining how rights operate is the consideration of defining the state. When thinking of the state, the most obvious manifestation is government, and more specifically parliament, where government ministers create laws and policies. In most democratic countries, the state is underpinned by a doctrine of the separation of powers, which requires that the principal institutions or 'branches' of the state – legislative, executive, and judicial – be clearly divided in order to help keep these three branches independent, accountable and ultimately safeguard people from undemocratic rule and forms of tyranny (e.g. a dictatorship). The legislative branch (e.g. congress or parliament) makes laws through debating, amending and passing bills, the Executive branch (e.g. President or Prime Minister and cabinet ministers) executes laws, develops and implements policies and runs the day-to-day operations of the government and the Judicial branch (e.g. courts of law) administers justice by interpreting the law, especially when its meaning is in dispute, and ensuring it is upheld. This offers a useful starting point in disentangling what we mean by the state.

Another obvious way of thinking of the state (for geographers at least) is the 'bricks and mortar' associated with state or 'public' land and buildings, expressing the materiality of the state (see Chapter 4 on social housing). Such infrastructure can include central government buildings, local authority offices, prisons and schools and geographers have indeed mapped and analysed how the various locations (and positioning vis-à-vis local populations) can shape citizen's experiences. Where public health care is available, geographers have also sought to examine the location of primary health care facilities, hospitals and community health facilities (Gattrell and Elliot, 2014). They have more broadly considered health as a centralised system like the UK (although private health care is available), or mostly privatised like the US, or a semi-state/semi-privatised entity like Ireland and Australia.

The state also encompasses the people who work in roles that have statutory duties, for example local state commissioners, the armed forces, police officers and teachers. Medical general practitioners and social workers are also subject to professional and statutory obligations in relation to safeguarding. Even staff working in public libraries are subject to working practices and professional obligations defined by the state, e.g. to provide a comprehensive and efficient library service for all persons (Hitchen, 2021). In Hitchen's study on libraries in the UK, she explores the emotional labour of library staff in meeting the requirements of the state while simultaneously being affected by statutory cutbacks to services. Despite these tensions, each of these actors operate as duty holders and gatekeepers, and are afforded discretion about dealing with the public. From the 'commanding highs, dispersed field offices, and trenches' (Migdal, 2001), power operates in the everyday, through these various role holders. Sometimes, this power is directed at the state itself, or mediated in ways which minimise the impact of more punitive policies for citizens.

It is helpful, therefore, to think of the state *relationally*, as a sector involving a myriad of different actors and institutions, which form various material entities like statutory agencies and spaces. The division between what we consider as the state sector and other sectors (i.e. the private sector and voluntary sector) is not set in stone. The relational boundaries have blurred considerably over time. In the past few decades alone, statutory responsibilities as well as public funding (although often not to the same extent) have been increasingly outsourced to the private and voluntary sectors. Public sector contracts are sometimes put out to tender, and both private and voluntary organisations can bid and secure work in the delivery of health, education, social care and other social services (Power et al., 2021). Outsourcing of government contracts to other sectors has become increasingly common, as explored in Chapters 4 and 5, and these settlements have a bearing on how social justice can be achieved. In the instance of the 2017 Grenfell Tower fire in London, it was difficult to actually pinpoint who was responsible for what. Moreover, as Pike (2023) explores, the local state can also become involved in the financialisation of its assets, by investing public money in private financing, raising concerns over the liabilities placed on the state (echoing concerns about the 2011 bailouts to Greece, Portugal and Ireland following the global financial crisis).

A key context to the current state is *neoliberalism*. Before we can discuss *neo*liberalism (neo denoting 'new'), we should explain that its roots can be found in an earlier phase of *classical* liberalism, which developed from the late 17th to early 19th century. Classical liberalism was a period where limits were placed on the power of monarchs, who had been more autocratic than their contemporaries. As explored in Section 3.4, classical liberalism emphasised individual freedoms with a limited government under the rule of law, private property and unregulated economic policy. It led to the birth of the industrial order by freeing up landowners and industrialists to set the conditions of production without any undue interference from the state. However, this resulted in the exploitation of the working classes, which gave rise to a subsequent appetite for later rights in the mid-20th century, as we shall see below.

Neoliberalism emerged as an ideological hegemonic project in the late 1970s and informs much of the political landscape in the present day (even in the Global South), led by governments which seek to enshrine some of the core principles of its classical predecessor. Hackworth (2007: 9) characterises it as rolling back the previous expansion of government in the mid-20th century, while rolling out new modes of regulation and governance (primarily around the management of the poor). Neoliberalism rejects the previous approach to governance that expanded opportunities and minimised inequalities, by dismantling and replacing its former policies and ways of governing, and fostering a new, more self-sufficient individual in terms of lessening their relationship to the state. On this last point, we develop and illustrate a concept of neoliberal citizenship later in the chapter. Further, we understand neoliberalism as (1) more a class project that favours the already powerful than a recipe for economic growth, and (2) an intensely urban one, albeit incomplete and uneven (DeVerteuil, 2015). As Brenner and Theodore (2002: 2) contended, 'cities have become strategically crucial geographical arenas in which a variety of neoliberal initiatives – along with closely intertwined strategies of crisis displacement and crisis management – have been articulated', as well as exported (Massey, 2007). Neoliberalism involves the active critique, repudiation and dismantlement of the public domain more generally, and previous welfare state arrangements more specifically. This is the essence of 'roll-back' neoliberalism, centrally concerned with hollowing out the state through deregulation, devolution, privatisation and austerity measures (Peck and Tickell, 2002).

Nonetheless, we treat this all-encompassing narrative with a certain scepticism, underlining the incomplete and uneven realities of current neoliberalism, as May and Cloke (2014: 894) argued:

> neoliberalism itself is best understood as a dynamic and complex process; a messy and precarious fabrication of forms and practices which always remains open to different kinds of contestation and resistance, and that researchers need to be careful not to assume that the locally contextualised practices of providing welfare ... will simply mirror the neoliberal environment in which those contexts are staged.

To take this one step further, geographers such as and Peck and Theodore (2015) and Peck (2011) have encouraged thinking about the state as not solely actors and institutions, but also as a series of networks and flows – of power, beliefs, ideas, discourses, funding and responsibilities – and how these manifest in the mobility of policies (e.g. 'flagship cities'). One example from Peck and Theodore (2001) is the mobility of the policy idea of 'workfare' (a term we explore further in Chapter 4) from Mexico to the US. This offers a useful lens for understanding the geographies of social justice issues, as it allows geographers to unpick how policies are moved and re-purposed from one place to another, and how they get translated, hybridised and re-worked. The networks involved in shaping these mobilities can also provide a platform for shaping ideologies about social issues. Ideologies refer to the system of ideas and ideals, which can form the basis of economic or political theory and policy. In some cases, governments use slogans to frame and gain consent for their political ideological positions, e.g. 'we're all in this together' in the UK to signal a commitment to austerity to address their financial deficit, or Trump's call to 'drain the swamp' in the US to seek to reduce the influence of special interests and lobbyists. Here, policies are politically framed by discourses and characterisations.

Finally, geographers are interested in modes of *governance*, and how in turn this can shape social (in)justice. Governance refers to the act of governing. The most obvious ways a state governs is through its laws and policies and their enforcement through the criminal justice system including the police (and then separately through the courts). Another visible way is through its taxation system and financing of public services – including the extent to which university fees are subsidised by the state or borne solely by you as a student. Geographers have been well attuned to making critical appraisals of forms of statutory governance, and their effects on specific populations, and this forms a large part of this book. More basic are the various tools that the state uses to monitor the population, such as CCTV (Closed Circuit Television), which is especially popular in democracies like the UK, but also more autocratic societies like China and Russia. CCTV is used for many things, including the surveillance of anti-social behaviour in social housing, green spaces and city centres. More on these tactics will be covered in Chapter 6 on youth.

Less sinister are the efforts by the state to regulate and compartmentalise various resources that humans need, especially the planning of shelter and land. More specifically, Huxley (2009: 193) defines planning as:

> state-related policies and programs for neighborhood, local and metropolitan areas, aiming to: effect broad-scale allocation of land uses to areas; order boundaries between them; manage ongoing uses of land, the spatial aspects of economic and social activities and connections between them; and ensure the optimal functioning of urban economic processes and social interactions.

These activities theoretically ensure enough space and amenities for the population. This is especially important in cramped spaces of the city, where land is valuable and much-competed over.

3.3 Citizenship and the Evolution of Rights

Our relationship with the state at the most fundamental level is bound up by our citizenship status. The term citizenship and its related language of 'rights' have also become very common in popular discourses within mainstream society. Citizenship refers to the position or status of being a citizen of a particular country, which also implicates a series of state-sanctioned articulations of power: passport, borders, controls and so forth. In a formal sense, people can typically apply for citizenship if they were born in or have a parent from a host country, or if they marry to have a civil partnership with a citizen from that country. Beyond that, citizenship is gained by meeting the legal requirements of a national, state or local government. A nation grants certain rights and privileges to its citizens. In return, citizens are expected to fulfil obligations outlined by the state. Contemporary geographies of social justice are thus often concerned with the blend and quality of rights and obligations accorded to citizens at a range of scales.

Beyond this formal definition, Susan Smith (1989: 147) is credited with arguing that 'citizenship refers to the relationship between individuals and the community (or State) which impinges on their lives because of *who* they are and *where* they live'. The concept of citizenship is thus about who we are – our characteristics such as our place of birth, age, gender, relative wealth and so on – and how these characteristics determine our entitlements and positions of power in relation to the state. This process of seeking and gaining recognition by the state is described as the production of citizenship (Isin, 2008).

A person's citizenship is not fixed in stone but continually evolves and is often renegotiated over the life course. As people grow older, their relationship with the state is likely to change, as they may become entitled to state benefits such as a travel pass and a state pension, for example, as well as require more medical or social care. The image directly below from the late 1990s shows the fluidity around citizenship in the popular imagination in Los Angeles, where Geoff was living at the time.

The image speaks to the increased scrutiny around non-citizens (as there is now, at the time of writing in 2025, amid the generalised hysteria around immigration in many countries including the US), and so the everyday language shifted. More specifically, the blacking out of 'citizen' and inclusion of 'over 65' in the sign illustrates a fear expressed that some clients would worry that getting a haircut required American citizenship, rather than simply being over 65.

Physical barriers and walls expose the materiality of the state, and constitute a more forceful dividing point between different citizenships, for example 100 miles to the south, there is the hard border between the US and Mexico (Figure 3.2). These walls, built in the 1980s and 1990s between Mexico and California, are designed to keep certain people out. And yet, a very significant amount

FIGURE 3.1 Los Angeles, 1997 © G DeVerteuil

FIGURE 3.2 Tijuana, Mexico, 1998 © G DeVerteuil

of goods and services flows across the border every day, to the point where the two places are effectively fused economically but certainly not politically or socially, and less so under the current Trump administration.

Smith's quote also reminds us that as well as citizenship hinging on who we are, this relationship between individuals and the state is further shaped by our geographical location. Where we live can reinforce these dynamics, as the relationship with the state is bound by local variation in customs and levels of political investment in a region. For example, for people growing up in a deprived neighbourhood, where there are limited jobs and opportunities, their prospects may be contingent on securing private or public investment in education or public assistance. In federal countries, like the US and Canada, these variations are more acute as local states/provinces can govern police powers, health, education, and welfare. In some countries, people seeking public services may also be bound by their geographical region of origin. In Mainland China for example, public services are conditional on resident registration, known as 'Hu-Kou', where individuals are only allowed to apply for welfare benefits in their registered place of origin, typically their birthplace or family residence. This maintains civil control by ensuring people exercise their permanent familial welfare responsibility by looking after their relatives in their local communities. For homeless people, this means that they must either return to their Hu-Kou region or rent an apartment initially to re-register their address in the new city before seeking any homeless welfare entitlements (an irony for homeless people).

Understood in this sense, social welfare policies and practices are determined by what has been called the *citizenship regime* (Evans, 2011).

A citizenship regime according to Evans is a temporarily stabilised set of narratives and institutional practices that define the boundaries of who belongs and who does not in a society – what we call social citizenship. These regimes differ according to the levels of responsibility they attribute to the state, market, family and community for social welfare; the extent of state recognition of rights and duties; the types of democratic rules and mechanisms for making political claims; and definitions of identity and membership that underpin feelings of belonging (Evans, 2011). Welfare practices reflect norms that prescribe who is to be protected, how and by whom, norms established by the prevailing citizenship regime.

Specific groups have historically been excluded from having a sense of full citizenship, including women, Black or ethnic minoritised people, LGBTQ+ people and disabled people. We explore some of these populations in Part II of this book. Geographers have examined access to citizenship from a variety of different scales. Some have explored the implementation of international conventions, e.g. the UN Convention for the Rights of Persons with Disabilities (Power, Lord and DeFranco, 2013). Others have sought to map rights across a country, e.g. transgender rights in the US (Nolan et al., 2023). Geoff and Andy have also mapped and explored how rights are mediated across sectors, e.g. the role of the voluntary sector in offering a safety net in lieu of social rights (DeVerteuil et al., 2020). Finally, some have examined rights in our everyday life (Staeheli, 2012). At this scale, questions of social justice hinge on the distinction between formal citizenship and informal citizenship. Although legislation can be enshrined to support equality, there is often a gap between the aspirations of legal rights (de jure citizenship) and how these manifest in daily life (de facto citizenship) (Smith, 1989). Citizenship can be assessed relative to the sets of rights and obligations conferred through 'national constitutions and realised or denied through the operation of markets, institutions and informal social relations' (Ibid: 152). We return to this theme of everyday lived citizenship in Section 3.6.

3.4 Geographies of Rights

Box 3.1 Engagement Questions

What **rights** can you think of?
What purpose do rights serve?
Can rights be grouped in any way?
As well as rights *from* the state, can you think of any **responsibilities** we have *to* the state and society?

Citizenship rights are typically grouped into five types which have evolved over time: human rights, civil liberties, political rights, social rights and cultural/civil rights. This section covers the first four of these, with the fifth covered separately in the next section. This loose classification and evolution of rights was first articulated by sociologist, Thomas Humphrey (T.H.) Marshall, an English sociologist who is best known for his essay 'Citizenship and Social Class' (1950), and subsequently expanded upon by Susan Smith (1989) and others. This section provides an account of the historical evolution of these rights from the 1800s to the present day. As noted, a key take-away from this section is that each strand of rights has been hard fought for and for specific reasons – often emerging as an outcome of calls to recognise individual or more collective interests.

Human rights are often regarded as the basic needs that humans need to survive, for example shelter, food and freedom from torture. These are often protected in international treaties, such as the United Nations Convention on Human Rights, which recognise the inherent dignity and autonomy of all people. International organisations such as the United Nations and non-governmental agencies such as Amnesty International seek to hold nations accountable for their human rights legislation. Labour or 'workers' rights emerged as a subset to the corpus of human rights, encompassing a large array of rights from the right to work, prohibition of all forms of forced labour (e.g. human trafficking) and the right to organise (e.g. join trade unions and negotiate with employers collectively). One of the first of these was the Factory Act 1833 passed by the British Parliament, which stated that children under the age of 9 could not work, children aged 9–13 could only work 8 hours a day, and children aged 14–18 could only work for 12 hours a day.

Civil liberties comprise a second corpus of rights that include the freedom of speech, freedom of religion, the right to privacy, the right to a fair trial and the right to protest. We dwell on this corpus of rights a bit longer as they are regarded as the prototypical form of *individual* rights (serving as the foundations of classical liberalism as noted in Section 3.2), and as we shall consider later in this section, are often held in opposition to social rights. Civil liberties are freedoms guaranteed to us by constitutional or common law to safeguard our autonomy and protect us from undue interference by the state. Collectively, these rights are concerned with the freedom from arbitrary interference in one's pursuits by other individuals or by government. These rights initially emerged as a direct response to previous unrepresentative modes of governing, which infringed on individual freedoms. A key example is the early forms of (autocratic) monarchy, briefly noted in Section 3.2, found across Europe and pre-Soviet Russia, which could block what they perceived as unacceptable politically-elected officials. Across Europe, the doctrine of the divine right of kings and queens asserted that a monarch is not accountable to any earthly authority such as a parliament (or the Pope in countries where Catholic leadership was dominant) because their right to rule is derived directly from God.

Civil liberties began to evolve as an outcome from the challenges to this earlier form of governing. The idea of a social contract between individuals and the state began to be articulated by early proponents of civil liberties.

One of the early vocal contributors was Thomas Hobbes (1651). Hobbes developed some of the fundamentals of (European) liberal thought: the right of the individual; the natural equality of all men (sic); and the artificial character of the political order which led to the later distinction between civil society and the state. Hobbes also believed that all legitimate political power must be 'representative'. However, Hobbes maintained his belief that we should subject ourselves to an absolute sovereign, preferably a monarch, to gain security in return. His justification was that one absolute and central authority is the best method of maintaining social order and avoiding discord and factionalism. In this foundational text, Hobbes imagines a time prior to the creation of social institutions, when humans were motivated solely by satisfying their own desires, which he argued had contributed to conflict and a state of perpetual war with people fighting to obtain what they need. Present-day examples are when oppressive regimes fall and are replaced by political instability and periods of lawlessness such as in Libya in 2011.

John Locke is credited with taking civil liberties one step further in his publication, *Two Treatises of Government* (1689). Locke's First Treatise was a flat rejection of the divine right of kings. His second was that each person is naturally free and equal under the law of nature, with common consent given to a civil government with the power to enforce the law of nature among the citizens of that state. This provided a timely endorsement of the Glorious Revolution of 1688 in Britain which created a 'Bill of Rights' which established a constitutional monarchy (a figurehead with limited powers), also known as a parliamentary monarchy. This proved a key moment which underpinned the extension of civil liberties. Naturally, this took time to implement, but the sovereign's influence on the choice of government gradually declined from this period. It also had a significant impact across Europe and America. For this reason, Locke is regarded as the father of classical liberalism due to his role in championing civil liberties. A key takeaway here is that civil liberties were justified on the basis of challenging earlier oppressive governance. Moreover, these earlier historical periods are often invoked when modern-day proponents fear that these rights are being curtailed, in sometimes excessive ways which disregard other collective rights or responsibilities, as we shall see below.

Political rights arose as an extension of civil liberties, where people sought to expect representation from government but to also allow democratic participation in the political system, either through the right to vote, including by referendum, to stand for election or to sit in parliament. This became known as Suffrage, from the Latin word *suffragium*, which means 'a vote', or 'the right to vote'. Another common term associated with voting is *franchise* which comes from the French word *franchir*, which means 'to free' or 'to cross'. Among the first nations that extended suffrage included France (1792; men only), Brazil (1824) and South Australia (1855). The Bill of Rights 1689 in England provided limited voting rights to landowning free men over 25 years of age. As explored in Chapter 12, women's right to vote was only extended much later in many countries and again was hard-fought. Other intersectional restrictions included race, religion and nationality.

Social rights emerged as a different breed of rights to those covered above. They largely emerged in the early to mid-20th century, although their roots can be found in earlier social movements that sought to challenge the looming casualties of the major industrial advancements and related urbanisation that took place from the 1800s up to the early 1900s. This was a period where 'laissez faire' government policy had taken precedence. Laissez-faire policy meant leaving things to take their own course, without interfering; this gave industrial developers, factory and mill owners complete autonomy in determining pay and conditions, and left individual workers to fend for themselves. It also marked the end of an era of public assistance, and gave rise to a harsher system of workhouses. Consequently, in many emerging industrial nations, this left working-class people under-educated, under-paid, frequently living in deprivation and squalor and facing poor physical and mental health. In 1840, for example, 57% of children living in Manchester died before they were 5 years old (National Archives, n.d.). The industrial development and political vacuum also shaped the geographies of cities, which experienced rapid urbanisation, and became characterised by overcrowded and poorly built housing, piles of rubbish and waste on streets, and poor drains and sewers.

The social rights which emerged were largely focused on addressing these inequalities underpinned by stark class divisions in society – and to address the 'nuisance' of urban poverty. Social rights were first enshrined by social reformers of the day, largely informed by Marxist thought. Political parties began to emerge out of the trade union movement and urban proletariat parties of the 19th century. Examples of social rights include mass education in the way we would recognise it today (from 1902 in England) and included later efforts to widen participation to compensate for disadvantages due to inequities in the social system (e.g. the US Elementary and Secondary Education Act of 1965 sought to equalise the education opportunities for all children). It also included mid-20th century public health and social security/welfare provision which sought to address the economic insecurity and housing and health care needs of the population. The products of these rights are explored in more detail in the next chapter.

T.H. Marshall defined social rights as being 'the whole range from the right to a modicum of welfare and security to the right to share in the full social heritage and to live the life of a civilised being according to the standards prevailing in the society' (Marshall, 1950: 10–11). The state no longer merely attempted to 'abate the obvious nuisance of destitution in the lowest ranks of society. It assumed the responsibility of modifying the whole pattern of social inequality' (Marshall, 1977, cited in Isin and Wood, 1999: 29).

Social rights helped to ameliorate the worst excesses of laissez-faire classical liberalism and to provide a basic safety net for citizens. However, social rights are argued by Drewett 'not to be "rights" at all, but instead are akin to party manifestos' (Drewett, 1999: 116–117), a statement that warrants some further consideration. By party manifestos, Drewett meant the political promises made by presidential or government party candidates during periods of election. During these periods, some candidates eschew the value of social rights and herald aspirational principles in proposed social policies, but do not necessarily adhere to these during their governance. Moreover, some populist right governments seek to undermine social rights in their manifestos and zealously protect their civil liberties over other collective rights or responsibilities.

Box 3.2 Engagement Questions

What did Alison Drewett mean by the statement 'social rights are argued by some not to be 'rights' at all, but instead are akin to party manifestos' (1999: 116–117)?

Why might it be easier to protect civil liberties than social rights?

Drewett's justification for this statement centres on the underlying premise that social rights are depicted as involving high levels of state expenditure and intervention, thus political promises are often constrained by budgetary concerns and curtailed due to prioritising of individual freedoms. Subsidising health care, unemployment assistance, education and long-term care to citizens does require the state to endorse taxation and invest in public services, and can be a relatively costly exercise. Social rights are also framed by different interpretations of social justice and perceptions of fairness over the distribution of wealth, with varying positions between individual self-sufficiency and collective consumption. By contrast, human, political rights and civil liberties cost the state relatively little money, beyond investments in police, prisons and elections. As the next section explores, these in-built tensions have continued to play out since, and from the 1960s, a call for a fifth type of rights began to emerge: '**civil rights**' (as opposed to 'civil liberties').

3.5 Civil Rights Emerge Amid Contested Geographies of Citizenship

While social rights were largely focused on addressing the inequalities associated with class divisions in society during the early 20th century, in this section, we explore the geographies of citizenship which have unfolded since the late 1960s and early 1970s, when large-scale protest and social movements began to call for a fifth type of right, referred to as civil rights. These also became known as 'group rights' or 'cultural rights'. These rights emerged as part of a more discernible shift in how specific disenfranchised population groups began to challenge many of the tradition-bound social norms of the 1940s and 1950s. In particular, different groups called for the right to have access to and occupy space on an equal basis with others. As this section explores, some geographers developed the concept of the *right to the city*, to highlight this spatial aspect of group rights.

Civil rights are the legal rights that protect individuals from discrimination (e.g., employment discrimination, barriers to public accessibility). Civil rights concern the basic right to be free from unequal treatment based on certain 'protected characteristics' (race [including colour, nationality, ethnic or national origin], gender, sex, disability, age, marriage or civil partnership [in employment only], sexual orientation, pregnancy and maternity and religion or belief) in settings such as employment, education, housing, access to public facilities and voting. In the US, most of these rights were enshrined with the Civil Rights Act of 1964. A year later, the 1965 Voting Rights Act became a landmark piece of federal legislation that prohibits racial discrimination in voting. This was a time in the US, when segregation based on race was still legal; it was also illegal to have gay intercourse, and when women were routinely bound to the home.

These rights were also driven by growing uncertainty with some of the ideals of modernity, specifically the belief that 'modern' government techniques, rational thought and science could improve humanity. This had captured the public's imagination for much of the earlier 20th century and underpinned the advance of nuclear power, the 'space race', urbanisation, nation state building and a belief in progress through science. However, following the Second World War, the Cold War and the subsequent decline in the long boom of Fordism (e.g. a system based in mass production that underpinned a general rise in living standards, a decline in inequality, and a more socially interventionist state), the public mood was shifting away from modernism, with a sense that it had strayed from its idealistic origins and had developed darker undertones. This scepticism led to a rejection of the grand narratives of modernity and a growing recognition of the influence of ideology in maintaining political power. It also led to a scepticism of romantic capitalism which had characterised the 1950s – the belief that society could be improved through the aestheticisation of consumption (Knox and Pinch, 2010), although this has arguably returned in a new form with online consumerism.

This shift in public mood was characterised by an increase in questioning of authority and for those affected, the institutional racism it endorsed. This sparked a rise in civil unrest and protest in the Global North among those who felt they had never gained full citizenship. Examples included the civil rights movement calling for the rights of Black Americans, the disabled people's Independent Living Movement and the women's movement. There were also wider questions over the capitalist system in some countries and how it was undermining workers' calls for better working conditions, higher salaries, earlier retirement and union representatives within the factories. This was seen in the Paris students' uprising and a general strike which involved some 10 million workers in 1968. This was in response to powerful forces in Paris which were forcing the working classes to vacate the central city to live in suburban estates.

While this uprising was playing out in Paris, an influential French sociologist Henri Lefebvre published a book that same year, *Le Droit à la Ville* (1968), translated as *The Right to the City*. Concurrent to Harvey, but having impacts on social geography only later in the 1990s, Lefebvre's notion of the *right to the city* claimed that all groups who live in the city ought to have access to, if not the right to occupy, urban space on an everyday basis. For Lefebvre, the right to the city was a spatial expression of ordinary, everyday citizenship, explored in Section 3.6. More to the point, social justice was both a distributional and procedural matter, of who gets what but also who could participate in those decisions. The endpoint for Lefebvre would be a socially just place-making. So for Lefebvre, (urban) social and spatial justice were intimately linked – the call for social justice cannot proceed if there is no space from which to make it, no space to sustain it: 'the right to the city manifests itself as a superior form of rights: right to freedom, to individualization in socialization, to habitat and to inhabit. The right...to participation and appropriation (clearly distinct from the right to property), are implied in the right to the city' (cited in Mitchell, 2003: 18). This spatial aspect to social justice resonated with many radical geographers, especially Don Mitchell. To Mitchell, creation of and access to public space was essential for the rights to the city, as it anchored an alternative public sphere distinct from the pervasive privatisation of urban space. More specifically, Mitchell chronicled how anti-homeless campaigns against public urination, sleeping on the street and panhandling (begging) have made it virtually impossible for the homeless to exist in prime urban spaces.

The following images, both in French (Paris and Montréal respectively), speak to the need of local inhabitants to the right to the city, appropriating the city. This could be done through *autogestion*, which is a French word that translates to 'workers' self-management in English. It refers to the control and management of an enterprise by its workers, often implying a collective and democratic approach to decision-making. It has been used as a term in French-speaking cities to characterise efforts by local neighbourhood groups to reclaim some control perceived to be lost through gentrification (see Chapter 11 for more information on this process). Crucially, both images speak to the importance of the local right to the city, yet still open to and embedded within the larger city. Both of these images speak to the importance of how the right to the city is *practiced* in terms of the production of space.

Translation from the French in Figure 3.3:

> *LET'S DISCUSS TOGETHER ABOUT OUR CITY, OF OUR NEIGHBORHOOD. LET'S CREATE CONNECTIONS THAT WILL ENABLE US TO LIVE TOGETHER. Build on what exists already. Protect*

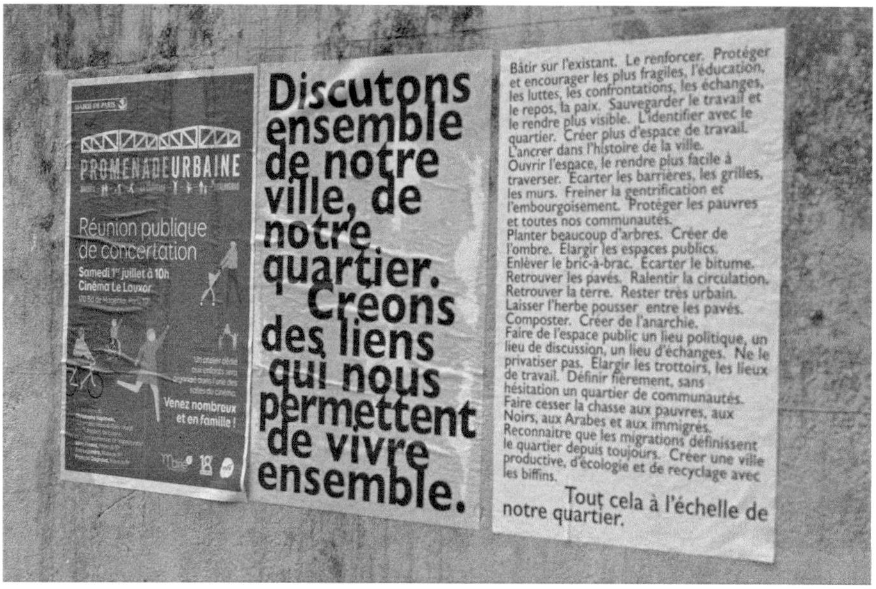

FIGURE 3.3 Paris, 2017 © G DeVerteuil

> *and encourage the most fragile, education, struggles, confrontations, exchanges, rest, peace. Safeguard work and make it more visible, while connecting it to the neighbourhood. Create more space for work, and anchor it to the history of the city. Open up space to make it easier to cross. Move aside barriers, fences, walls. Put a brake on gentrification. Protect the poor and all in the community. Plant more trees. Make shade. Expand public spaces. Remove odds and ends. Take out asphalt. Rediscover the sidewalk. Slow down traffic. Rediscover the ground. Stay urban. Let the grass grow on the pavement. Compost. Create anarchy. Make public space into a political space, a site of discussion and exchange. Do not privatise them. Widen the sidewalks, spaces of work. Define neighbourhoods as communities. Stop hunting the poor, Blacks, Arabs and immigrants. Recognise that migration helps define neighbourhoods, and since a long time. Create a productive, ecological city. And all this at the scale of the neighbourhood.*

Translation from the French in Figure 3.4 (overleaf): *The library is ours. Gentrification versus self-management.*

The two images help convey a wider point – that the grounds of citizenship are contested and must be continually negotiated amid wider structural forces. This became clearly evident soon after the 1960s' wave of civil rights that were enshrined. Despite constitutional and legal measures aimed at protecting the rights of racial minorities in the US, the private beliefs and practices of many Americans took longer to change and meant racism persisted. This was illustrated soon after the 1964 Civil Rights Act

FIGURE 3.4 Montréal, 2019 © G DeVerteuil

by the Detroit Uprising in 1967, which was sparked by a police raid of an unlicensed, after-hours bar, known as a 'blind pig', on the city's Near West Side. The bar was the site of a welcome-home party for two returning Vietnam War veterans. The police arrested all patrons in attendance, including 82 African-Americans. It exploded into one of the most destructive social insurgences in American history, lasting five days and surpassing the scale of Detroit's 1943 race uprising 24 years earlier. After the initial punitive response, President Johnson established the National Advisory Commission on Civil Disorders to investigate the deeper causes of the unrest. They identified high levels of frustration, resentment and anger that had been created among African-Americans by unemployment and under-employment, persistent and extreme poverty, racism and racial segregation, police brutality and lack of economic and educational opportunities (Locke, 2017).

The timing of the unrest so soon after securing civil rights serves as an important reminder that rights do not operate as a rust solvent to quickly dissolve social injustices. On the one hand, they do provide the basis for holding the state accountable and in this case, they instigated the process of desegregation that got underway. On the other hand, like other rights that have come before, the maintenance and enforcing of civil rights has required constant vigilance. Sometimes the language of 'rights' gets (ab)used for other purposes (e.g. private property rights, corporate rights, political power). This became particularly apparent when the extension of civil rights coincided with a global fiscal crisis at the beginning of the 1970s which tempered how these rights were applied. As noted in Chapter 2, the geographer David Harvey (1973) saw an unfolding crisis emerge from deindustrialisation in Baltimore where he was based, and the uprooting of capital investment from the region. This gave rise to increased socio-economic polarisation and was echoed in many other industrial centres globally. He argued that the capitalist system (as it was being implemented in the mid-20th century) was denying social groups access to social justice.

This paved the way for a new form of **neoliberal citizenship**, which began to supplant the post-war social citizenship regime of universal inclusion. Neoliberal citizenship refers to how the process of neoliberal expansion shifted the boundaries of social inclusion to be defined in relation to the social obligations of citizens rather than social rights afforded by the state (Brodie, 2008). These social obligations – autonomy, self-sufficiency and personal responsibility – privilege a 'neoliberal' subject (Moore and Fraser, 2006). People who can meet these social obligations are socially included (i.e. those that are deemed to be healthy, self-sufficient, not dependent, and exhibiting behaviour that fits with social norms). Those that are deemed not to meet these obligations (as many Black Americans were being positioned, despite their fledgling civil rights) are excluded from the realm of full social citizenship, although we would argue that this is not absolute but is more akin to a gradation of differential citizenship and inclusion.

More generally, *all* rights suffer from some degree of fallibility, as Tushnet (1984) outlines: Rights suffer from *instability*, that is, they are not universal and abstract, as the discourse about them often claims, but rather exist only as products of particular political and social moments. As these moments change, so too do the content of rights. Second, rights suffer from *indeterminancy*. That is, the language of rights is so open and indeterminate that opposing parties can use the same language to express their positions. Third, rights suffer from *reification*, the act of treating abstract ideals of rights as if they were concrete, material entities, that can by themselves automatically change practices; and fourth, rights suffer from *disutility*, where rights are used for other motives to protect powerful interests and privilege (e.g. commercial speech or economic rights of high-income people), instead of the oppressed and minorities. This informs understandings of 'us' and 'them', leading to new spaces of exception, which we turn to next, where groups are not recognised and lack political status in 'ordinary' everyday life, and forms of boundary work have served to reproduce existing social divisions and arguably create new ones.

3.6 Tracing the Spaces of Exception in 'Ordinary' Everyday Lived Citizenship

While the last section outlined a new generation of civil rights and its spatial expression through the right to the city which resulted in several landmark pieces of legislation enacted into law, we saw that a gap between formal de jure rights and the de facto citizenship nonetheless persisted. Citizens may in theory possess civil liberties, political, social or civil rights, but do not possess the means to fully access them (Higgins, 1999). Although the formal rights and regulations concerning the position or status of being a citizen of a particular country are important, so too are the everyday negotiated practices that constitute belonging to a particular state or community. In this final section, we consider how people experience **lived citizenship**, often within this gap. We explore how lived citizenship is about recognising not just the rights of people but the relational nature of their 'ordinary' everyday citizenship, practised in social relations (Isin and Turner 2002; Jæger Fjetland et al., 2022). At this scale, we can appreciate how relationality and encounter are managed through the social interactions between people.

In the wake of the historical contexts outlined in the previous section, it is important to consider that as well as thinking about what the state offers, in terms of protection from coercion and rights and entitlements, it is also important to consider the *spaces of exception* (Ong, 2006), which can be created when the state (sometimes inadvertently) makes exceptions for some populations. Spaces of exception can be created through less-overt acts such as de-regulation of outsourced public services for example, which creates exclusions for people being supported by these services. In such spaces, protections from the state are replaced by ambiguity and contingency in provision which can indirectly make those reliant on such rights being abandoned or given up on (pp. 5–6). According to Ong (2006: 6), the elements that typically come together to create citizenship – rights, entitlements, territoriality, a nation – become disarticulated, and rearticulated with forces set in motion by market forces. Ong only briefly considers the welfare state (which we explore in Chapter 4) in her work, focusing primarily on the citizenship rights of workers where the spaces of exception are more absolute. We are aware that ambiguity and contingency are more normative features of welfare provision and historically have always been present. However, Evans (2011) provides a potent account of the spaces of exception which emerge for homeless people. In his account, this space of exception is where the political status of homeless people becomes suspended – mirroring the experiences of other groups or individuals that are not easily assimilated or integrated into the political system (often due to structural inequalities embedded in the system!). This state of exception creates a unique relationship between the state and these un-integrated groups.

These spaces of exception have arguably expanded and colonised other groups. For those that occupy the 'gray zones' of citizenship (Evans, 2011), marginal to the perceived centre, or at the borderlands of citizenship, voluntary sector organisations typically operate in helping people negotiate this space. Trudeau (2012) argues that the contributions of local community-based nonprofit organisations for migrants help construct alternative, subnational visions of citizenship, through constructing locale-specific discourses about migrant belonging. These local groups thus seek to secure rights for recognition and for political status to either escape *from* or create spaces of belonging *within* these spaces of exception. In the same spirit, scholars have sought to highlight the *lived* experiences of citizenship, which we turn to next.

Isin and Turner (2002) proposed a social model of citizenship that seeks to explore what lived citizenship means to people whose claim to citizenship may be difficult to realise. This is particularly relevant when acknowledging and challenging the exclusionary nature of citizenship, as we saw above. This idea of a social model of citizenship focuses less on individual rights and the relationship between the individual and the state and more on *social participation*. The model seeks to expose the limitations of frameworks that focus solely on rights or on identity and difference. It thus challenges Smith's definition of social citizenship, understood simply in terms of the formal relationship between the individual and the state. It its place, the model privileges the local contexts and the importance of personal networks, where an individual's sense of identity is produced through their social relationships (MacIntyre, 2021).

Lived citizenship is thus understood as being practiced and experienced in 'ordinary' spaces that people use and occupy in their everyday lives; through interactions in schools, homes, shops, nurseries and community groups. These interactions can help to establish different relational networks for individuals and their families, and to be recognised as visible and valuable citizens in wider society. For some people, they may face obstacles to making these connections and gaining access to opportunities available to the population in general (Ponce and Rowe, 2018).

Even with civil rights protections to ensure equality, as we saw, some people may face contested spaces of citizenship through their everyday relational encounters. Staeheli et al. (2012) makes the case for understanding everyday ordinary lived citizenship as being simultaneously constituted through encounters with law and daily life. One example is at the intersection of race and gender. While anti-discrimination legislation is enshrined to protect both Black people and women, the British artist and photographer Ingrid Pollard has been credited for highlighting the subtlety of encounters faced by Black women which may undermine their lived citizenship, such as stares, condescension, and feelings of being *out of place*. In an exhibition, *Pastoral Interludes* in 1987, she curated a series of self-portrait photographs in the Lake District in the North of England to portray these feelings. Her exhibition (V&A, 1987) sought to challenge the stereotypically White rendering of the idealised rural landscape of Britain and the traditional representation of Black people's 'space' as urban space. This is captured in her caption:

> It's as if the Black experience is only lived within an urban environment. I thought I liked the Lake District; where I wandered lonely as a Black face in a sea of white. A visit to the countryside is always accompanied by a feeling of unease; dread. (Pollard, caption in V&A, 1987)

Ingrid's provocation serves as a powerful tool in de-familiarising the familiar, in making visible the spatial codes which can go unnoticed by the majority and challenging the ordinary, everyday barriers to being a full citizen. According to Painter and Philo (1995: 115):

> If citizenship is to mean anything in an everyday sense, it should mean the ability of individuals to occupy public spaces in a manner that does not compromise their self-identity, let alone obstruct, threaten or even harm them more materially.

How different groups occupy alternative spaces of citizenship is relevant when we consider the pervasive presence of neoliberal citizenship and the spaces of exception which exist within it. Consequently, the politics of recognition is now as important as ever in our current online digital age where visual representation is omnipresent, to recognise the lived experiences of people at the fringes of citizenship and to make rights violations more visible.

3.7 Conclusion

Issues of citizenship, rights and responsibilities and the role of the state have been at the forefront of geographic enquiry since the 1970s. Their significance in the geographies of the social, inequality and social justice cannot be overstated. As we saw, rights first emerged to prevent groups from imposing their will oppressively on others but over time, certain principles of exclusion have persisted and have been constantly open to (re)negotiation. Despite having common foundational principles, social justice has many different faces, depending on how these key concepts have been employed. In more recent times, there have been very real concerns over the nature of contemporary citizenship and governance, that have become dominated by a neoliberal framing. This has meant individual responsibilities have overshadowed rights under de jure citizenship. However, different social movements have equally made significant gains in achieving greater visibility and recognition for their respective constituencies, thus reshaping their de facto citizenship – or at least carving out spaces of alternative citizenship in which they can belong. This does not mean that groups have abandoned efforts to secure and enforce rights.

Despite their flaws, rights nonetheless provide a worthwhile pursuit for protecting vulnerable groups from abuse and provide a standard that holds governments accountable, particularly when it comes to distributional and procedural justice. Just the same, this is balanced with an awareness of the importance of the second strand of social justice – recognition – where efforts to cultivate more positive narratives and representations of identities often hold equal sway. It is here that the socio-spatial relationality of citizenship arguably plays out and are lived and felt in people's 'ordinary', everyday materiality and everyday lives. These encounters between state and citizen keep these issues at the forefront of geographical inquiry and serve to shape the role of the state in public services, which we turn in the next chapter.

References

Brenner, N. and Theodore, N. (2002) Cities and the geographies of 'actually existing neoliberalism'. *Antipode*, 34, 349–379. https://doi.org/10.1111/1467-8330.00246

Brodie, J. (2008) The social in social citizenship. In E. Isin (Ed.), *Recasting the Social in Citizenship.* Toronto, ON: University of Toronto Press.

DeVerteuil, G. (2015) *Resilience in the Post-Welfare Inner City, Voluntary Sector Geographies in London, Los Angeles and Sydney*. Bristol: Policy Press.

DeVerteuil, G., Power, A. and Trudeau, D. (2020) The relational geographies of the voluntary sector: Disentangling the ballast of strangers. *Progress in Human Geography*, 44(5), 919–937. https://doi.org/10.1177/0309132519869461

Drewitt, A. (1999) Social rights and disability: The language of 'rights' in community care policies. *Disability & Society*, 14(1), 115–128.

Evans, J. (2011) Exploring the (bio)political dimensions of voluntarism and care in the city: The case of a 'low barrier' emergency shelter, *Health & Place*, 17(1), 24–32.

Fjetland, K.J., Gjermestad, A. and Lid, I.M. (2022). Introduction. In *Lived Citizenship for Persons in Vulnerable Life Situations - Theories and Practices* (pp. 9–28). 1 June 2022. https://doi.org/10.18261/9788215053790-2022-01

Gatrell, A.C. and Elliott, S.J. (2014) *Geographies of Health: An Introduction.* 3rd edition. London: Wiley-Blackwell.

Hackworth J. (2007) *The Neoliberal City: Governance, Ideology, and Development in American Urbanism*. Ithaca: Cornell University Press.

Harvey, D. (1973) *Social Justice and the City*. Baltimore: Johns Hopkins University.

Higgins, J.W. (1999) Citizenship and empowerment: A remedy for citizenship participation. *Community Development Journal*, 34(3), 287–307. https://doi.org/10.1093/cdj/34.4.287

Hitchen, E. (2019) The affective life of austerity: Uncanny atmospheres and paranoid temporalities. *Social & Cultural Geography*, 22(3), 295–318.

Hobbes, T. (1651) *Leviathan*. London: Andrew Crooke.

Isin, E. F. (2008) Theorizing acts of citizenship. In E.F. Isin, and G.M. Nielsen (Eds.), *Acts of Citizenship* (pp. 15–43). London, UK: Palgrave Macmillan.

Isin, E. F. and Turner, Bryan S. (2002) Citizenship studies An introduction. In E. F. Isin, and B. S. Turner (Eds.), *Handbook of Citizenship Studies* (pp. 1–10). London, UK: Sage. https://doi.org/10.1080/13621020601099773

Isin, E. and Wood, P. (1999) *Citizenship and Identity*. London: Sage https://doi.org/10.4135/9781446218129

Kline, N.S., Webb, N.J., Johnson, K.C., Yording, H.D., Griner, S.B. and Brunell, D.J. (2023) Mapping transgender policies in the US 2017–2021: The role of geography and implications for health equity. *Health & Place*, 80. https://doi.org/10.1016/j.healthplace.2023.102985

Knox, P. and Pinch, S. (2010) *Urban Social Geography*. London. Routledge. https://doi.org/10.4324/9781315847238

Locke, H.G. (2017) *The Detroit Riot of 1967*. Detroit: Wayne State University Press.

McCann, M. (2006) *Law and Social Movements*. New York: Routledge.

MacIntyre, G., Cogan, N., Stewart, A., Quinn, N., O'Connell, M. and Rowe, M. (2021) Citizens defining citizenship: A model grounded in lived experience and its implications for research, policy and practice. *Health and Social Care in the Community*. https://doi.org/10.1111/hsc.13440

Massey, D. (2007) *World City*. UK: Polity Press.

Marshall, T.H. (1950) *Citizenship and Social Class*. Cambridge, UK: Cambridge University Press.

May, J. and Cloke, P. (2014) Modes of attentiveness: Reading for difference in geographies of homelessness. *Antipode*, 46, 894–920. https://doi.org/10.1111/anti.12043

Migdal, J. (2001) *State in Society: Studying How States and Societies Transform and Constitute Each One Another*. Cambridge, UK: Cambridge University Press.

Mitchell, D. (2003) *The Right to the City, Social Justice and the Fight for Public Space*. NY: Guilford Press. DOI: 10.1177/1078087404269422

Moore, D. and Fraser, S. (2006) Putting at risk what we know: Reflecting on the drug using subject in harm reduction and its political implications. *Social Science & Medicine*, 62, 3035–3047.

National Archives (n.d.) Victorian Industrial Towns. London. www.nationalarchives.gov.uk/education/resources/victorian-industrial-towns/

Ong, A. (2006) *Neoliberalism as Exception: Mutations in Citizenship and Sovereignty*. Durham, NC: Duke University Press. doi:10.1215/9780822387879

Painter, J. and Philo, C. (1995) Spaces of citizenship: An introduction. *Political Geography*, 14(2), 107–120.

Painter, J. and Jeffrey, A. (2009) *Political Geography: An Introduction to Space and Power*. 2nd edition. London: Sage.

Peck, J. (2011) Geographies of policy: From transfer-diffusion to mobility-mutation. *Progress in Human Geography*, 35(6), 773–797.

Peck, J. and Theodore, N. (2001) Exporting **workfare**/importing welfare-to-work: Exploring the politics of Third Way **policy** transfer. *Political Geography*, 20(4), 427–460.

Peck, J. and Theodore, N. (2015) *Fast Policy: Experimental Statecraft at the Thresholds of Neoliberalism*. Minneapolis: University of Minnesota Press.

Peck, J. and Tickell, A. (2002), Neoliberalizing space. *Antipode*, 34, 380–404. https://doi.org/10.1111/1467-8330.00247

Pike, A. (2023) *Financialization and Local State Craft*. Oxford: Oxford University Press.

Ponce, A.N. and Rowe, M. (2018) Citizenship and community mental health care. *American Journal of Community Psychology*, 61, 22–31. https://doi.org/10.1002/ajcp.12218

Power, A., Lord, J. and DeFranco, A. (2013) *Active Citizenship and Disability: Implementing the Personalisation of Support*. Cambridge, NY: Cambridge University Press.

Power, A., Hall, E., Kaley, A. and Macpherson, H. (2021) Voluntary support in a post-welfare state: Experiences and challenges of precarity. *Geoforum,* 125, 87–95, https://doi.org/10.1016/j.geoforum.2021.07.003.

Smith, S.J. (1989) Society, space and citizenship: A human geography for the 'New Times'? *Transactions of the Institute of British Geographers*, 14(2), 144–156.

Staeheli, L.A., Ehrkamp, P., Leitner, H. and Nagel, C.R. (2012) Dreaming the ordinary: Daily life and the complex geographies of citizenship. *Progress in Human Geography*, 36(5), 628–644. https://doi.org/10.1177/0309132511435001

Trudeau, D. (2012) Constructing citizenship in the shadow state. *Geoforum*, 43(3), 442–452. https://doi.org/10.1016/j.geoforum.2011.10.001.

V&A (1987) *Ingrid Pollard Pastoral Interlude collection.* V&A Museum. https://collections.vam.ac.uk/item/O107865/pastoral-interludeits-as-if-the-photograph-pollard-ingrid/ (Accessed 23 April, 2025).

Wilkinson, E. and Ortega-Alcázar, I. (2019) The right to be weary? Endurance and exhaustion in austere times. *Transactions of the Institute of British Geographers,* 44, 155–167. https://doi.org/10.1111/tran.12266

4 GEOGRAPHIES OF STATE SUPPORT AND WELFARE

4.1 Social Justice Theories of State Support

Before considering the welfare state, it is important to return to the conversation between social geography and social justice, briefly signposted in Chapter 1. This will be connected to the three strands of social justice – distribution, recognition and procedure. For social geography, the original focus on social justice in the context of state support and welfare was embedded in a liberal approach based on John Rawls' *A Theory of Justice* (1971). Rawls is considered a liberal theorist of justice for two reasons. First, he outlined that everyone ought to have an equal right to enjoy equal basic liberties (synonymous with classical liberalism outlined in Chapter 3) such as individual rights, equality and fairness, while also aiming to justify a social contract that secures both liberty and economic justice for all. Second, he argued that social and economic inequalities should only exist if they advantage the worst-off (known as the Difference Principle). By this, he meant that society could allow certain people to earn more money or hold more power as he accepted that some inequality incentivises productivity, innovation and economic growth. However, this was based on a condition that this arrangement needs to improve the lives of the least advantaged more than any alternative system would. Rawls is not arguing for strict equality of income or status. But Rawls insists that the benefits accrued by some must 'trickle down' in a meaningful way to those who are worse off – not just symbolically, but materially (e.g., if more jobs are created or services improve, the inequality is justified under Rawls' framework). It is worth noting that Rawls' Difference Principle has been extensively co-opted to rationalise neoliberal, trickle-down economic policies, where wealth concentrates at the top, with the vague promise that 'everyone will eventually benefit'. Consequently, it has been critiqued for placing too weak a constraint on inequality and failing to address structural injustice. It also fails to address global economic injustice, by letting rich nations off the hook and not accounting for colonial legacies which harm the world's poor.

Nonetheless, Rawls influenced a subset of human geography in the 1970s to focus on universal, distributive issues of 'who gets what, where, and how'. Welfare geography and its liberal foundations, however, would

DOI: 10.4324/9781003414889-5

also have to contend with a more explicitly radical challenge in the form of Marxist formulations of justice, themselves heavily based on David Harvey's *Social Justice and the City* (1973). In his book, universalistic and eternal characterisations of justice were presented and then decisively replaced by more focused attempts at understanding, and striving for, notions of justice embedded in the actual inequalities embedded in the socio-economic structures of capitalist production. As such, Harvey's focus was twofold: the first part of the book was squarely rooted in the liberal conception of justice, prioritising the least advantaged, while the second part of the book was a Marxist analysis that revealed the inherently unjust nature of the capitalist city and the need for more radical strategies. Set within a context of increasing societal inequality in the 1970s, the influence of Harvey's radical approach on the nascent Marxist geography was immediate and profound.

In contrast, post-structuralist formulations recognised the distributive concerns of liberal formulations, as well as the Marxist focus on class, the production system and economic relationships, as necessary but not entirely sufficient in explaining the full range of injustice in the social world. Post-structuralists sought to extend the context-based in/justice of the Marxist approach, retaining the emphasis on economic marginalization, exploitation and powerlessness but also moving beyond to emphasise oppression and domination based on a variety of group-based differences: gender, race, disability, sexual orientation and so forth. To post-structuralists, the economically-based notion of justice is simply too narrow since oppression would persist even if all economic injustices were abolished.

This broadened perspective was largely influenced by Iris Marion Young's Justice and the Politics of Difference (1990), as signposted in Chapter 2 on inequality. Young's central concern is how social institutions subject some groups to domination and oppression. To her, 'domination' means: 'structural or systematic phenomena which exclude people from participating in determining their actions or the conditions of their actions' (1990: 12), constraining self-determination. 'Oppression' means 'structural phenomena that immobilise or diminish a group' (1990: 53), constraining self-development. Oppression is systemic rather than tyrannical, not the result of a few people's choices or policies, but rather embedded in unquestioned norms, habits and symbols, institutional rules. There is no need for a clear oppressor for oppression to exist; rather, oppression is structural and built in to the system. Finally, by 'group' she means a collective of persons differentiated by cultural forms, practices or way of life. Groups range widely, including men and women; various ages; different races; different sexuality and gender identities. Recognition now becomes the key strand over and above distribution, in that even an economic system bent on economic equality such as socialism would still leave some groups behind.

In the 2000s, there were several attempts to reconcile some of the tensions across the three aforementioned formulations, or at least to retrieve some important commonalities that move beyond the universalist/relativist divide.

David M. Smith was a leading thinker in this regard, seeking to update liberal formulations of justice in an 'age of difference' brought on by post-structural critiques. In his 2000 article entitled 'Social justice revisited' (Environment and Planning A), he sought to align the appreciation of difference with a larger framework of human sameness and universal needs that transcend, in his words, 'the here and now of specific individual, group or local practices' (1151). Without abandoning the need to recognise different groups and their different needs, Smith argues that our minimal needs for survival constitute universal human rights – an approach not entirely dissonant with the capabilities literature spearheaded by Martha Nussbaum and Amartya Sen. Procedure now becomes important, in terms of how social justice is done with/on various groups across various places. The main challenge to the equality of these rights is that resources are geographically distributed in an entirely arbitrary and accidental fashion. The place of good fortune can only be transcended if an ethics of care is present that promotes the lives of 'distant strangers' and territorial social justice. Gerry Kearns and Simon Reid-Henry (2009) added to this perspective by examining the role of 'geographical luck' of birth and residence in determining longevity, and how massive inequalities in longevity across the world necessarily raise issues of justice. These issues require a recognition of the interconnectedness of human life that stretch beyond the everyday.

At its essence then, social justice is always the following: contested; spatial; practical and utopian; linked to social injustice; and especially, though not exclusively, urban. In the next section, we define the welfare state as framed by the social justice strands of distribution and, to a lesser extent, recognition. The welfare state frames a variety of relationships and encounters with those who depend on it. This dependency is obligatory in many cases where citizens rely on the support of the state. These obligatory relationships come in various forms. We can talk about mutually beneficial relationships, where both sides benefit, but also more parasitic relationships, where one side benefits by harming the other, or commensal, where one side benefits while the other side is not affected or does not seem to mind. Think of these relationships in the following sections, in terms of the transactions between the welfare state and citizens.

4.2 Defining the Welfare State

Given the inherently political nature of the welfare state – of who gets what and where – issues of social justice and power are unavoidable, as are questions about the role of the state in the everyday lives of citizens. This includes not just who counts in society, but also the 'bricks and mortar' manifestations of the state in terms of social housing, as well as 'bread and butter' issues of access to state resources, and who deserves what in society. As such, we consider the welfare state as both a complicated set of relations between the state and citizens, but also as a physical space of encounter where resources are transacted between welfare offices, social housing estates and local government on one side, and the various clientele on the other. These meet in various contact zones as we shall see later in the chapter. The welfare state in essence serves to smooth out the rough edges of societal inequality, which involves redistributing from the well-off to the less well-off, usually via a system of taxation and benefits. Recognitional and procedural justice are part of this redistribution, although it can seem secondary to the distributional elements of who gets what, and where.

What is the welfare state then, and how does it relate to social and economic inequality, concepts that were introduced in Chapter 2? For some, the welfare state is very narrowly associated with benefit payments directly to people in need, generally those who cannot work. These are people who have recently lost their jobs or do not work because of disability, old age or who live in areas with few employment opportunities, people who are carers (especially for young children), people who have arrived from other nations without the legal ability to work. In the US, it has become associated with government programmes that have become maligned by negative associations with benefit dependency. But we can cast it as something wider in its application. Basically, the welfare state is anything that the state spends on to benefit the wider population – this can run the gamut of public lighting, roads, prisons to schools, public assistance, pensions and the like. Each of these material provisions create very specific relationships with citizens as well. With the rise of an incomplete neoliberalism, some have argued that it is harder to discern what the state does anymore, as it now shares governance with community groups, the voluntary sector (Chapter 5) and the private sector. The 'welfare state' is not a single entity, and while political, ideological and economic shifts matter (as we explore below), a diversity of spaces, practices and relationalities may co-exist (Newman, 2012). Newman uses the idea of 'border work' for such arrangements and practices that are difficult to discern whether they are acting within or outside a welfare state. Yet even in this age of muddled state provision, there remain clear lines of provision.

So the welfare state ranges widely, and when it does intervene it necessarily stratifies the population into deserving of assistance, and undeserving – or even ignores whole layers who fall through the cracks (asylum seekers, immigrants). To those deserving, welfare can be a lifeline and a response to a lack of mainstream options for employment. To those undeserving, support must be sought elsewhere – from family, the voluntary sector or through informality, criminality or out-migration. With these stratifying tendencies, the welfare state can also be understood as a form of poverty management. Within a management frame, the state – alongside other institutions and elites – articulates and employs techniques to maintain the social order, including the supportive (for example, affordable housing), the ambivalent (for example, ceding the homeless agenda to the voluntary sector) and the punitive (for example, anti-homeless laws). These techniques translate into sites of management over social marginality that vary over time – from 19th-century asylums to the contemporary homeless shelter (Wolch and DeVerteuil, 2001); more on this in Chapter 10.

Even for those considered 'deserving' of welfare assistance, there may be a degree of losing out in other ways from this transaction. A key criterion for determining 'deservingness' is a person or group's perceived vulnerability. Certain people are often deemed to be inherently 'vulnerable', e.g. disabled or older people. The term suggests a seemingly protective and therapeutic emphasis, based on a perceived ethical, moral duty to protect vulnerable people. This is linked to a well-meaning approach to those 'less well off'. However, this casting of certain groups as inherently vulnerable has been critiqued by Brown (2012) who argues that this emphasis on vulnerability has more stigmatising and exclusive effects. In policy and social welfare practice, it has been used to classify those who are less accountable for their circumstances or actions and those who have less autonomy in the development of perceived difficulties in their lives. In effect, it has become a disempowering term, a static label: e.g. 'vulnerable groups in society', 'dependent', 'non-threatening' and those who must be 'protected'. Typically, this distinction is presented as a binary between focusing on a person's vulnerability (deserving), and expecting an individual to be held responsible for their actions and contexts (undeserving). This gives rise to an important point of contention over whether the likely cause of vulnerability is in the inherited traits of people or due to external factors which may disempower people and erode their security.

As a result of this binary and the overlooking of external factors, certain groups may fall between the cracks. For instance, it can lead to exclusions of other deprived groups who may benefit from public services, e.g. young poor Black men who in certain contexts have been narrowly perceived as a threat to community security rather than deserving of support. Wilton (2004) explores how people with mental health issues are also increasingly held responsible for their own material well-being and public conduct, yet are less able to exercise control within everyday life. Returning to Brown (2012), she presents an alternative view in her 'vulnerability thesis': that all individuals experience vulnerability at different points in their lives. In this reading of vulnerability, she argues that we should emphasise the structural causes of people's varying degrees of fragility and need (Brown, 2012), and how different people's 'place' in the world can affect their vulnerability. This is the rationale used by proponents of a living wage for all citizens, in recognition of the rising level of in-work poverty around the world, which can be linked to Young's (1990) exploitation and powerlessness.

4.3 The Welfare State Across the World

There is no doubt that the welfare state's direct payments – for housing, cash benefits, medical coverage, disability coverage, education, pensions and so on – play an outsized role in how inequality is directly managed via (re)distribution. Esping-Anderson (1990) measured the welfare state by its degree of de-commodification: how many necessities are provided by the state without direct costs, compared to what must be paid for through the market. This is commonly referred to as collective consumption which refers to the shared use of goods and services, particularly those provided by the government or with significant state involvement, where the benefit is enjoyed by the community as a whole or by a large segment of society. Further, Esping-Andersen conceptualised various national welfare state regimes that could be used to compare the levels of universalism, stratification and de-commodification in Western nations. He posited that the ideal welfare state would be one that maximised de-commodification and universality, while minimising stratification and commodification. Of course, Esping-Andersen only focused on established welfare states in the Global North, comparing the robust ones of Scandinavia and Western Europe with the less-comprehensive models in the UK and the US.

The principles of universalism, stratification and de-commodification can be illustrated with the origins story of the British welfare state in the immediate post-war period. In 2022, the Beveridge Report marked its 80th anniversary This report was the founding blueprint of the UK's modern welfare state, by Sir William Beveridge, British economist and Liberal politician, titled 'Social Insurance and Allied Services' (1942), in which he outlined his proposals for a new welfare state. This blueprint provided a wide range of suggestions aimed at eradicating what he called the five 'giant evils': want (e.g., poverty), disease (e.g., health, care), ignorance (e.g., education), squalor (e.g., housing) and idleness (e.g., employment). Beveridge thought that by taking the burden of healthcare and pension costs away from corporations and individuals, and onto the government (de-commodification) to provide them to British citizens (universalism), would increase the competitiveness of British industry while producing healthier, wealthier, more motivated and more productive workers keen to buy British goods. Its timing was fortuitous as there was a healthy public appetite to turn away from the harsh, punitive forms of pre-war welfare found in the poor houses, which many people feared.

For the first 40 years, these ideals proved useful. The sustained post-Second World War period of economic growth and near full employment that characterised the Fordist mode of production lasted until the late 1970s. This period saw falling poverty, slum clearance, the founding of a free health service and education system alongside rising real incomes and falling inequality – which, in turn, led to higher tax revenues and helped the UK pay off its war debts. While this period can justifiably be celebrated, it is worth adding a side note that while the welfare state was *ahead* of its time, it was also *of* its time, in terms of being built on gender norms that prevailed in the 1940s and 1950s – that men would be the sole breadwinners earning the family wage while women would stay at home and contribute to social reproduction (stratification).

The UK model was underpinned by Keynesian economics, developed by John Maynard Keynes, which was particularly influential in the post-Depression era in the United States (covered later in Section 4.4). Sweden was also an early adopter of Keynesian policies in the 1930s. Keynesianism offered a counter to the classical liberal view outlined in Chapter 3 that free markets would naturally lead to full employment. It emphasised the role of government intervention to address unemployment and economic instability. It provided the theoretical framework for government intervention to stabilise economies and combat unemployment. Later, countries like Ireland, Canada, and even India adopted this model post their independence.

However, both the UK and US in the 1980s under Margarat Thatcher and Ronald Reagan respectively, proceeded to dismantle, devolve and privatise the state, with important consequences for the welfare state. These involved the following:

-> *Dismantling* involves the wholesale reduction in or withdrawal of national funding, or alteration of programme guidelines which significantly transforms existing government programmes at all levels. An example would be social housing, which we will use as an example of material welfare provision (in the form of bricks and mortar) in Chapter 12.

-> *Devolution* involves internal transformations that promote responsibility to a different governmental tier, usually lower. Think of devolving various programmes from the national to local level (e.g. homeless programmes). Frequently, devolution involves downloading responsibility without the necessary resources – this is the case when the national government placed increased responsibility for provinces/states/counties to take care of health and education, including universities.

-> *Privatisation* involves the transfer of traditional government functions to external commercial or nonprofit entities. Privatisation is more than simply returning services to the private sphere. Privatisation holds the promise of replicating the quality of public provision, while at the same time representing an underlying shift of claims in a society and casting into doubt the desirability of public provision. This is the essence of neoliberalism. Think of the UK government's privatisation of British Airways, as well as the rail system.

The impacts of this change are covered in Section 4.4, but before we explore these further, we must extend our focus beyond UK shores. Since we are geographers, we are well-aware that welfare states differ greatly to in different parts of the world. In the Global South, the welfare state remains effectively the family – if you lose your job, your housing and so forth, you will turn to your close ones first for survival. In the Global East, the welfare state has begun to emerge more robustly in various nations, especially those with ageing populations. For instance, in Japan there has been an expansion in the welfare provisions for elderly people, homeless people and children. This was done through the 'Livelihood Assistance program', which began in 1998 and was strongly expanded in the wake of the 2008 financial crisis. The focus of the programme is on elderly households who receive no, or only a low, pension because of insufficient payments in the past, single mother households, and households with handicapped, injured or sick people (Japanese Ministry of Health, Labour and Welfare, 2022). Underpinning this was a strong sense of social inclusion and social coherency, which runs against the neoliberalisation of the welfare state in other nations. This becomes evident in a statement by the Japanese national government in 2000:

> During the period after WWII, Japan transformed itself from a poor society into a wealthy one. We should surely appreciate the contribution made by social welfare, which sought to escape poverty during that time. However, we cannot deny that following urbanization, the trend towards nuclear families, industrialization, and internationalization have caused people to lose their common ties. Moreover, even though society has become economically wealthy, we must acknowledge that new challenges have presented themselves, and respect for one another has been weakening ... From the start, Japan's social welfare has been supported by the social solidarity of its citizens. Our welfare state must now shift away from a remedial system towards a system that respects user choices, the promotion of community welfare, and to reconnect ties among communities, a social backup.
>
> (Mizuuchi, 2020: 72–73)

In other words, social welfare should promote some kind of universality and social coherency, rather than further stratification and stigmatisation. Similar welfare state regimes have emerged in South Korea and Taiwan, albeit blended with strong Confucian ideals over the role of the family, with welfare being seen more as an aid to employment. The social programmes in places like Hong Kong and Singapore ultimately have the same goals (see the next subsection), and are heavily dependent on subsidised, decommodified (social) housing as a way to assist the general population, upwards of half the housing stock.

Much of the discussion around the welfare state is implicitly around distribution, or more specifically *re*distribution of resources from the haves to the have-nots, usually via taxation (on income, sales, and less usually on wealth itself). However, we want to bring in the strand of recognition, of how the welfare state can struggle to recognise the special needs of particular populations, and how those populations demand their own institutions from the grassroots to the state. In settler nations such as Australia, New Zealand, the United States and Canada, the Indigenous populations remain marginalised economically, spatially and socially. Over the past 40 years, a movement among Indigenous peoples – in the form of *indigeneity* – seeks to counter the colonial erasures of identity, and reclaim respect for knowledge of and connection to the land that was once theirs. The colonisation, containment and exclusion of Indigenous peoples in most settler societies – in the form of dispossession from traditional territories, suppression of cultural beliefs and practices, the creation of isolated reserves, chaining benefits to reservation status (in Canada for instance) and high levels of segregation in most cities – have resulted in a host of persistent social, economic and health inequalities (Kearns et al., 2009). This has also resulted in the alienation of Indigenous cultural conceptions, spirituality and spaces from mainstream society, a fact reflected in discriminatory institutional practices across health, educational, legal and criminal justice systems.

Within the Canadian context, over 1.8 million people self-identified as Indigenous in the 2021 census, which is close to 4% of the total population (Canada, 2022). Since the 1960s, the Canadian Aboriginal population has extensively urbanised. This resonates with Indigenous urbanisation in other settler societies, including New Zealand and Australia. Despite this urbanisation, most local authorities have not developed Indigenous-specific policies and programmes in Canadian cities. As a result, the urban Indigenous population is frequently compelled to use the non-Aboriginal social service system where their special status, cultural beliefs and traditions may not be acknowledged.

Winnipeg makes a good case study, given its Indigenous population is over 15% and growing. In that city, a set of institutions has emerged to respond to these needs, making demands on the local government while promoting Indigenous-heavy areas as the natural location for new sites of collective consumption (DeVerteuil and Wilson, 2010). These have included Indigenous-focused housing programmes, treatment centres and drop-ins. The scale and scope of these organisations has greatly expanded over the last 20 years, to the point now where an Indigenous group now owns The Bay, the former flagship department store in Winnipeg, and an Indigenous man now sits as premier of Manitoba. Still, the need for culturally appropriate services is outstripping current capacity in the cities with growing Indigenous populations, especially given the currently disproportionate level of social problems assailing the urban Aboriginal community that have their origins in colonisation, racism and the deep legacy of residential schools that stripped individuals of their indigeneity.

4.4 A Welfare State Divided and 'Reformed'

The examples in Section 4.3 brought up intersectionality, including class, race, indigeneity and citizenship, which belie the ideal of a universalistic welfare state but certainly supports the idea of a stratifying one. Increasingly, welfare states are moving towards a more targeted strategy that recognises differences among various groups in need. Of course, this could just be seen as deepening the division between the deserving and undeserving, and of how the purported universality of the welfare state was never actually implemented.

The ambitions outlined in the Beveridge Report are now in grave danger of being undone. The 'five giants' are creeping back into the mainstream of daily UK life. The Rowntree Foundation *UK Poverty Report 2024* found that more than 1 in 5 people in the UK (22%) were in poverty in 2021/22 – 14.4 million people. This included 8.1 million (or around 2 in 10) working-age adults, 4.2 million (or nearly 3 in 10) children and 2.1 million (or around 1 in 6) pensioners. Poverty rates have returned to around their pre-pandemic levels. Some of this poverty is a holdover of extremely high levels of inequality in the 1980s, when the economy boomed at the same time that the welfare state was successively cut back. This has not been reversed, meaning current levels of poverty are around 50% higher than they were in the 1970s. Some groups of people face particularly high levels of poverty. This includes larger families, single-parent families, certain minority ethic groups (Black and Asian), disabled people, informal carers, low-wage earners, part-time workers and persons in rented accommodation.

Along somewhat similar lines, the current American welfare state was essentially 'reformed' in 1996 with the Personal Responsibility and Work Opportunities Reconciliation Act (PRWORA), which overturned many of the ideals set forth in the 1930s New Deal and the 1960s Great Society. While the 1935 and 1965 welfare acts served to expand the American welfare state, the 1996 Act fundamentally reduced it. The 1935 Social Security Act created a system of social insurance programmes and more adequate provision of welfare for different groups in society. For example, the Aid to Dependent Children (ADC) program. However, the Act specifically excluded domestic work and agricultural labour – industries that relied heavily on Black men and women – from Social Security eligibility. States also had substantial discretion in determining eligibility for welfare – flexibility that southern states sometimes used to restrict access to benefits during harvesting seasons. This was a clearly racialised policy, and therefore the roots of contemporary stigma associated with welfare in the US were hard wired into the original values of welfare from the outset. By 1939, the ADC caseload was 80% White, despite the disproportionate burden of poverty on Black families. This effectively coerced poor, Black families into working in fields and domestic labour at whatever wages were offered. Hence, from the outset, the die was cast with an assumption that those deserving of state support were White American families.

By the late 1970s, the US was, like many other Global North nations, facing a global fiscal crisis. From 1980 onwards, President Ronald Reagan was looking for a way to cut welfare spending. He sought to ramp up anti-government spending rhetoric and anti-poor resentment. His welfare 'reform' hinged on finding a scapegoat. This was found in a news story by the Chicago

Tribune of a Black woman, Linda Taylor, who was convicted of welfare fraud. This was later picked up by Reagan who coined the expression the 'welfare queen' in his presidential speeches to describe everything he believed was wrong about 'welfare' – that people were too lazy to work, instead relying on public benefits to get by, paid for by the rest of 'us' upstanding citizens. While Taylor was a troubled character, the wider racialised narrative that was woven was always a myth – White people had always made up the majority of those receiving government checks, and fraud was not endemic as Reagan suggested. Nonetheless, the 'welfare queen' myth cemented the stigma associated with welfare in the US as a racialised dependency trap at odds with American values of independence from the state and justified the first major overhaul of the nation's welfare system since it was created in 1935. Yet, the irony is that middle-class White people receive a significant share of total welfare benefits, especially through universal programs like Social Security and Medicare (see Appelbaum and Gebeloff, 2012). Nonetheless, in 1981 and 1982, Reagan made more than $22 billion in cuts to social welfare programs: Aid to Families with Dependent Children (renamed from the earlier ADC), Food Stamps, Rent supplements and Medicaid (Ojala, 2021). He later signed into law the Family Support Act 1988, and claimed in his own words, 'It is fitting that the word "family" figures prominently in the title of this legislation. For too long the Federal Government, with the best of intentions, has usurped responsibilities that appropriately lie with parents'. The welfare queen myth has continued to persist in American lexicon ever since and has often been used by governments as a short hand rationale for welfare reform. This paved the way for the beginnings of the neoliberal (or post-Fordist) welfare state, which as we described earlier is characterised by dismantlement, devolution and privatisation.

These very hallmarks would also underpin the 1996 PRWORA: devolution from federal to the state, local and voluntary sector levels, lifetime restrictions on eligibility (5 years) and non-eligibility for legal immigrants. It also entrenched a moral/individualistic rather than structural understanding of poverty (Henriksen et al., 2012). This Act firmly established a two-tier system of recipients, between the deserving (the job ready, elderly) and the non-deserving (long-term unemployed, single mothers, immigrants). The American welfare state has not fundamentally changed since 1996, although at the state levels there is now much geographical diversity, between more generous states like California and New York and those that have almost no safety net at all (e.g. Florida, Texas). This growing unevenness has become a political matter, with each tendency attracting criticism or praise.

PWORA also sought to galvanise a system called workfare. Workfare is a government programme where individuals must undertake work searching activities and learn new skills to receive welfare benefits or risk losing them. At the time, this was regarded as a novel policy, but its enforcement has become ever more punitive over time. Workfare has become part of what we deemed, in Chapter 3, as neoliberal citizenship. By being socially obliged to seek employment rather than making demands on the welfare state, the individual no longer takes advantage of the social rights afforded by the state. The obligation to be autonomous, self-sufficient and personally responsible epitomises the neoliberal subject. For those who can meet these obligations, the reward is social inclusion. For those who cannot, due to disability, mental illness, caring responsibilities or living in areas with perpetually few employment opportunities, are excluded from full social citizenship. Moreover, in good economic times workfare is potentially beneficial to those who can work, but seeking employment during an economic downturn can prove difficult. It is also related to the minimum wage: some places have a high wage, others not so much. For the latter, this means that work might actually pay less than welfare benefits. The minimum wage itself is a political matter, with some places raising it to make it a living wage, which is enough to live on based on the costs of a particular locality.

The final examples of welfare state 'reform' revolve around once-deeply Catholic nations of the Global North (e.g. Ireland, Spain, Portugal, but also the Canadian province of Québec). Up until the 1960s, and even beyond, it was the Catholic Church that largely embodied the welfare state in those places – orphanages, institutions for people with mental health and disabilities, homes for single female mothers, housing for elderly people, assistance to the unemployed and in the Canadian case, residential schools for the Indigenous population. This institutional context was an inheritance from the pre-welfare state time when, according to Jessop (2002: 118), 'arrangements should be instituted to encourage family, neighbourhood, informal, or market-based and market-sustaining solutions to the problems of social reproduction', essentially reverting to a state of affairs in which the individual, family, community, and God were the primary support systems (DeVerteuil, 2015). Beginning slowly in the 1960s in Ireland and Québec, the Catholic Church was forced to relinquish its responsibilities in the face of a drive to modernise and expand the state, as well as persistent scandals into the way the church had ran its social welfare institutions, particularly around orphaned children and single mothers, and Indigenous populations across Canada.

Now that we have established a range of national and sub national contexts, it is time to consider the inevitably local manifestations of the welfare state in terms of a variety of material frontlines, including social housing, street-level bureaucracy and contact zones. This builds on the Winnipeg case study, which suggested that the welfare state is necessarily embedded in local conditions, with a focus on the materialisation of the (welfare) state in concrete forms in the city.

4.5 Localising and Materialising Encounters with the Welfare State: Street-Level Bureaucracy and Contact Zones

In this section, we seek to transition from the national welfare state to more *local* and *material* manifestations, bringing it down to the everyday and the mundane but still focused on distribution and recognition. An initial way to localise the welfare state is through examining local forms of collective consumption, whereby the local state provides the basics of life without any strings attached. In 1983, Manuel Castells wrote *The Grassroots and the City*, which studied how citizens can participate in the creation and maintenance of cities. The book must be considered one of the most important works ever published on urban social movements, defined as 'organizations formed by residents to protest about, or make demands for changes in, the urban environment and urban services' (1983: 2). To Castells, cities were the result of an endless historical struggle over the definition of urban meaning by antagonistic social actors. Urban social movements combine struggles over (1) collective consumption with the (2) defence of cultural-territorial identity (e.g. recognising different populations via their particular neighbourhoods), making (3) specific demands on the local government. Castells suggested that urban social movements could be at the forefront in laying the foundation for a new democratic and just city, a city in which quality of life, preservation of local and cultural identities and community participation are taken seriously in decision-making. This collective consumption is essentially a right to the city through services that treats citizens as clients, a redistribution but not a radical revisioning of the city that will be revisited in Chapter 13.

We can further localise and materialise the welfare state in at least two other ways. The first is through the provision of public (or social) housing. Public housing is considered a social right, falling under the broader category of the right to shelter, which is recognised as a basic human right. While the right to housing is fundamental, the means of achieving it, like public housing, are considered social rights because they are not absolute and depend on government policies and resources. Public housing represents a foundational element of the welfare state, aimed at ensuring access to secure, affordable living conditions for individuals and families who are unable to meet housing needs through the private market. Social housing schemes are designed not just to alleviate homelessness or overcrowding, but to reduce inequality and promote social stability. Most forms of social housing are provided via subsidised rental housing – often managed by local authorities or non profit organisations. Here, the welfare state intervenes directly in the housing market to support those most vulnerable to market failures, including low-income families, the elderly and disabled people. In doing so, public housing embodies a broader commitment to collective responsibility and social citizenship, asserting that the well-being of individuals is a public concern and not merely a private burden. However, as we shall see in Chapter 11, social housing provision has greatly diminished over the past 50 years, due to policies and political rhetoric that have promoted homeownership and viewing public housing as a 'last resort'. Consequently, there has been an erosion of older stock levels and a reduced rate of new housing provision. As it has become more residual, typically only the poorest and most needy individuals and families are now eligible in many countries. This negative political framing, coupled with unfavourable media representation has meant that public housing has become more stigmatised for those living in such housing, mirroring the broader stigma surrounding welfare.

The second is through the relationship between those who serve on behalf of the welfare state, and those who are served by the welfare state. One way to understand this relationship is through *street-level bureaucracy,* whereby the servers have some agency at the local level (e.g. welfare office, police officer, case manager) to give or deny services or support to those who seek them. As a conceptual tool, it frames how individuals working in frontline public agencies use their individual discretion and ingenuity to spread limited resources more thinly and ration services when demand far outstrips supply. Rather than vehicles to transmit policies emanating from higher-tier governmental agencies, these street-level bureaucrats have autonomy to limit outlays in contexts of 'chronically inadequate resources in circumstances where the demand [would] always increase to meet the supply of services' (Lipsky, 1980: 81). Street-level bureaucracy is one way to frame the very real encounters between clients seeking services with those who dole out the services, exercising a certain discretion, and even agency.

Examples of street-level bureaucracy abound. The police officer who patrols a neighbourhood has extensive leeway regarding who they arrest, for what, and why. Think also of the governmental guard at an airport or even at the border – while they will all insist on showing identification, they do have some discretion as to who to question. In terms of the welfare state itself, this can take the form of applying rules in various ways, enabling, or denying benefits to specific people. This street-level bureaucracy becomes crucial for those neighbourhoods where welfare has strongly insinuated itself into the everyday rhythms, and has become symbiotic. In these so-called 'welfare neighbourhoods' (DeVerteuil, 2005), the attachment between welfare and place can be profound, including how many local merchants come to rely on welfare payments (e.g. grocers).

Second, the local welfare office becomes a contact zone between staff/managers and recipients, a space of encounter to the welfare state. Lawson and Elwood (2014: 214) define contact zones as 'interactions in which people grapple with social difference in ways that are neither a celebratory appreciation of difference ... nor a disciplining and defensive position that seeks to exclude or assimilate difference'. Contact zones can make boundaries between clients and staff/managers, but can also break them. The welfare system tends to feature both. First, boundaries are erected because demand far exceeds supply, but also because of the widespread notion that individuals are primarily responsible for their own fates, thus leading to a lot of emphasis on autonomy and self-improvement. These kinds of boundaries have been increasingly popular ways to shift the responsibility onto individuals, in terms of managing their own relationship with the welfare state. But second, clients can band together to make specific demands on the welfare state, sometimes through the voluntary sector (see Chapter 5), and sometimes as activists. This can occur in locally specific ways. A good example is Skid Row Los Angeles, where a large proportion of people are homeless and rely on some kind of welfare benefit – local relief for single people, housing subsidies and disability payments for people with long-term mental or physical impediments. We will revisit Skid Row in much more detail in Chapter 8, but suffice it to say that most of its inhabitants are unable to join the labour market for a variety of reasons (especially disability). This overall lack of opportunities has created an expansive contact zone between needy clients and an array of welfare services, one that largely works in the favour of the former. Further, the large proportion of people on welfare reduces the stigma associated with receiving public benefits and provides a measure of social acceptance and emotional/material support.

In other words, welfare offices, welfare websites, and the street-level bureaucrats become the contact zone, a place where the two sides meet. In the next section, we contemplate the more recent tendencies of the welfare state – that is has been so thoroughly 'reformed' that it is no longer recognisable, and that the 2020 pandemic revitalised the generosity of the welfare state, albeit briefly.

4.6 The Post-Welfare State Alongside the Pandemic Expansion

The high-water mark of the welfare state was arguably in the early 1980s in most nations of the Global North, and certainly the US and UK. Since then, the talk among social scientists and geographers has been around an emerging post-welfare state, one less focused on redistribution, universalism, and de-commodification. This drift has been framed by the idea of a 'hollowed out' welfare state where social policy moves 'away from redistributive concerns based on expanding welfare rights in a nation state towards more productivist and cost-saving concerns in an open economy' (Jessop, 1994: 17–18). Within certain nations, like the US, the impetus for the national welfare state has been to devolve responsibilities to the lowest rungs of the state, even to the voluntary sector (see Chapter 5), privatising traditional government functions, and withdrawing programmes entirely. This purported post-welfare state is characterised by 'a period of intensely coordinated activity to dismantle the liberal welfare state in accordance with the new ideological and political imperatives of market liberalism' (Fairbanks, 2009: 17). A post-welfare state has been represented as a sure sign of social breakdown, of capitalism run amok, of every person for themselves, of unfettered neoliberalism.

And yet, the promise of the post-welfare state must bump up against the reality that parts of it persist, or even re-emerge, sometimes with tacit state approval. These could be emerging sites of collective consumption, such as in Winnipeg, or the residual yet persistent nature of social housing in places like London. Further, it can re-emerge in the form of a fully-fledged, generous entitlement programme such as in the National Disability Insurance

Scheme (NDIS) in Australia. The NDIS is one of the largest, most costly and most rapidly implemented social policy reforms in Australia's history, effectively doubling federal spending on services for disabled people (Millar and Hayward, 2016), albeit via a hyper-marketised model. Re-emergence also included the radical spurt in the interventions of the (welfare) state to avoid mass unemployment and destitution during the 2020–2022 pandemic. During that time in many nations of the Global North, various welfare states mobilised to fill the gaps, (re)building a social safety net in record time. In the USA, the CARES Act, passed in March 2020, included the largest increase in benefits – and the widest eligibility – since the 1960s, if ever. Between 2019 and 2020, unemployment insurance went from 500,000 to over 5.5 million (Covert, 2024), signalling a full-throttle embrace of the (welfare) state as a way to protect workers from destitution. This destitution was through no fault of the worker, of course – entire industries basically shut down, from cruise liners to air travel to hotels, clubs and live music. Just as the first lockdown was met with overwhelming approval among Americans, so too did this first round of dramatic generosity.

Similarly, in the UK, the expansion of benefits during the pandemic was a welcome relief from a decade of austerity that effectively gutted every area of government, beginning in 2010 with the decree that every department would need to cut its budget by an average of 25%. In many cases, the government could not make these cuts via efficiency savings, and simply resorted to wholesale cuts. Regarding direct welfare payments, local housing allowances were capped, payments to disabled people were slashed with enormous new conditions imposed, a two-child benefit cap was instituted to ostensibly deter parents from having more children while on benefits, and so forth. Not surprisingly, homelessness soared, alongside the use of food banks and children in poverty. It should come as no surprise then that despite a pandemic-related spurt, especially around supporting workers who had lost their jobs, the UK welfare state remained insufficiently robust to fully counter the impacts of the pandemic (MacLeavy, 2024). Within this context, work by Sarah M. Hall (2018) and Eleanor Jupp (2022) have helped advance a relational account of the geographies of family, friendship and intimacy to better understand everyday survival before and during the pandemic. Their focus on the everyday brings the institutionalised and formalised spaces of austerity and the pandemic – the home especially – into everyday life, where personal and lived experiences revealed substantial evidence of mutual aid when the welfare state remained insufficient to offset pre-pandemic austerity (see also MacLeavy, 2024).

During the pandemic, there was even talk about a guaranteed wage for all citizens, including those who do not work in remunerated employment. In California, this was implemented in small batches in various cities, including Los Angeles. An article in the *Los Angeles Times* (July 31, 2024) by Rebecca Plevin and Dakota Smith focused on The Basic Income Guaranteed: Los Angeles Economic Assistance Pilot, or BIG: LEAP. This disbursed $38.4 million in city funds to 3,200 residents who were pregnant or had at least one child, lived at or below the federal poverty level and experienced hardship related to COVID-19. Participants were randomly selected from about 50,000 applicants and received the payments for 12 months starting in 2022. The cash assistance was 1,000 US Dollars per month and there were no strings attached, so the recipients could use the money as they saw fit. This unconditionality is quite rare. The *Los Angeles Times* article reported on a policy study which found that the city-funded programme was enormously beneficial to recipients. Beyond immediate everyday needs, the study also found that the initiative gave people the time and space to make deeper changes in their lives. That included landing better jobs, leaving unsafe living conditions and escaping abusive relationships. This became especially apparent in comparison to those applicants who did not receive the monthly payouts. Further, in a city with sky-high rents, participants reported that the guaranteed income functioned as 'a preventative measure against homelessness', according to the report, helping them offset rental costs and serving as a buffer while they waited for other housing support. Interestingly, the funding for the programme came out of the former Police Department budget, in response to nationwide protests after the George Floyd murder in May 2020 (see Chapter 13).

4.7 Conclusions

For some citizens, the welfare state operates very much in the background, ensuring that society works smoothly. But for others, as explored, the welfare state is an obligatory relationship, a matter of survival, of life and death. Being tied to the welfare state through poverty, disability or old age means that the relationships are very much symbiotic. In terms of social justice, the welfare state is evidently about distribution, of who gets what. This relationship is frequently measured between the citizen and non-citizen, and between the deserving and the underserving. In terms of procedure, the welfare state's impact on localities and individuals is the proverbial devil in the details. To those dependent on services, street-level bureaucracy interferes and sometimes seemingly operates on the whims of the bureaucrats themselves. Contact zones could be understood as shams or could actually create real connections to a welfare state that is sometimes reluctant to give clients that which they are actually entitled.

Recognition of difference is difficult to implement when the welfare state operates on universalistic grounds, i.e. everyone is equally entitled to certain benefits. But as the residual welfare state becomes ever more fragmented and targeted, the likelihood that the different needs of different groups are recognised can only grow, especially if they are able to organise along lines of collective consumption. But even this is rather precarious, in that the supposed post-welfare state is even more likely to stratify between deserving and undeserving along neoliberal lines.

And so the relationalities and encounters now seem more confusing, perhaps less fair, and certainly less universal and ambitious than they used to be, although we acknowledge that true universality is exceedingly rare. A running theme of this chapter is how a receding welfare state (or even a post-welfare state) is steadily giving ground to more individual and community-level engagement and bottom-up, grassroots collective consumption and activities to ensure daily survival in the 2010s and 2020s. The move away from, alongside or even beyond the (welfare) state implicates a new set of relationalities and encounters around 'care' and 'voluntarism', which we will consider in the next chapter.

References

Appelbaum, B. and Gebeloff, R. (2012) Even critics of safety net increasingly depend on it. *New York Times*, February 11. www.nytimes.com/2012/02/12/us/even-critics-of-safety-net-increasingly-depend-on-it.html?pagewanted=all&_r=0

Brown, K. (2012) Re-moralising 'Vulnerability'. *People, Place and Policy*, 6(1), 41–53. https://doi.org/10.3351/ppp.0006.0001.0005

Castells, M. (1983) *The City and the Grassroots*. Berkeley and Los Angeles, CA: University of California Press.

Covert, B. (2024) Is this what happens when you build a real social safety net, then take it away? *New York Times*, 12 March, 2024. Opinion I What Happened to Our Covid Social Safety Net? – The New York Times (nytimes.com).

DeVerteuil, G. (2005) Welfare neighborhoods: Anatomy of a concept. *Journal of Poverty*, 9(2), 23–41.

DeVerteuil, G. (2015) *Resilience in the Post-Welfare Inner City: Voluntary Sector Geographies in London, Los Angeles and Sydney*. Bristol: Policy Press.

DeVerteuil, G. and Wilson, K. (2010) Reconciling indigenous need with the urban welfare state? Evidence of immigrant-serving nonprofit sector services and spaces for Aboriginals in Winnipeg, Canada. *Geoforum*, 41, 498–507.

Esping-Andersen, G. (1990) *The Three Worlds of Welfare Capitalism*. Cambridge, UK: Polity Press.

Fairbanks, R. (2009) *How it Works*. Chicago: University of Chicago Press.

Jessop, B. (2002) Liberalism, neoliberalism, and urban governance. In N. Brenner, & N. Theodore (Eds.), *Spaces of Neoliberalism* (pp. 105–25). Oxford, UK: Blackwell Press.

Jupp, E. (2022) *Care, Crisis and Activism: The Politics of Everyday Life*. Bristol: Policy Press.

Hall, S.M. (2018) Everyday austerity: Towards relational geographies of family, friendship and intimacy. *Progress in Human Geography*, 43(5), 769–789.

Harvey, D. (1973) *Social Justice and the City*. Baltimore, MD: Johns Hopkins University Press.

Henriksen, L., Smith, L.R. and Zimmer, A. (2012) At the eve of convergence? Transformations of social service provision in Denmark, Germany and the United States. *Voluntas*, 23(2), 458–450

Japanese Ministry of Health, Labour and Welfare (2022) Annual Report on the Ageing Society [Summary] FY2022 June 2022 Cabinet Office Japan.

Jessop, B. (1994) Towards a Schumpeterian workfare state? Preliminary remarks on Post-Fordist political economy. *Studies in Political Economy*, 40, 7–40.

Kearns, R., Moewaka-Barnes, H. and McCreanor, T. (2009) Placing racism in public health: a perspective from Aotearoa/New Zealand, *GeoJournal*, 74, 123–129.

Lawson, V. and Elwood, S. (2014) Encountering poverty: Space, class and poverty politics. *Antipode*, 46(1), 209–228.

Lipsky, M. (1980) *Street-level Bureaucracy: Dilemmas of the Individual in Public Services*. New York: Russell Sage Foundation.

McLeavy, J. (2024) *Enduring Austerity: The Uneven Geographies of the Post-Welfare State*. Bristol: Bristol University Press.

Miller, P. and Hayward, D. (2016) Social policy 'generosity' at a time of fiscal austerity: The strange case of Australia's National Disability Insurance Scheme. *Critical Social Policy*, 37(1), 128–147. https://doi.org/10.1177/0261018316664463 (Original work published 2017).

Mizuuchi, T. (2020) Development of social inclusion in Japan. In T. Fukuda (Ed.), *Materiality, People's Experience and Making Geographical Knowledge* (pp. 69–103). Japan: Osaka Prefecture University.

Newman, J. (2012) *Working the Spaces of Power: Activism, Neoliberalism and Gendered Labour*. London: A&C Black.

Ojala, E. (2021) The Shadow of Ronald Reagan is costing us dearly. *The New York Times*, NY.

Rawls J. (1971) *A Theory of Justice*. New York: Harvard University Press.

Rowntree Foundation (2024) *UK Poverty Report 2023*.

Statistics Canada (2022) *Canadian Census of Population 2021*. Ottawa.

Wilton, R.D. (2004) More responsibility, less control: Psychiatrie survivors and Welfare State restructuring. *Disability & Society*, 19(4), 371–385. https://doi.org/10.1080/09687590410001689476

Wolch, J. and DeVerteuil, G. (2001) Landscapes of the new poverty management. In J. May and N. Thrift (Eds.), *TimeSpace* (pp. 149–68). London: Routledge.

Young I.M. (1990) *Justice and the Politics of Difference*. New York: Princeton University Press.

5 GEOGRAPHIES OF CARE AND VOLUNTARISM

5.1 Geographies of Care

We first explore the question of 'who cares' as part of the foundations of a more relational social geography, in that care creates relationships among and across various individuals and social groups. First, we move away from the state to embrace informal care, as part of civil society, that is the un-coerced and un-structured relationships provided by family, neighbours and the wider community. Second, we see how care is sometimes formalised into voluntary action and the voluntary sector. Care has become a crucial way to bridge the gaps in the welfare state (Chapter 4), but of course burden some people and places more than others. The uneven and incomplete post-welfare system allows (or has even come to rely on) these informal communities in civil society to take up some of the slack, especially when the market is not interested in stepping up (unless generous state procurement is involved), and when state austerity (e.g. cutbacks in government spending) is the main mode of governance, leading to a post-welfare state.

Conradson (2003) defined care as both a disposition ('caring about') and a practice ('caring for' or 'caregiving'); spaces of care constitute therapeutic environments that are fashioned by, for and through one person's interest in the well-being of another. Care is not present to the same degree everywhere, and it takes place in particular spaces. Sophie Bowlby (2012) and colleagues (Bowlby and McKie, 2019) seek to address this via the notion of *caringscapes*, a multifaceted approach to the everyday contexts, resources and relationships via which care is achieved in people's lives. More broadly, place-based approaches to the landscapes of care (for example Milligan and Wiles, 2010; Power and Hall, 2018) offer opportunities to consider these dynamics of care within specific contexts, which encourage questions about how space is constructed, experienced, bounded and scaled through caring activities.

There is a longstanding and deep connection between care and feminist geographical thought, in that caring for people is an essential component of social life but a responsibility that falls particularly on women (Hall, 2022). Caring for, and about people and places, and care as both a disposition and practice, relates to an older concept of social reproduction, a topic intimately

DOI: 10.4324/9781003414889-6

tied to women's unpaid labour, first raised in Chapter 2. The intricacies of raising children, for example, is rarely paid for directly by the welfare state, but rather in time and lost employment. The same goes for taking care of elderly parents, especially where it is mandated in the Global East, and in some cultures this feminised burden is undertaken in the place of directly-remunerated pay in the labour market. In these ways, care becomes social justice, particularly around the recognition that certain populations need care, and the procedures by which care is delivered and by whom.

This is a prominent theme of Jennie Middleton and Richard Yarwood's (2015) work on how care is delivered and received in the austere city. They explore the growing presence of Street Pastors in relation to their role as Christian volunteers who patrol urban places at night to provide practical support and care for the users and providers of the night-time economy. Attention is placed on the ways in which religious faith and ethics are performed to create liminal spaces of understanding and care in urban areas. Building on these insights, Middleton and Farhan Samanani (2021: 35) argue that care does not simply refer to practices of social reproduction or emotional attachment, but is a distinct mode of ethics, where 'everyday worlds are replete with matters of concern', visible in efforts to help people navigate and remake social worlds. Care is thus a matter of social justice (Gesser et al., 2022). Inherent in this work are considerations of how care is made possible, or indeed remade, across different spatial relations. So, rather than solely considering care as a virtue, it is also a process and an everyday practice, such that while good care is desirable, care happens regardless of its quality (Kittay, 2019), and can even be coercive, forcing certain vulnerable populations (especially the homeless) to seek it as a way to obscure their presence in public spaces.

Unlike the welfare state, which is territorially bound, care must reconcile social and geographical distance – in other words, is it possible to care about distant others or is care always localised? This relates to an ethics of care, of which some feminists have made much in recent years: 'humans are not fully autonomous, but must always be understood in a condition of ... interdependence ... all humans need care' (Lawson, 2007: 1). For vulnerable populations such as the homeless and those suffering from mental ill-health, community-based drop-in centres or homeless shelters can function as a therapeutic environment offering sanctuary, moral support, and essential sustenance. In their book *Swept Up Lives?*, Paul Cloke et al. (2010) investigated the multiple service spaces that homeless people use in the UK, including foodbanks/soup runs (outdoor relief), day centres and hostels (indoor relief). Further, they demonstrated how these spaces were intertwined with an urge to care among the staff that cannot simply be seen to hide the abrasive presence of those using their services. These contact zones are highly gendered, with women more likely to be carers and volunteers. This runs counter to the Marxist-inspired views (Mitchell, 2003) that emphasise the strongly unjust, regulated and uncaring nature of current political responses to visibly vulnerable populations. More forcefully, it remains to be seen whether the notion of care, both localised and for distant others, can effectively challenge deep-seated injustices, or if they are more about surviving in the meantime, a form of resilience (see Chapter 13).

Outside of urban areas, the point about care needing to reconcile geographical distance is of particular relevance in rural areas. Here, considerations must be made about how to best support long-term care in place, particularly for older people living in geographically remote areas (Skinner et al., 2017). Mark Skinner's work reveals the many factors influencing older people's ability to remain at home in these rural communities. Housing is one issue, as the accessibility of the home becomes paramount as people age in place. Older people may require housing adaptations to remain living at home. More generally, healthcare and long-term care services and reliable transportation options are more limited in rural areas. Finally, opportunities for maintaining social interaction may become more challenging, as people's mobility and social networks may shrink. These challenges are further exacerbated as geographical remoteness increases, disproportionately affecting older people in the most isolated regions. Naturally, the extent of remoteness differs by county, so rural areas in Canada where Mark Skinner is writing from, are extremely remote by global standards due to the country's vast geography, low population density and limited infrastructure outside urban centres. Rural restructuring has also prompted an imbalance between older residents who wish to age in place and younger family members' who wish to move to urban areas, affecting their ability to support them. While defining 'rural' is context-dependent then, at the very least there is consensus that rural locales present different challenges and opportunities to health and care delivery to those evident in urban settings.

Returning to Vicky Lawson (2007), quoted above, she talks about how the decrease in state-provided support systems pushes care to the fore, especially its ethical implications. To her, 'care ethics focuses our attention on the social and how it is constructed through unequal power relationships, but it also moves us beyond critique and toward the construction of new forms of relationships, institutions, and action that enhance mutuality and well-being' (2007: 3). An ethics of care demands that citizens build greater connections and mutualities in daily life. In turn, this means taking seriously the various contact zones of care: between the served and the servers, between the needy and the privileged and especially between strangers. Bolstering these claims is the fact that everyone, particularly early on and later in their lives, need care to survive. This is universal; no person is an island, especially when they are children and when they are elderly. Even those people who purport to live 'off the grid' have benefited from care in their early years, including schooling and parenting. But at an even larger scale, we can talk about the existence of a contract between citizens and society that involves the obligatory reliance on key infrastructure (such as housing, water, electricity, roads), which is collectively supplied and used.

As we have seen in Chapters 1 and 4, this symbiotic relationship between individuals and society can be captured in a variety of ways, although in this chapter the focus is primarily on the mutually beneficial.

In these respects, care overlaps with ideas around reciprocity and hospitality, which in relational terms is all about mutual benefits (both sides win). Reciprocity is about exchange – in this case care – with others for mutual benefit, even if they are not always in equal proportions. Care can be immediate but can also be stockpiled for later use. Another form of mutuality is hospitality, that is the welcome extended to strangers (Ince and Bryant, 2019). This welcome can be conditional, i.e. based on certain expectations of reciprocity, for example, or it can be truly unconditional, without strings attached and potentially overlapping with the commensal or even the parasitic. Let us take an example of care given to unhoused individuals. At first glance, this care is only tangentially about reciprocity – there is really no expectation that the unhoused individual will 'give' something back, at least not something tangible. But the care given mutually benefits both sides, in that the giver obtains a sense of gratitude and 'doing good'. This giving is usually with some strings attached, some expectation that the unhoused individual will behave in certain ways so as to ensure mutual benefit. But there are cases where the unhoused individual might not express gratitude, or that the giver expects nothing in return – this is approaching what we have deemed unconditional hospitality. As previously noted, care given to the individual might have so many strings attached that it becomes coercive, that the individual *must* form an (obligatory) relationship with the care-giver in order to eat, sleep, even exist. This obligatory nature will be explored in more depth in Chapter 9, which is concerned with the relationship between disabled individuals and their caregivers, and especially their various space of encounter.

5.2 Care and Social Infrastructure

One way to geographically ground care, to localise it, to put it in relation with everyday material reality, is through the emerging concept of *social infrastructure*. The urban sociologist Eric Klinenberg (2013; 2018) understood social infrastructure as 'informal, incremental, peopled…infrastructure that supports social reproduction in cities' Jennie Middleton (2021: 777) in a special issue on the concept, outlines how a Chicago heatwave in 1995 and the varied responses by neighbourhood groups in the city provided a striking illustration for Klinenberg of the value of social infrastructure:

> During the heatwave – an event that claimed 739 lives – the pattern of deaths largely followed the city's geography of poverty and racial segregation. But there were exceptions. Englewood and Auburn Gresham, two adjacent neighborhoods […] both ninety-nine per cent African-American, with similar proportions of elderly residents. Both had high

> rates of poverty, unemployment, and violent crime. Yet, Englewood experienced one of the city's highest death rates, Auburn Gresham, one of the lowest. The difference, argues Klinenberg, was Auburn Gresham had comparatively robust social infrastructure – 'the people, places, and institutions that foster cohesion and support' (Klinenberg, 2013). He quotes Betty Swanson, a long-time resident, who recalls: 'During the heat wave, we were doing wellness checks, asking neighbors to knock on each other's doors [...] The presidents of our block clubs usually know who's alone, who's aging, who's sick. It's what we always do when it's very hot or very cold here.'

Klinenberg (2018: 1) underlines that social infrastructures comprise 'physical places and organisations that shape the way people interact', not social capital (see Chapter 10) but the physical conditions that determine whether social capital develops. This takes us into an analytical frame which grapples with unequal and shifting contexts of care, and a need for a critical approach to the practices, politics and ethics of care in relation to specific sites, beyond interpersonal caring dynamics (Jupp, 2022). When social infrastructure is robust, 'it fosters contact, mutual support, and collaboration among friends and neighbours; when degraded, it inhibits social activity, leaving families and individuals to fend for themselves' (2018: 5). Social infrastructure frames and grounds care by bringing the spatial and social together in particular places, such as libraries, care homes and food markets. As such, social infrastructure is designed to be highly accessible; as Klinenberg (2018: 124) argues, vulnerable populations 'need an environment that's not like every other environment they've ever known, that judges them, that takes advantage of them, that doesn't want anything to do with them, doesn't understand their role in society'. Alan Latham and Jack Layton (2019) saw the crucially public nature of social infrastructure – that the state, or some other collectivity (such as the voluntary sector, in the next section) can provide social infrastructure for public and private use. As such, social infrastructure is particularly important to vulnerable populations such as the homeless, the elderly, and refugees, all of whom have been particularly forgotten in the current crisis, or worse, a target for stigma through places such as care homes, homeless shelters and prisons.

Yet social infrastructure is never monolithic – one person's sociality and care is another person's exclusion, given that hospitality is rarely extended in completely unconditional ways, what Ash Amin (2008) called a cultivation of 'urban surplus' that encourages trust, civility, encounter and common purpose (Middleton and Samanani, 2022). This urban surplus runs against the austerity that is the hallmark of the post-welfare state (and street-level bureaucracy) more specifically and neoliberalism more generally. In this regard, the following image gives a sense of social infrastructure in the city, its materiality. The image directly below shows a commercial street in Hackney, London. A variety of private and voluntary uses sit across just three short

FIGURE 5.1 London, 2014 © G DeVerteuil

storefronts: two religious centres (Christian, Hindu) as well as a combined hair salon/international money transfer/furniture store.

None of these storefronts are part of global chains; rather, they are social infrastructure in terms of being locally-focused places of interaction and certainly caring, at least for those who avail themselves of the services on offer.

Social infrastructure can also extend beyond the confines of local neighbourhoods to embrace *the city as social infrastructure*. Caring for the city can connect to a sense of belonging, the multifaceted and unstable process whereby some people belong to certain places (more in Chapter 10), and how this feeds into the politics of identity and the politics of place. Belonging is a key aspect of cities, involving notions of hometown, rootedness, place-based attachment and place-based identities, but always at the risk of being insular, of being wary of strangers and those who do not fit it. The following image reminds passersby that the city – in this case Buenos Aires – is yours, take care of it.

Care is thus crucial in the practice of making cities liveable on an everyday basis – a relationship that takes great effort, is very much daily and ordinary, and thus usually unsung.

One way to frame social infrastructure and care for the city is through the idea of the '15-minute city', which had by 2025 become an important local policy in Paris. This concept brings together a more parochial focus around the immediate neighbourhood, where more activities should occur within a tighter circle rather than citywide, i.e. living in one area, commuting to another, shopping in yet another, strongly segregated by class and cut through with automobile-oriented infrastructure. Rather, at least in Paris,

FIGURE 5.2 Buenos Aires, 2015. Translation: *The city is yours, take care of it* © G DeVerteuil

local shops are subsidised, and new housing is mixed according to income. The images from Paris on 56 in Chapter 3 speak to an emerging 15-minute city that keeps everyday life to the scale of the neighbourhood as much as possible.

Box 5.1 Engagement Questions

How close does the current area you are living in fit with the 15-minute city model, and how might it act as social infrastructure?

For those living away from their family home, how does the area where you grew up differ (many university towns are frequently modelled on the 15-minute city, without actually saying so)?

Tight-knit Puerto Rican *barrios* (neighbourhoods) in New York City, whether in Spanish Harlem, the South Bronx, the Lower East Side or South Williamsburg, are renowned for their memorial murals, part of a longer Latin American tradition of *memento mori* (reminders of mortality). These murals operate as a kind of material and place-based social infrastructure, indicating that people who have passed on where once part of the neighbourhood, were loved, were important and are now missed.

The murals speak, in very obvious ways, to the very strong connection between neighbourhood pride and the people who live there. In Figure 5.5, a strong link is made between the recent rebuilding of the South Bronx and the efforts of Father Louis Gigante in channelling funds to housing

FIGURE 5.3 Brooklyn, 2020 © G DeVerteuil

FIGURE 5.4 Lower East Side, Manhattan, 2020 © G DeVerteuil

programmes. This message seeks to reverse decades of stigmatisation to an area once supremely associated with urban decay and abandonment in the 1970s, 1980s, and 1990s.

But the image also speaks to mutual aid, reciprocity and taking care of one own's community-in-place, care again literally as city-making.

FIGURE 5.5 South Bronx, 2019 © G DeVerteuil

This community care can be done in a variety of ways, including through community-based murals. Murals tend to be marginal (the authors, but also the location), anonymous, spontaneous, precarious and fleeting. Critically, to many outsiders, murals are cryptic and senseless; they are understood by only a fraction of those who encounter them. Usually read as a sign of urban decay and as a loss of control over the physical environment, murals in places like the South Bronx can be understood more as a voice for the voiceless, and a means of promoting community solidarity, from 'way back'. Suffice it to say that the stigmatisation associated with these neighbourhoods led to particular spatial expressions like graffiti, which in turn sought to challenge those stigmas.

As such, murals can present a different side to a community, one that counters negative portrayals. Of course, focusing on one own's neighbourhood, community and people is a potential drawback to social infrastructure – that it tends to focus on the 'near and dear', with little regard for other places and communities, parochial even. Parochial is a loaded term. It can capture a sense of localised belonging, which is positive, or a wariness and outright hostility towards outsiders, which is negative. It can sustain social infrastructure, but at the expense of others. It is, therefore, a messy concept that embraces some and leaves others out entirely.

Yet care for some neighbourhoods can do the opposite, and extend lines of support across different places. A good example of this is through *remittances*, which usually involves immigrants sending money from the Global North or East to the Global South to support their sending communities, a sort of bottom-up globalisation and everyday transnationalism. This is

FIGURE 5.6 South Bronx, 2019 © G DeVerteuil

very much part of the geographies of the elsewhere, as part of an extended urbanisation. Remittances add more to the Global South economy than foreign aid from the Global North, constituting a crucial bridge between richer nations and poorer ones, a geopolitics of care that links places such as: Los Angeles and Mexico; Hong Kong and the Philippines; Dubai and Pakistan; Madrid and Morocco; London and Ghana; Miami and Haiti; Tel Aviv and Ethiopia; and so forth. In terms of surviving on the margins, more about this relationship will follow in Chapter 8.

During the pandemic, the utility of social infrastructure was severely curtailed given social distancing, but simultaneously greater demand was placed on facilities that catered to populations who sought to avoid indoor spaces. More broadly, the pandemic was a crisis of care, especially in terms of burdening some populations more than others. Those who could not work from home were burdened with commutes and the danger of catching COVID-19, when compared to those who had the luxury of working from home. In this sense, immobility became a privilege (Marvin et al., 2023). But even those who did work from home had to deal with the lockdowns on schools (if they had school-age children), which meant that their children also had to be educated from home, alongside work.

As with social infrastructure and other place-based modes of care in the city, similar pressures were put upon the voluntary sector (which we explore in our next section) during the pandemic. Care in the community is always geographically and socially limited, and there comes a time when more collective formalised agents, in the form of the voluntary sector, must intervene on behalf of those in need, especially within a context of a receding welfare state and for those, due to the sometimes parochial nature of social infrastructure, who are left out entirely. Our focus thus turns next to the voluntary sector.

5.3 The Voluntary Sector

The uneven withdrawal of the state from welfare and care provision, paralleled by the growth in vulnerable groups without familial, informal, state or market ties increasingly implicates the voluntary sector to fill the gaps. The voluntary sector comprises nonprofit organisations, charities, community groups and volunteers that operate independently from government and for-profit businesses. It forms an important component of social infrastructure, alongside physical and organisational structures that support the quality of life in a community, such as schools, libraries, parks, health services and community centres. Indeed, voluntary organisations often deliver essential social services (e.g., food banks, shelters, mental health support, elder care). These services fill gaps in government or market provision, and acts as a backbone of social infrastructure, particularly in underserved areas. Once again, the contact zones created by the voluntary sector are highly gendered, with women more likely to be volunteers. In many cases, social infrastructure would be incomplete – or significantly weakened – without the contributions of the voluntary sector.

This obligation of managing strangers sets the voluntary sector apart from the informal community-based care via family and friendship (Hall, 2018). Put differently, the 'voluntary sector is distinguishable from the state by its independence; from the market by its emphasis on the non-profit principle, mutualism, and altruism; and from the family/community by its formality' (DeVerteuil, 2015: 41). And relating back to care, voluntary sector organisations may be conceived as 'all in one way or another caring organisations, providing services or goods with a "dual" public (collective) and private (individual) nature. Care for others on a voluntary basis, directed at a more or less defined and exclusive "other," is regarded as their common denominator' (Brandsen et al., 2005: 750). Salamon and Anheier's more detailed definition (1998: 219) also followed this approach to foreground their own extensive, cross-national study of the voluntary sector. First, the voluntary sector organisations are organised, i.e. they possess some institutional reality. They are private (although not to be confused with private sector which denotes for-profit entities), i.e., institutionally separate from government. They are non profit-distributing, i.e. not returning any profits generated to their owners or directors. They are *self-governing*, i.e. equipped to control their own activities. They are *voluntary*, at least in part, i.e., they involve some meaningful degree of voluntary participation, either in the actual conduct of the agency's activities or in the management of its affairs.

The latter point serves as a crucial distinction between the voluntary sector and voluntarism. While they are closely related, they are distinct concepts. The above paragraph defines the voluntary sector, whereas voluntarism is the principle or practice of giving time and effort freely, without financial compensation. This can involve individual or community action and can happen within or outside of formal organisations. It is motivated by altruism, civic duty, religious beliefs, etc. Voluntary sector organisations often rely on voluntarism, i.e. the involvement of volunteers, but they can also include paid workers, sometimes in very specialist roles.

The voluntary sector therefore emerges as a boundary institution, joining and formalising the urge to care with the larger welfare state. But the voluntary sector must negotiate with both of these sides, occupying a space between the state and citizens (or those at the margins of citizenship), absorbing but also deflecting intentions of care, mutuality and reciprocity on the one side and state impositions on the other (Trudeau, 2008). Indeed, since the 1980s and the onset of the decline of the welfare state, the exact role of the voluntary sector remains a point of vigorous debate. In the wake of the 1980s fiscal crises, the voluntary sector was regarded as a panacea by governments in the Global North for helping to resolve these crises by delegated responsibilities for social welfare to nonprofit and voluntary organisations (Skinner and Power, 2019). Jennifer Wolch (1990) was enormously influential in identifying a new partnership model between the state and the voluntary sector, one she defined as a *shadow state*, through which the welfare state could safely devolve risk and responsibility to non-state actors to fill the missing gaps. Wolch argues that, particularly in the United States and the United Kingdom at this time, this delegation of responsibilities meant that these organisations, while operating independently, effectively functioned as extensions of the state. As a result, voluntary organisations became primary providers of services like housing, healthcare and welfare support. In turn, the voluntary sector became subservient to the (receding) welfare state, dependent on its funding and captured by its rhetoric of a leaner set of services. Their reliance on government funding and contracts often limited their autonomy and capacity to advocate for systemic change. Wolch contends that this arrangement allowed the state to co-opt the sector and maintain control over service provision while minimising direct accountability. Moreover, it meant that the voluntary sector needed to implement more punitive elements, to control limited resources and manage their clients.

Yet the voluntary sector can also forge its own path, exercising a certain latitude to engage with vulnerable communities in ways that diverge from a strictly co-opted model. And so the voluntary sector is caught between doing the welfare state's 'dirty work' and providing genuine sites of care, help and sustenance for vulnerable populations, a paradox as much as a panacea. The example of food banks is apt – they provide immediate resources for everyday survival for, and among, complete strangers, but they also act as spaces of care, introducing alternate values of generosity that run against the increasingly stringent and bare-bones nature of the welfare state. Food banks are also spaces of voluntary labour, in that most of the work is done by volunteers on an everyday basis. Food banks can, therefore, operate not just against but also alongside this welfare state. This connects to a common desire on the part of voluntary sector organisations to serve, though not exclusively, dependent and vulnerable people, usually those ill-served by other societal structures, including the failure of the market and/or state to absorb their needs, as well as a lack of connection to informal communities such as family and neighbourhood. As such, these populations see the voluntary sector as a safety net for everyday survival (Clifford et al., 2012) and 'voluntary sector organisations maintain a critical layer of social protection that … can mean the difference between life and death' (Evans, 2011: 24).

So, there is a larger debate around the role of the voluntary sector in the management of people who are obliged to rely on it, revolving around the punitive versus the caring/supportive (DeVerteuil et al., 2020). The former punitive framing sees the voluntary sector as a key component to coerce vulnerable populations off the street and into services, obscuring their visibility in prime urban areas and chaining them to dependent relationships in the service of neoliberalism and shadow state constructs. In other words, the voluntary sector is enlisted to micro-manage vulnerable populations – part and parcel of 'the criminalization of urban poverty' (Wilson and Keil, 2008; Gowan, 2010). By contrast, the latter more caring framing sees the voluntary sector as at least partially redeemable, functioning as crucial and potentially alternative spaces of citizenship by taking in and tolerating/sustaining those excluded by neoliberalism as non-producing and non-consuming (Trudeau, 2008; Evans, 2011). As such, the voluntary sector also provides a vehicle for social movements, alternative citizenship and survival, despite its notoriously uneven nature and its continued dependency on the state (Cloke et al., 2010; Williams et al., 2012). Dan Trudeau, in particular, is crucial in showing how the voluntary sector can carve out alternative citizenships for people who are otherwise excluded from the national citizenship regime (Chapter 3).

Another way to illustrate this delicate balance and the alternative citizenship model is through a focus on the voluntary sector that caters to specific immigrant communities. At least within immigrant-receiving nations such as the US, this specific aspect of the voluntary sector emerged in the 19th century, as part of a paternalistic welfarism characterised by locally connected self-help and mutual aid organisations embedded within immigrant communities (Hasson and Ley, 1997; Milligan and Conradson, 2006). Places such as Little Italy and Chinatowns became the sites of immigrant-serving organisations. But with the rise of a more organised, national-level welfare state, the prominent components of (immigrant) paternalistic welfarism were absorbed into a more universalistic welfare state, including hospitals, care for neglected children and education, articulated through a normative 'model of citizenship based on collective rights, distributive justice, and representational democracy' (Amin et al., 2002: 3).

Present-day immigrant hubs are one way to illustrate the reverse trend: the transition from a centralised, universalistic welfare state back to an increasingly patchy, community-based and DIY (do-it-yourself) one. The case here is that of Central American immigrants in Los Angeles, who over the past 40 years have created a home-grown voluntary sector that sprang from grassroots beginnings and mutual aid, duplicating survival skills developed in the Global South. For this community, the grassroots origins were strongly linked to the Sanctuary movement (Landolt, 2008: 61), in which 'churches ... declared their places of worship as refugee safe havens'. This was set within a period of what Popkin (1999) deemed 'reactive ethnicity', of rising solidarity around national origin, bolstering the role of community organisations and consolidating (if not expanding) links to the homeland. The geography of the Central American immigrant hub remains remarkably concentrated within a small area west of Downtown LA. For such a car-dependent city, this concentration, if not co-location, seems almost quaint. This immigrant hub was also, for most of its history, a largely DIY affair. In other words, activists created the organisations themselves, without any government aid. It was only 15 or 20 years later that some organisations began seeking governmental funding. To this day, many of the organisations refuse any funding, given that it restricts their ability to cater to certain subgroups, such as undocumented arrivals. In this respect, the current immigrant-serving organisations revive and align themselves with the remnants of older forms of assistance from the pre-welfare state, when immigrants had to create their own institutions from scratch.

The Central American example shows how the voluntary sector can blur the Global North and Global South welfare models. In this respect, we also need to consider the emergence of the NGO (non-governmental organisation). NGOs are voluntary sector organisations that focus on some kind of social or humanitarian mission, usually in the Global South, and usually in the service of building local capacities in water, shelter, food but with a view to empower the local population rather than encourage permanent reliance. NGOs can operate on an everyday basis in some nations, but also can intervene during disasters. In this respect, well-known organisations such as the Red Cross, Islamic Relief, and Médecins sans Frontières are all NGOs that operate in disaster-mode in particular. While many NGOs are well-meaning and can be crucial during disasters, they suffer from the same issue as the voluntary sector in the Global North – their existence enables (sometimes very corrupt) states to ignore the needs of the population, as well as absolve nations in the Global North in their obligation to directly help exploited places in the Global South.

5.4 Philanthropy

The growth of the voluntary sector at the global scale has paralleled the growth of large-scale donations, otherwise known as *philanthropy*. While some voluntary sector organisations are heavily dependent on donations, philanthropy itself is not limited to the voluntary sector. Rather, money can be given to a wide array of organisations, from sports teams to colleges, opera companies, orchestras and museums. Philanthropy tends to be focused on cities, but can also range to the global level. Over the past 20 years, the reach of philanthropic organisations has increasingly dramatically, in lockstep with the obscene growth in millionaires and billionaires globally, many of whom see philanthropy as a way to bolster their image. Philanthropy also very much pre-dates the welfare state outlined in Chapter 4, burgeoning in the era where the poor and needy depended on handouts and charity from powerful donors, religious-based orders or even companies.

Nowadays, we need to present a critical yet balanced perspective on philanthropy, some of which clearly bolsters 'good causes' and some of which clearly makes the world a more socially unjust place (Fuentenebro and Acuto, 2022; Hay and Muller, 2014). If we take the latter, then we could say that philanthropy is largely done by the rich for the rich, and that these large donations take away from taxable income and wealth that could be used by the government in more democratic ways, and that many very large donors have made their money with the help of tax loopholes in the first place that may well have worsened social inequalities. So in this view, philanthropy is less about redistribution, and more about promoting elite causes that only reinforce the inequalities of the world. More specifically, the biggest donations tend to go to elite universities and schools that the rich themselves had attended. In the UK, in the 10-year period to 2017, more than two-thirds of all millionaire donations – £4.79bn – went to higher education, and half of these went to just two universities: Oxford and Cambridge (Vallely, 2020). In the United States, the playing field becomes even more lopsided, with some universities (e.g. Harvard, Yale, Stanford, Princeton) having larger endowments than the GDPs of some poorer nations.

Conversely, certain large foundations – such as the Gates Foundation – invest more than the foreign aid budget of many nations which aspires to go to benevolent causes (e.g. global health and poverty reduction). Philanthropy is also connected to funding research on socially just causes. Indeed, Andy's PhD and first academic job on family caregiving and researching disability rights respectively, were both underpinned by an Irish-American philanthropy, *The Atlantic Philanthropies*. Finally, philanthropy is connected to social investment, in that donations can help fund social enterprises that, unlike the voluntary sector, are expected to generate profits (that are then rolled back into the enterprise itself). This has been deemed *philanthrocapitalism*, in that donations become part of a business model that privileges the return on investments.

However, philanthropy need not always be based on the personal whims of the super-rich – it can be a source of positive change and support. An early example was Andrew Carnegie, a 19th-century steel magnate. He came penniless from Scotland, but then emerged as the world's richest

FIGURE 5.7 Atlantic City, April 2019 © G DeVerteuil

individual. Carnegie built a network of nearly 3,000 libraries like the one pictured in Figure 5.7 and other institutions to help the poor elevate their aspirations.

Interestingly, Carnegie himself never questioned the built-in inequalities of the system that had made him and others so rich. We would have to wait until the mid-20th century before the welfare state emerged as a strong competitor to philanthropy, at least in the Global North.

In contemporary times, many philanthropists are well-known entertainers who use their money and fame to attract attention to good causes. For instance, venerable musicians such as Elton John and Dolly Parton donate funds to help medical research, social justice issues around homophobia and women's rights, and access to university (as in the case of Stormzy's programme to help under-represented students get into Cambridge University).

For instance, here are the mission statements from Elton John's AIDS Foundation, from Our Mission I Elton John AIDS Foundation:

> The Elton John AIDS Foundation was established in 1992 and is one of the leading independent AIDS organisations in the world. Our mission is simple: to be a powerful force in ending the AIDS epidemic.
>
> We believe AIDS can be beaten and we're committed to making it happen. That means everyone at risk of HIV – no matter who they are or where they are – must get compassionate support and care that allows them to stay healthy and safe, and live with dignity.
>
> No more discrimination. No more HIV infections. No more AIDS deaths.

- We focus on communities who are most at risk of HIV – including LGBTQ+ people, young people, and people who use drugs – to make sure no one is left behind. We know how to reach communities who have been marginalised.
- We aren't afraid to fail, to take risks and make a stand.
- We work in some of the most challenging countries and contexts, and we work at all levels to influence change – with local partners for local solutions and, amplified by Elton's voice, at national and international levels.

Philanthropy is a good example of how social geography blurs with economic geography – of how certain social issues attract resources while others do not. But, like all issues of social geography, there is a discussion to be had around the social justice aspects of philanthropy.

5.5 Conclusions

In this chapter, the divides have been quite close to the ground, more so than the last chapter, and involve several obligatory relationships. They include the care that is provided as mutual aid within communities, but also how the voluntary sector can formalise some of the care as a boundary institution. The voluntary sector becomes yet another contact zone alongside the welfare state, between total strangers (e.g. clients, volunteers, professionals), each with their own agendas. As we saw, not surprisingly, these contact zones are highly gendered, with women more likely to be carers and volunteers. Returning to a symbiotic approach, care and the voluntary sector can be mutually beneficial yet also coercive, lopsided, parasitic. From the case studies, the relationships tended to be mutually beneficial (care, hospitality, reciprocity) but occasionally coercive via the voluntary sector. In other words, the contact zones around care and voluntarism can be positive and emancipatory, or just as disempowering as the older welfare state that they replaced. Equally, social infrastructure can be welcoming or parochially closed to only the insiders.

Care can relate to the three strands of social justice: procedure, in terms of an ethics of care that is inclusive (or not); distribution, in terms of more resources (but uneven and very localised); and recognition, in terms of catering to specific groups and their needs in a way that a universal welfare state could not or will not (e.g. immigrant hubs). Compared to Chapter 4, the scale of our perspective has become very localised, to specific neighbourhoods, cities and groups.

Our outline of the various relationships that mark society and social geography across five chapters included the social, social justice, inequality, difference, intersectionality, the state, citizenship, the welfare state, the voluntary sector and the community. These relationships can in turn frame encounters with various sub groups, the subject of Part II of this book. In the next chapter then, we begin to consider these relationships and spaces of encounter for a particularly ignored group – the young.

Further Reading

Baillie Smith, M. and Laurie, N. (2011) International volunteering and development: Global citizenship and neoliberal professionalisation today. *Transactions of the Institute of British Geographers*, 36, 545–559. https://doi.org/10.1111/j.1475-5661.2011.00436.x

Fuentenebro, P. and Acuto, M. (2022) The gifted city. Setting a research agenda for philanthropy and urban governance. *Urban Studies*, 59(9), 1944–1955.

Heng, T. (2021) Photographing absence in deathscapes. *Area*, 53(2), 219–228.

Leiter, J. (2008) Non-profit isomorphism: An Australian–US comparison. *Voluntas*, 19, 67–91.

Montero, S. (2020) Leveraging Bogotá: Sustainable development, global philanthropy and the rise of urban solutionism. *Urban Studies*, 57(11), 2263–2281.

Nielsen, A.B. and Papin, M. (2021) The hybrid governance of environmental transnational municipal networks: Lessons from 100 Resilient Cities. *Environment and Planning C: Politics and Space*, 39(4), 667–685.

Parnell, S. (2016) Defining a global urban development agenda. *World Development*, 78, 529–540.

Phelps, N.A. and Miao, J.T. (2020) Varieties of urban entrepreneurialism. *Dialogues in Human Geography*, 10(3), 304–321.

Piper, K. (2019) *Why this billion-dollar foundation is becoming a corporation*. Vox, February 7, 2019.

Puttick, R. (2023) The influence of philanthropic foundations on city government innovation. *International Journal of Urban and Regional Research*, 47(5), 689–875.

Reich, R. (2018) *Just Giving: Why Philanthropy Is Failing Democracy and How It Can Do Better.* Princeton: Princeton University Press.

Roelofs, J. (2015) How foundations exercise power. *American Journal of Economics and Sociology*, 74(4), 654–675.

Smith, D.M. (1994) *Geography and Social Justice*. Oxford: Blackwell.

Vallely, P. (2020) How philanthropy benefits the super-rich. *The Guardian* 8 September. Philanthropy.

References

Amin, A. (2008) Collective culture and urban public space. *City*, 12(1), 5–24.

Amin, A. and Thrift, N. (2002) *Re-imagining the Urban*. Cambridge UK: Polity.

Bowlby, S. (2012) Recognising the time – space dimensions of care: Caringscapes and carescapes. *Environment and Planning A: Economy and Space*, 44(9), 2101–2118. https://doi.org/10.1068/a44492

Bowlby, S. and McKie, L. (2019) Care and caring: An ecological framework. *Area*, 51(3), 532–539. https://doi.org/10.1111/area.12511

Brandsen, T., van de Donk, W. and Putters, K. (2005) Griffins of chameleons? Hybridity as a permanent and inevitable characteristic of the third sector. *International Journal of Public Administration*, 28, 749–765.

Clifford, D. (2012) Voluntary Sector Organisations working at the neighbourhood level in England: Patterns by Local Area Deprivation. *Environment and Planning A*, 44. https://doi.org/10.1068/a44446

Cloke, P., May, J. and Johnsen, S. (2010) *Swept up Lives? Re-envisioning the Homeless City*. London: Wiley RGS-IBG Series.

Conradson, D. (2003) Spaces of care in the city: The place of a community drop-in centre. *Social and Cultural Geography*, 4(4), 507–525.

DeVerteuil, G. (2015) *Resilience in the Post-Welfare Inner City: Voluntary Sector Geographies in London, Los Angeles and Sydney*. Bristol: Policy Press.

DeVerteuil, G., Power, A. and Trudeau, D. (2020) The relational geographies of the voluntary sector: Disentangling the ballast of strangers. *Progress in Human Geography*, 44(5), 919–937.

Evans, J. (2011) Exploring the (bio) political dimensions of voluntarism and care in the city: The case of a 'low barrier' emergency shelter. *Health & Place*, 17(1), 24–32.

Elton John AIDS Foundation (2024) Our Mission I Elton John AIDS Foundation. eltonjohnaidsfoundation.org

Gesser, M., Zirbel, I. and Luiz, K.G. (2022) Care for disabled people with complex dependency. *Estudos Feministas*, 30(2), 1–15.

Gowan, T. (2010) *Hobos, Hustlers, and Backsliders: Homeless in San Francisco*. Minneapolis: University of Minnesota Press.

Hall, S.M. (2018) Everyday austerity: Towards relational geographies of family, friendship and intimacy. *Progress in Human Geography*, 43(5), 769–789.

Hall, S.M. (2022) For feminist geographies of austerity. *Progress in Human Geography*, 46(2), 299–318. https://doi.org/10.1177/03091325211065118

Hay, I. and Muller, S. (2014) Questioning generosity in the golden age of philanthropy: Towards critical geographies of super-philanthropy. *Progress in Human Geography*, 38(5), 635–653.

Hasson, S. and Ley, D. (1997) Neighborhood organisations, the welfare state, and citizenship rights, *Urban Affairs Review*, 33, 28–58.

Ince, A. and Bryant, H. (2018) Reading hospitality mutually. *Environment and Planning D: Society and Space*, 37(2), 216–235.

Jupp, E. (2022) *Care, Crisis and Activism: The Politics of Everyday Life*. Bristol: Policy Press.

Kittay, E. (2019) *Learning from My Daughter: The Value and Care of Disabled Minds*. Oxford: Oxford University Press.

Klinenberg, E. (2013) Adaptation. *The New Yorker*. www.newyorker.com/magazine/2013/01/07/adaptation-eric-klinenberg

Klinenberg, E. (2018) *Palaces for the People: How to Build a more Equal and United Society*. New York: Bodley Head.

Landolt, P. (2008) The transnational geographies of immigrant politics. *Sociological Quarterly*, 49, 53–77.

Latham, A. and Layton, J. (2019) Social infrastructure and the public life of cities: Studying urban sociality and public space. *Geography Compass.* https://doi.org/10.1111/gec3.12444

Lawson, V. (2007) Geographies of care and responsibility. *Annals of the Association of American Geographers*, 97, 1–11.

Marvin, S., McFarlane, C., Guma, P., Hodson, M., Lockhart, A., McGuirk, P., McMeekin, A., Ortiz, C., Simone, A. and Wiig, A. (2023) Post-pandemic cities: An urban lexicon of accelerations/decelerations. *Transactions of the Institute of British Geographers*, 48, 452–473.

McIlwain, C. (2007) Teaching and learning guide for: From local to global to transnational civil society: Re- Framing development perspectives on the non-state sector. *Geography Compass*, 2(5), 1729–1735. doi: 10.1111/j.1749-8198.2008.00134.x

Middleton, J. and Samanani, F. (2021) Accounting for care within human geography. *Transactions of the Institute of British Geographers,* 46, 29–43. https://doi.org/10.1111/tran.12403

Middleton, J. and Samanani, F. (2022) Whose city? Which sociality? *Urban Geography*, 43(5), 777–783. https://doi.org/10.1080/02723638.2021.2007665

Middleton, J. and Yarwood, R. (2015) 'Christians, out here?' Encountering Street-Pastors in the post-secular spaces of the UK's night-time economy. *Urban Studies*, 52(3), 501–516.

Milligan, C. and Conradson, D. (2006) Contemporary landscapes of welfare: The "voluntary turn"? In C. Milligan and D. Conradson (Eds.), *Landscapes of Voluntarism* (pp. 1–14). Bristol, UK: Policy Press.

Milligan, C. and Wiles, J. (2010) Landscapes of care. *Progress in Human Geography*, 34(6), 736–754. https://doi.org/10.1177/0309132510364556

Mitchell, D. (2003) *The Right to the City: Social Justice and the Fight for Public Space*. London: Guildford Press.

Olen, H. (2017) Is the new way to give a better way to give? Donor-advised funds are gaining popularity, but charities may be losing out. *The Atlantic*, December 13, 2017.

Popkin, E. (1999) Guatemalan Mayan migration to Los Angeles. *Ethnic and Racial Studies*, 22, 267–89.

Power, A. and Hall, E. (2018) Placing care in times of austerity. *Social & Cultural Geography*, 19(3), 303–313. https://doi.org/10.1080/14649365.2017.1327612.

Power, A. and Skinner, M. (2019) Voluntary sector and urban health systems. In I. Vojnovic, A.L. Pearson, G. Asiki, G. DeVerteuil and A. Allen, (Eds.), *Handbook of Global Urban Health* (The Metropolis and Modern Life) (1st ed., pp. 134–148) London: Routledge. doi: 10.4324/9781315465456-8

Salamon, L. and Anheier, H. (1998) Social origins of civil society: Explaining the non-profit sector cross-nationally. *Voluntas*, 9, 213–48.

Skinner, M.W. and Winterton, R. (2017) Rural ageing: Contested spaces, dynamic places. In M.W. Skinner, G.J. Andrews and M.P. Cutchin (Eds.), *Geographical Gerontology: Perspectives, Concepts, Approaches*. London: Routledge.

Trudeau, D. (2008) Towards a relational view of the shadow state. *Political Geography*, 27, 669–690.

Wilson, D. and Keil, R. (2008) The real creative class. *Social and Cultural Geography,* 9(8), 841–847.

Williams, A., Cloke, P. and Thomas, S. (2012) Co-constituting neoliberalism: Faith-based organisations, co-option, and resistance in the UK. *Environment and Planning A*, 44(6), 1479–1501.

Wolch, J. (1990) *The Shadow State: Government and Voluntary Sector in Transition*. Ann Arbor, MI: Foundation Center.

PART II

THE SOCIAL GEOGRAPHIES OF SPACES OF ENCOUNTER

In Part I of this book, we laid the foundations by exploring key concepts in social geography and the significant relationships shaping citizenship, social justice, welfare, care and so forth, via the state, the voluntary sector, local infrastructures of care and the wider publics. Now the key relationalities have been covered in detail, we focus particularly on the second cross-cutting thread of the book; that of spaces of encounter at the interface of the everyday. We now turn to explore how the three themes around social geography – the social, inequality and social justice – are encountered, experienced, lived and felt by people across some of the key intersectional social identifiers within society. Our attention is placed on *lived space*, which is mutually constituted by our ordinary daily practices and relations of power, difference, and identity (Worth, 2015).

We cast quite a wide net by including a comprehensive range of different groups, from those clearly defined, such as young people, disabled people and family carers, to more loosely defined groupings such as people seeking access to housing or mental health care. More formally, we consider the five key social identifiers (see Chapter 2) and their intersectional qualities across several chapters – class, race, disability, age and gender – as well as carving out specific chapters for each (Chapters 8, 7, 9, 10, 6

DOI: 10.4324/9781003414889-7

and 12 respectively). These identifiers serve to highlight how each group's encounters, depending on the spaces they occupy, are gendered, racialised and understood through the prism of class, disability and age. Through the intersectional characteristics that people embody or face, we seek to understand their everyday ordinary citizenship. Through enacting this citizenship, we also explore how certain groups seek to shape or resist their political identities in various ways, ranging from distancing to conviviality.

These efforts to shape identities play out across the three strands of social justice (distribution, recognition, procedure). The strands vary in their presence; some are more visible, depending on the characteristics and power dynamics associated with each group, but each are interrelated in complex ways. The recognition of a group typically shapes the distribution of entitlements, and the procedures that endorse or deny their standing in society. The wider structures, including the geographies of production and social reproduction, are also at play in the background in shaping the various edges of social injustice these groups encounter.

Part II is not about 'social problems' (Section 1.4) per se, but about how specific groups of people experience the world and encounter various social problems in their everyday lives that are often derived from how the group is perceived. Some chapters are focused on more clearly defined stigmatisations, among for instance young people featured in the next chapter. Others are more about how society at large experience social problems such as inequitable levels of poverty and marginality, and limited and inequitable access to appropriate housing.

Encounters across the various social groups are often with the state, the voluntary sector, civil society and the market. These encounters implicate a variety of actors, including landlords, welfare officers, the police, but also peers, family members and neighbours. Sometimes these encounters are equally framed by the built environment, which for some come to define their everyday spatial experiences (e.g. wheelchair users, parents with strollers). These spaces of encounter can also be – and frequently are – more progressive, empowering and mutualistic, such as with carers, volunteers and social workers. Moreover, there can be a blurring of the boundaries in some cases (i.e., 'servers are people too'), which can enhance their mutual aims. It is through the interrelationship between these various encounters that group identities are formed and access to social justice is determined.

Reference

Worth, N. (2015) Youth, relationality, and space: Conceptual resources for youth studies from critical human geography. In J. Wyn and H. Cahill (Eds.), *Handbook of Children and Youth Studies*. Singapore: Springer.

6 GEOGRAPHIES OF YOUTH AND INTERSECTIONALITY

6.1 The Social Construction of Youth

> Children now love luxury; they have bad manners, contempt for authority; they show disrespect for elders and love chatter in place of exercise. Children are now tyrants, not the servants of their households. They no longer rise when elders enter the room. They contradict their parents, chatter before company, gobble up dainties at the table, cross their legs, and tyrannize their teachers.

The above quote reflects many of the negative stereotypes that are commonly levelled at young people. Its author is, in fact, the ancient Greek scholar Socrates, which provides a potent insight into how long standing negative representations of young people have persisted. Similar sentiments have been reported in more recent accounts, such as the quote below, from a mother quoted in Alison Clarke's book, *Childhoods in Context* (2013: viii):

> I just think there's a whole different, you know, they're a lot less frightened of authority than we were, we'd have never done certain things that they do now I'm sure of that ... I don't think that they do have much respect for authority or older people now at all ... But like I say when ... they're all hanging around I find it really intimidating and to be, to feel like that about kids really, it's not nice to think that, not nice.

Whether you are a current young student or at a later life stage, this chapter encourages you to think about your own experiences of youth, and, by extension, *age*. We also encourage you below to think about how your experiences have been shaped by intersectionality with other characteristics or contexts, and how it might compare (or have compared) to the experiences of other young people that have different backgrounds and characteristics to your own.

DOI: 10.4324/9781003414889-8

Box 6.1 Engagement Questions

1. How would you characterise *your* experience of 'Youth':
   - What places do/did you use and occupy?
   - How free are/were you to go places of your choosing free from parental control?
2. How would you characterise how youth is perceived *by society* today?

Additional Questions

- What are the features of 'youth culture' in your country?
- Do you think there are enough things for young people to do and places to go?
- Do you think that young people are allowed to be themselves?
- Would you say that young people are 'marginalised' as a group?

As noted, young people have endured negative attitudes towards them. This has included moral panics about the ungovernability of children and young people. One notable period was in the mid-1800s, when poverty was more acute internationally due to laissez-faire politics which framed the widening incorporation into the capitalist world-system and led to a decline in wages to below subsistence, a deterioration in human stature and an upturn in premature mortality (Sullivan and Hickel, 2023). It was generally expected at the time that young people should work and were effectively considered as mini units of industrial production. The geography of production was, therefore, the main determining factor in shaping young people's lives. Children and young people worked in gas works, factories, construction sites, shipyards and chimney sweeping. They became easy casualties of labour market exploitation and faced in-work poverty and high mortality rates. Alongside this context, many countries sought to tackle the perceived 'nuisance' of youth delinquency. The Malicious Trespass Act of 1839 in Victorian London, for example, criminalised 'nuisance' behaviour including 'picking fruit hanging over the walls of orchards'. Meanwhile, the Metropolitan Police Act 1839 prohibited flying any kite or playing at any game to the annoyance of the inhabitants or passengers. This offers a potent reminder of the role of the state, building on Chapter 3, in the lives of young people.

Recovery from this prolonged period of crisis occurred relatively recently: the late 19th century in Northwest Europe and the mid-20th century for many nations beyond, as detailed in Chapter 4. From the post-Second World War period, the social construction of youth entered a new phase, as young people born in the mid-1940s and early 1950s defined a new generation ('baby boomers'), and carved out a counter-cultural movement in the 1960s. This movement sought to expose social tensions across generational lines, sexuality, women's rights, traditional modes of authority and the end of racial segregation, particularly in the anglosphere.

While young people gained considerable rights since then, from the turn of the 21st century, efforts to remove young people from public space (as signposted in Chapter 1) have again become increasingly pervasive, against a backdrop of widespread panic about children's safety and the unruliness of teenagers (Collins and Kearns, 2001). Here the recognition of young people is typically characterised by questions over the degree of tolerance extended by wider (adult) publics.

The social construction of a deviant youth is a theme explored by Valentine (2004) in *Childhood in Crisis?* In it, she compares two contrasting tropes: first, children as innocent and dependent, and children as menacing and dangerous. Part of the issue inherent in these contrasting positions is that the terms 'children' and 'young people' are often used collectively or interchangeably. The frequency of this conflation has been recently raised as a concern; with the implications being that the unique needs and characteristics of teenagers and adolescents are not being sufficiently recognised (Weller 2006).

Tracey Skelton (2007) also raises this concern, arguing that this conflation of identities under a label of 'youth' undermines young people's agency – and consequently affects their well-being. In discussing the UN Convention on the Rights of the Child (UNCRC), Skelton (2007: 166) worries that young people are not very often mentioned per se in the document: 'Young people who do not consider themselves to be children anymore, and are in many ways not perceived as such by wider society, may feel that the rights defined for children do not apply to them'.

According to Gill Valentine (2003), the age range 16–25 years is commonly used to define 'youth'. Valentine uses a metaphor of boundary crossing to discuss the transition between childhood and adulthood, yet argues it is very difficult to define. Notably, it is blurred by the ambiguous period of 'youth', which is contested, fluid and bears little resemblance to legal definitions of youth, e.g. for buying alcohol and cigarettes, being criminally responsible, taking part in jury service, working and getting married. While each country differs, one common trend is that the age for each entitlement (or obligation) itself differs. Here, we see the hand of the state directly shaping experiences of youth. For your country, take a moment to look these up and consider the rationale for each of these being different.

Box 6.2 Engagement Questions

What age can you do the following?

- Buy cigarettes
- Be held criminally responsible
- Be responsible for jury service
- Take a part-time job
- Leave education or training

- Marry (with/out parental consent)
- Undertake military service (obligatory in some countries for males and sometimes females too)
- Buy a gun

To illustrate the differences, in England, for example, the age of being criminally responsible is 10. This is the lowest minimum age compared to any state in the European Union. Germany by contrast, sets the age at 14 years, but treats young people as juveniles if they show a lack of maturity up to the age of 21. These geographic differences between countries in the age of being criminally responsible reveal politically ideological influences over perceptions of youth crime and protecting the public. Meanwhile, the age for being permitted to take part in jury service is 18. The rationale is that jurors have immense responsibility and are asked to make important decisions in sometimes traumatic cases. Jury service can also sometimes take a long time. While these decisions are pragmatic and reasonable, the discrepancy between ages deemed old enough for being criminally responsible and for judging if others are (if found guilty), nonetheless raises interesting questions over how we define the age of autonomy and consent.

Beyond these examples, ages of entitlement or obligation are shaped by wider political contexts in countries. For example, in the US, under the Gun Control Act of 1968, shotguns and rifles, referred to as long guns, and ammunition for both, can only be sold to individuals 18 years old or older. Meanwhile handguns, including their ammunition, can be sold only to individuals 21 years old or older. However, states typically implement their own age restrictions. Texas, Wyoming, Ohio, New Hampshire, Montana and Maine have no age restrictions when it comes to who can be in possession of either a handgun or a long gun. Following the proliferation of domestic terrorism and the increase in gun violence across the country – where guns were the leading cause of death among children and adolescents in 2020 (Jones, 2024) – gun ownership has understandably been a contentious political debate. Meanwhile, some countries also have mandatory military service. Service in the Israel Defense Force and in the South Korean army is obligatory for their respective citizens at 18 years of age both men and women. Meanwhile, in the East Asian country of Laos, the minimum age for compulsory military service is 15 years.

The problems in the fixing of age boundaries also extends to the social construction of other youth identities such as 'adolescence' and even 'teenagers'. While the term 'teenager' seems on the surface to be an innocuous descriptor of those between the age of 13–19, it has become loaded with wider social identity politics. Geographers have examined the spatio-temporal specificity of the term 'teenager', which we return to in Section 6.3 (spaces of leisure). Historically it is rooted in the growth of a younger consumer market in Western capitalist societies in the 1950s (Valentine et al., 1998).

Representations of youth, at least in the Global North, are dominated by Western constructions of childhood, youth and adulthood. The former is typically characterised by play, frivolity, freedom, innocence, dependency and a lack of responsibility, and the latter by work, seriousness, independence and responsibility (Evans, 2008). Geographers have challenged the idea of youth in this binary fashion. Reality is far more complex than can be captured in a simple division between dependency (parental home) and independency (own home). Youth sits in the middle or at the 'social-chronological margins' (Pain et al., 2001), and is often understood as a period of transition between childhood and adulthood, marked predominantly by increasing responsibility and independence.

Skelton (2002) argues that widespread adoption of individualisation in Western (neoliberal) society has led to the construction of 'failed transitions' where young people who are deemed not to follow expected life paths are blamed for their own failures (Skelton 2002). For those seeking to choose their own life path, the policing and blame for not adhering to these social norms is sometimes extended to parents, usually mothers. These perceived transitions are argued by Beck and Beck-Gernsheim (2002) to be informed by 'traditional' life course norms previously sanctioned and structured through institutions such as state schooling and religious and moral norms around sexual relations and family structures. However, generational differences are evident, based on housing affordability, career aspirations and childcare costs, which suggests that young people's transitions to adulthood are becoming elongated (Arnett, 2004). As a specific example, Tomanović and Ignjatović (2006) explore youth transitions within the specific geographical, post-socialist context of Serbia. As a result of a lack of resources stemming from unemployment and housing unavailability, and cultural constraints from paternalistic family relations, young people postpone the key life events that are perceived as reaching adulthood. As Kathy McIlwaine and Kavita Datta (2004: 485) note, 'the transition from childhood to adulthood is partial, inconsistent and contradictory' in both hemispheres. For these reasons, we advocate a contextual and relational approach to youth transitions (Punch, 2002), and what she calls 'negotiated interdependence', which might usefully be extended across the life course.

This individualisation of responsibility and placing of blame on young people is at odds with how youth is considered in societies which accept a wider social responsibility towards youth development. The African proverb 'It takes a village to raise a child' initially popularised by Margaret Mead's (1928) groundbreaking study of adolescence in Samoa, has gained greater approval in many nations which have sought to recognise the wider shared responsibilities for child rearing in larger kinship groups and communities, a sort of social infrastructure within civil society. It has since been recognised in other nations, including Indonesia and India (Seymour, 2013). Its recognition in anglosphere countries has been more mixed. While gaining some political currency among socially progressive politicians, including in the US, as illustrated by Hillary Clinton's (1995) book, *It Takes a Village*, yet persistent blame culture remains (Sear and Coall, 2011). Schissel (2021) argues that while this blame culture in public discourse has persisted, it has nonetheless changed in form. During the 1990s, the media was more blatant and obvious in its attempts to create folk devils out of young people in trouble. Since this time, in North America at least, when the media write about youth crime and misconduct they generally attack those who are marginalised (Schissel, 2021). Mothers who live below the poverty line or from ethnic minorities are portrayed as either inadvertent or deliberate offenders.

While this blame culture is problematic in its own right, it should also be considered in tandem with the long standing lack of social avenues and places for young people. In the 1980s, being an adolescent was described in Allison James' (1986: 155) study as: 'being nothing', having 'nowt to do' and 'nowhere to go'. 'Being nothing' was defined as 'not adults' and being undervalued. Having 'nowt to do' was a criticism of the lack of provision of facilities and activities for young people. Having 'nowhere to go' spoke to how young people were absent from consideration in the design and planning of public space. This chapter offers a glimpse into the enduring nature of these concerns for young people. We return to consider some of the more contemporary echoes of this sentiment later in the chapter.

6.2 Intersectionality and (Stigmatised) Youth

Understanding experiences of youth is largely shaped by its intersectionality with other identities, as signposted in Chapter 2. Geographers have helped identify the multiple spaces and structures which continue to affect young people's experiences, interwoven in complex ways with other categories such as gender, ethnicity and class. When taking an intersectional lens, we must be mindful that not all youth are necessarily stigmatised. Some young people have relatively privileged lives in terms of wealth, education and opportunities to take part in meaningful activities. We acknowledge a larger challenge with respect to social geography and the disproportionate focus on the margins. We accept that the abiding focus on those at the margins has overlooked the significance of privilege in perpetuating and maintaining the social reproduction of a 'social mobility gap', based on disparities in opportunity, and how different young people can be treated very differently. As university students yourselves, you are considered educated, certainly more than the average person. Given your education, you can participate in the knowledge or 'creative' economy (a theme we pick up in Chapter 11), and thereby enjoy some of the benefits of social mobilities afforded to you, such as greater opportunities for personal and professional growth, travelling the world and gaining graduate-level jobs. Of course, education is only one marker of difference, and can be inflected by other intersectional identities, as noted above, that can mean that opportunities are conditional on space and relational encounters and can be withdrawn in different contexts. For instance, in the case of an educated, young gay man who experiences homophobic violence.

While a focus on privilege is warranted, it must be acknowledged that youth in upwardly mobile, upper-middle class community contexts have also recently been identified as statistically 'at risk' (Luther et al., 2013) of experiencing issues from internalising problems associated with the social pressures of being individualised, neoliberal subjects with high expectations. Here Luther et al. (2013) refer to contemporary American culture which emphasises maximising personal status, and how this can threaten the well-being of young individuals. This can lead to directing feelings inwardly towards oneself or one's body which can lead to anxiety-depression among affluent adolescents as well as externalising problems which can lead to higher levels of drug and alcohol use. This should be countered by an acknowledgement of the likely social capital accrued by upper-middle class, in terms of the skills and capabilities gained through social networks and education. Nonetheless, Luther et al. conclude with suggestions for interventions at multiple levels, targeting youth, parents, educators, as well as policymakers, toward reducing pressures, and maximising positive adaptation among 'privileged but pressured' youth and their families.

We now consider two particular intersections of youth in more depth here – minority ethnic and working-class identities. Peter Hopkins' work (2010) explores the experiences of the embodiment of young Muslim men. His research is informed by interviews with young Muslim men aged 16–25 who live in Scotland. The research considers the multiple ways that young men negotiate their national young identities as Scottish Muslims. Here, intersectionality plays out in shaping a young person's links to a collective identity (i.e. being 'Scottish'), and how it aligns with cultural tropes such as drinking 'Irn-Bru' (a popular Scottish orange flavoured soft drink), playing football, having a Scottish accent and ascribing to a particular (alcoholic) drinking culture. Having an appreciation of Scotland's natural environment is also used as claims to Scottishness. Hopkins found that Muslim young men often limit themselves from belonging completely within these boundaries of Scottish identity, and thereby occupy a hybrid identity.

The research also focuses on exploring how signifiers of Muslim identity had gained prominence in the early 2000s, following the Twin Towers attack in New York September 11, 2001, and the July 7, 2005 bomb attack in the London Tube (underground metro). Hopkins (2010) found that this series of events resulted in the bodies of young Muslim men becoming intensified sites of social inscription, in terms of how their identities were being read, and responded to by society. Similarly, young Muslim women's bodies are contested sites for the construction of their religious, ethnic, and gendered identities. Dwyer's (1999) work has explored the experiences of young Muslim women in the West and how their everyday identities were influenced by ethnicised cultural and religious discourses. Wearing the hijab for instance, provoked debate, and sometimes disquiet, within Muslim households and among peers. According to Dwyer (2008), choosing to adopt this highly visible sign of Muslim identification was not taken lightly. Some young women found themselves positioned, implicitly, in opposition to this expression of a more self-conscious Muslim identity where others felt this was a big step which they might contemplate in the future. These illustrate some of the dilemmas of dress for young British Muslim women and the ways in which dress functions as one of the most powerful signifiers of identity.

A second intersection that we consider here is working-class male identities, and therefore, we bring three different core identifiers (gender, class, and age) into view. McDowell (2007) explores how working-class young boys have become subject to political narratives in the UK, framing them as disrespectful and disreputable. In this narrative, young people are being framed as having a lack of discipline in schools, loitering in shopping malls and exhibiting 'laddish' public behaviour, and by extension, terrorising their neighbourhoods with anti-social behaviour and committing crime. Here, the framing of working-class male youth identity has become 'over-determined', which Tyler (2008: 18) describes as having become publicly imagined in excessive, distorted, and caricatured ways, portrayed as disaffected vandals, 'Chavs' or 'yobs'.

In education and employment, Linda McDowell (2003) explores how boys deemed low-achieving and low-skilled – in the main from working class families – are seen as problems. Little or no acknowledgement is made in political accounts that many young people experience *in-work* poverty, as a result of seeking to make a living in the precarious Gig Economy. McDowell argues that the class-based expectations in their localities constitute an encounter which these young men must navigate, through which they enact their masculine identities. This space of encounter also determines their employment prospects and future trajectories. In response, education policy at the time of her study had become focused on ensuring young people could meet generally recognised standards and access to regular work. At the same time, however, the same boys were being represented as out of control and in need of surveillance and punishment. McDowell (2007) examines the policy ambivalence at the heart of the New Labour government's approach in the early 2000s, where working class boys were constructed as needing to continue education or earn despite having lower national pay rates, while simultaneously needing control and harsher punishment. This differs considerably to the geographer, Joe Painter's own account of receiving welfare support (being 'on the dole') in the 1980s.

This resonates with more recent work by Malcolm James (2015) on urban multiculture in Outer East London which explores the ramifications of this political context, where working-class boys are set up to fail. In his ethnographic study, James traces the multi-layered nature of the local context which frames the young people's experiences. This includes a long history of extensive migration, primarily by industrial labourers, a subsequent gradual reduction of job opportunities and intense poverty – synonymous with Baltimore in the US which David Harvey wrote about several decades before. Among these economic changes, James also notes an erosion of community and a detachment from local government's championing of cosmopolitan solidarity, typified by investment in 'shopping centres' as a form of urban regeneration. These legacies have paved the way for native territorialism, a theme we explore in the next section.

6.3 Contemporary Geographies of Youth and the Spaces of Encounter

Geographers have considered the various spaces that young people use and occupy, revealing how, and *where* youth is experienced and perceived. Youth geographers (Worth, 2015) have been especially interested in what's been called 'social space' or 'lived space'. It is at this scale that we can see how ordinary lived citizenship is lived and felt on an everyday basis. Here, the intersectionality of youth, explored above, is continually at play in shaping spatial practices and experiences. This is a useful lens to better understand how youth is shaped and experienced through different spaces like the street and neighbourhood, home, and 'youth spaces' of school, leisure and social media – recognising that identity is both read and performed differently across these spaces. Each space has different sets of encounters – with the state, wider publics, parents, teachers, youth workers and security staff – in terms of how they produce and reinforce relations of power (which often operate through broader neoliberal contexts), and shape experiences of difference and identity.

The Street and Neighbourhood

We use 'the street' as a metaphor for all public outdoor places where young people are found, including neighbourhoods, parks, high streets, suburban cul-de-sacs and derelict spaces. The street is often important to the formation of identities among young people (Hopkins, 2010). Notably, the space of the neighbourhood and city street becomes the primary domain for many, particularly after dark, when many adults have retreated to the sanctuary of the home. This is often the only autonomous space many teenagers can carve out an identity for themselves and is therefore an important social arena where young people can be together (Skelton, 2013). Yet, young people are acutely aware of adult attitudes of their presence in public space, sometimes as menacing and dangerous (Valentine, 2004). The visibility of youth and their competing use of street space positions them in the frontline of conflict over its use.

Perhaps for this reason, for many young people since the turn of the last century there has been a shift from public outdoor play to private extra-curricular activities. Dowling (2000) explores how the car has become crucial to 'good mothering'. This links back to our earlier discussion on privileged youth in the last section. More specifically, travelling by car to pre-defined (paid) activities has become synonymous with middle-class childhood and parenting alike, typified by a belief that a young person's time should be spent productively rather than 'wasted' on free play in public space. The following quote offers a critical take on this trend: 'These days our children are not so much free-ranged as battery-reared' (Sunday Times journalist, cited in McNeish and Roberts, 1995: 3). By contrast, working class young people have been found to experience more of their local neighbourhood and less participation in institutional activities. Hopkins (2010) argues that *territoriality* can vary between different young people, with working class young people placing great value on the street and local neighbourhood, although this is by no means universal.

Young people of course do not all come from urban areas or live in such immediately stimulating environments, where they can meet and encounter other young people to socialise, create shared identities and hang out. Andy certainly did not, growing up in a rural village in Ireland. As a counterpoint to the street and neighbourhood encounters, which are squarely urban, Leyshon (2008) explores how rural areas constitute a different kind of spatial construction of youth. He refers to accounts of rural youth of the countryside as 'dull' and a 'boring place where nothing ever happens'. The village can also be associated as a site of surveillance, with the presence of 'busybodies' who are perceived as nosy and intervene whether help is welcome or not. While Leyshon's account acknowledges these experiences as a part of rural life, he found that village spaces are nonetheless revered by local rural youth as the key site in the mediation and production of their rural identity. In this sense, rural youth seem to buy into the village—with mythic icons of cottages, pub, shop and church framed by the 'natural' beauty of the countryside. Importantly, the fields, woods, hedgerows, farms and lanes beyond the village also provide the context for acts of production, resistance and mediation in and around the village. So, for some young people this 'dullness' is a challenge to be overcome, by making their own entertainment. Rural youth, therefore, see themselves as imaginative and inventive problem-solvers – a welcome illustration of the social.

Young people in rural areas also affirmed each other's collective claim to ruralness and legitimated their position within their village through drawing upon symbolic representations of rural living in their accounts. Rural social spaces are often coded and categorised by a powerful set of ideological and symbolic representations of a pleasant, clean environment with a close affinity to nature, a lifestyle often romanticised in popular culture. Leyshon's study indicates that rural youth themselves draw upon these symbolic representations in their own accounts of rural living. Embodying these representations become the keys or the criteria for admission. In other words, the rural and the urban are imaginatively constructed in relation to each other. Conversely, these can reinforce exclusionary practices of *recognition* which seek to maintain spatial stability and purity of rural life as idyll, almost parochial. He therefore suggests that the lives of young people in the countryside are multi dimensional, complex and textured.

Returning to the urban sphere, youth in public space is often socially constructed as deviant, which offers a convenient tool to protect the adult world against the intrusion of young people (Valentine, 2004: 29). More broadly, it reinforces how the production of public space is increasingly being seen as an adult space. Increasingly urban renewal schemes involve state planners and corporate enterprises to co-produce carefully monitored and controlled privatised 'public' environments 'that are based on desires for security rather than interaction' (Michell 1995: 119). The increasingly privatised public domain (known as privately owned public spaces or 'POPS') is where control over access is actively enforced. A common assumption is that there is one shared set of 'public' values to which all members of the civil society subscribe (echoing Young's cultural imperialism), and which determines what is deviant and who is welcome (4). This undermines what Vidler (2002) describes as an urban life worth living, which is constituted by publicness itself. The tools to manage public space throw into sharp relief questions about the nature of public space for young people, how it is to be used, and under what conditions. It also raises more fundamental questions about the right to the city for young people.

Collins and Kearns (2001: 389) explore the use of curfews in many US, Canadian and Australian cities. These curfews are ostensibly motivated by a desire to reduce youth crime and victimisation. This is certainly the case in the US where juvenile curfews are most prevalent, despite rhetoric about the fundamental nature of individual freedoms. They argue that curfews reflect a contemporary preoccupation with achieving social control through the control of space. It is contended that these curfews are as much about enforcing a particular notion of parental responsibility as controlling young people themselves. Similar justifications are used for 'Mosquitos', which are electronic devices which emit high-pitched alarms to deter loitering and are commonly tuned to be heard primarily by younger people. These tactics offer powerful symbolic expressions of the adult control and purification of public space.

Public space is never entirely regulated solely by 'sophisticated policing mechanisms but through contingent, contextual and local processes exercised by the street users' (Malone, 2002). This reminds us that public space is to some extent self-regulated, through everyday encounters with public stares, condescension, passive aggressive comments or merely being ignored. These encounters serve as an important reminder of young people's everyday ordinary citizenship. Notions of *territoriality* in public space have also been explored through the lens of urban gangs and the use of graffiti. Italian for 'writing on the walls', graffiti tends to be marginal (both the artists/authors and the location), anonymous (rarely know the artists/authors), spontaneous/furtive, precarious (the use of cheap, easy to obtain materials) and fleeting (do not last long). To many outsiders, graffiti and murals are cryptic and senseless; they are understood by only a fraction of those who encounter them.

Usually read as a sign of urban decay and as a loss of control over the physical environment, graffiti has become one of the most potent cultural languages of youth. Yet as signposted in Chapter 5 in terms of murals in places like Puerto Rican New York, graffiti offers geographers an important means of understanding the story of urban change for young people. In the 1970s in the US, there was a significant rise in incidences of gang warfare in various urban areas including Los Angeles and Philadelphia. Ley and Cybriwsky (1974) traces how this unfolded in the Monroe district of Philadelphia. The gang's use of graffiti became a core part of their creation of territories and was linked with 'turf wars' between gangs of poor young males. In a similar vein, James (2015) provides a formidable account of the territoriality of young people in an inner-city area in London. As part of his ethnographic study, James spent time with local young working-class people experiencing multiple disadvantage and austerity. He explored how they had adopted nihilistic outlooks, meaning they were rejecting moral principles in the belief that life is meaningless and becoming despondent and isolated from wider society. This is often a source of inspiration for urban grime music, which James explores in his work. While such research provides a bleak appraisal of inner-city working-class youth, it nonetheless offers a forceful reminder of the importance of investing in youth spaces and generating opportunities for young people.

The Home

The home is another important place for young people and vital to their sense of ontological security (Easthope, 2004; Hiscock et al., 2001). It is a site of care, where parents or guardians care *for* and *about* young family members, drawing on our ideas around reciprocity and hospitality from the last chapter. Thus, the home can be understood as another important relational zone of encounter – between immediate kin and young people. Yet, it can be subject to another more difficult set of encounters, in this case from controls by parents over the use of space and time. Young people are also living at home with their parents for longer periods (White, 2002), hence their transitions and relational encounters through childhood, teenage-years, adolescence and young adulthood can be more prolonged and often must be mediated for longer, in close proximity with others in the home. Controls from parents are continually negotiated, and change over time as children grow. This was particularly acute during the lockdowns associated with the COVID-19 pandemic, which often meant these relations became more constrained and thus more intense.

The home is the primary domain where most young people attempt to carve out their own microspaces, and set their own times and routines. More to the point, the concept of home is associated with personal meanings of stability, privacy, identity, memory, domesticity. But it can also sometimes be a site of exploitation, and what Marxist geographers call 'use value', in which a dwelling assumes a symbolic investment that goes beyond the actual exchange value of the property. While Marx justifiably nudges us to think of the symbolic value of the home, we should also not forget the structural barriers facing young people in gaining access to the housing market – a point we explore further in Chapter 11 on housing. Yet, the domestic lives of young people have rarely been considered of significance either in studies of the family or in studies of youth.

A sense of home can be made or unmade, depending on where there is a perceived loss of independence, privacy and freedom of choice. This is often a fluid process, negotiated through everyday encounters. To illustrate this point, Collins (2015) draws on qualitative research with British teenagers to highlight how young people and their parents interact when managing household material consumption. Objects of consumption around the home can be a source of comfort for teenagers and help them manage the crystallisation of their identity. One of her participants, a young teenager called John, recounts a continual battle over his old trainers, which his mother continually tried to throw out.

John: They've survived repeated attempts of trying to get rid of them by my mum.
Int.: Why is it that your mum wants to get rid of them?
John: It's because they're old. And she doesn't ... like clutter and she wants to ... to get rid of them.

In John's account, he later shares how these contestations with his mother were less about emotional attachment to the objects in question; rather, they had become a battle of wills based upon his insistence that he should be able to decide upon the usefulness or uselessness of his possessions on his own terms. Often these negotiations involved playing off parents against each other, with participants invoking the attitude of one to legitimise to the other a particular course of action. John, for instance, later reminds his mum that his father's old trainers are kept for gardening which has the desired effect in legitimising their retention.

Another teenager in Collins' study, Tina, illustrated the same principle in reverse. Her wish to sell her no-longer-used keyboard was obstructed by her father's insistence that she keep it. Although she no longer identified with the younger self that had once played it, she suggested that her father's reluctance for her to part with it was based on his attachment to an image of his younger daughter enjoying an item he had bought for her. This reveals a frequent tension with parents over the positioning of 'modern childhoods' in contrast to remembered idyllic past childhoods. It also shows how relationships in the home can become intertwined with household hierarchies of power, for example related to intersections between gender, birth order, and cultural or social expectations.

Beyond the family home, Ortega-Alcázar and Wilkinson (2020) focus on the experiences of young women in rental properties and explore how limitations in housing access (Chapter 10) has typically meant shared accommodation with strangers. They explore how sharing with strangers creates increased vulnerabilities, and asks whether it is possible to feel at home when living with strangers? The paper outlines a set of difficult experiences of strangers and reveals how the young women feel fear, intimidation and loneliness. Here, both the material design of the shared house, and the social configuration of the household, were seen as a barrier to maintaining close relationships. Most living spaces had been turned into bedrooms, and many of the bedrooms had been subdivided to create small single rooms. The design of the shared house which often had little if any common shared space, limits the possibilities of finding connections amongst sharers or for tenants to invite guests or keep pets Ortega-Alcázar and Wilkinson (2020). These experiences can force people to carve out and occupy both physical and online spaces. As part of this next section, we consider the spaces of education, leisure, and social media.

Spaces of 'Youth': Education, Leisure, and Social Media

Between the street and the private space of the home, there are a wide array of spaces which young people occupy and that are directly associated with youth. We focus on three such spaces: spaces of education, spaces of leisure, and spaces of social media. We accept that these distinctions are arbitrary, and that social media can be regarded as a form of leisure. These spaces also frequently overlap, in that people can be online *in* physical spaces. We therefore emphasise the physical, in-person experiences of education and leisure to suitably distinguish the former two from the latter.

For spaces of education, most young people spend most of their day at school or in some form of education or training, so naturally this has been a central interest of youth geographers. Here, our core themes of the social, inequality and social justice are all evident in the school as a relational institution and education as a social-political process. One example is on the restructuring of educational provision in neoliberal economies. In these contexts, Sarah Holloway et al. (2010) outlines how inequalities in access to education are laid bare in advanced capitalist political economies, where many students from low-income families and minority ethnicities get a poor deal from state education. Considerations of power and social relationships are also central to Louise Holt's (2010) research on the intersection of disability and school spaces, where children with special educational needs and learning differences often face significant battles to secure resources for supporting their child's education. Differential access to education, and geographical and gendered variations and gaps in educational attainment (aggravated by the pandemic), have thus been important themes.

Noting these inequalities, schools and other educational spaces have themselves been viewed as 'precarious geographical achievements' (Philo and Parr, 2000, cited in Holloway et al., 2010: 587). Meanwhile, in more privileged contexts, research has shown how well-resourced schools are designed and perceived through adult interpretations of an ideal childhood. Peter Kraftl (2006) reveals the extent this is the case through a focus on a Steiner school in Wales (a cooperatively built and run school which falls outside mainstream state-funded education). He explores how 'alternative' ethical, nature-based views of childhood informed the design, construction and use of the school building, including the plastering, paint, door handles, nature tables and wall charts (Kraftl, 2006). In so doing, he explores how educational space can shape young people's practices as a form of 'performance art'.

Holloway et al. (2010) has also highlighted the importance of the curriculum in shaping different scales of identity and citizenship, whether regional, national or trans national, and their links with state and broader economic formations. English children for example are reportedly tested longer, harder and younger than anywhere else in the world, according to a UNICEF (2011) report that compared school standards in 22 countries, leading to critical commentary about British schools programming robots for exam factories (Roberts-Holmes, 2017). Living in a rich country does not guarantee equal access to quality education, according to UNICEF (2018). Children in less wealthy countries often perform better at school despite fewer national resources. For instance, Romania and Lithuania rank higher than the UK and US. Towards the end of compulsory schooling, Latvia, Ireland and Spain are the three most equal countries. Ireland also has the highest rate of people with Bachelor's degrees in the world – thanks to a generous state subsiding of higher education (of which Andy graciously benefited from, illustrating their different welfare model).

While the design and administration of the curriculum is important, so too are the informal lessons which students learn and enforce or reject in schools and post-compulsory education. Here the focus is not on the content of lessons, but the nature of socio-spatial relations within different learning spaces which shape young people's identities and experiences. Thomas (2005), for example, explores these ideas in her analysis of racial segregation in a southern US high school, concluding that the effects of girls' everyday sociability within schools (who they sit with at lunchtime, how they address one another, etc.) reproduces race as an identity positioning practice within schools.

Connections between education and inequality are perhaps most explicitly felt in the higher-education sector, with the spatial phenomenon of educational markets driven by league tables (Dorling, 2017). This is explicitly connected, in many cases, with strategies of internationalisation designed to improve universities' position in circuits of expertise, as explored by Jöns and Hoyler (2013). They trace how 'world-class' university status has become a widely shared aspiration for higher-education institutions internationally, which can bring direct dividends in terms of increased fee income and investment. This extends to increased job creation, innovation and creativity within the university's locality and region. This trend has also been bolstered by the much-cited work of Richard Florida (2002) which advocated the idea that cities possessing higher proportions of educated residents are more creative and productive (but also see Chapter 11 for more on Florida's theories). Yet, this trend has not been immune to tensions, with research exploring how students often outcompete other residents in local housing markets of university towns, and create a studentification of city areas (Hubbard, 2009) – a topic we explore in Chapter 11 on housing.

A second space associated with youth is the space of leisure, itself made up of multiple discrete spaces often carved out along intersectional lines. For spaces of leisure, these are often spaces associated with sociability and with weakly defined or open boundaries, characterised by social mixing and diversity, e.g. sporting venues, clubbing venues, carnivals and festivals. We consider two places in particular which are commonly used by young people – youth clubs and shopping malls. In Norway, as in the other Nordic countries, youth clubs are relatively prevalent, and Erikson and Seland (2020) traces how they can offer young people an alternative place for socialising outside often overcrowded urban flats and facilitating simple leisure activities like board games. Having no charge or only symbolic fees can place no demands on regular attendance or parents' time or effort. Therefore, youth clubs are often described as alternatives to organised leisure activities like sports. The prevention of problem behaviour is an important rationale for maintaining youth clubs in cities; to keep young people occupied and off the streets, ultimately to prevent them from turning to drugs, crime and gangs. Yet, Erikson and Seland argue that clubs are now important arenas for the prevention of mental health issues among youth (a theme we later explore in Chapter 10). Relationships with youth workers have been shown to be particularly important for giving young people academic and practical help, life skills and support (Rhodes, 2004), as well as for young people who are in difficult circumstances, for example, in the child welfare system.

By contrast, in the UK, youth clubs have been steadily disappearing due to cutbacks to public funding. Geographers have sought to turn the lens towards youth workers to better understand how these cutbacks have been managed. In study by John Horton (2016), a youth group manager in the English Midlands revealed the anxieties he felt 'in the current climate' characterised by austerity cutbacks to their youth club:

> ...we're just waiting for the inevitable to happen ...We're expecting our funding to be cut completely cut – in the next six months, maybe the

> next quarter. I wouldn't be surprised if we've only got a few weeks left ...We're keeping things going but ... it's a really anxious, heartbreaking atmosphere for everyone working in this sector. (p. 349)

These anticipatory anxieties, Horton argues, have manifold every day, lived consequences that pervades their working lives and the lives of young people. As Horton argues, young people are diversely affected by, and engaged in, the circulation of these anxieties and hopes. However, he also recognises that young people's geographies *go on* and sometimes offer hopeful ways to live 'in the current climate'.

A second physical space of leisure (and consumption) associated with youth that we consider is the shopping mall. The mall is seen as a popular, convenient meeting place for young people, echoing some of the relational characteristics of the street, neighbourhood or village. The mall arguably reached its zenith in pop cultural relevancy and growth in the United States in the late 20th century (Kumar, 2021). Indeed, it came to define teenagers, as a demographic that was named to complement the commercial culture that came to define such spaces. The term 'mallrats' also became used to define youth who habitually spend a lot of time in shopping malls. In this sense, mall culture and youth culture have become closely intertwined. Yet, despite malls being primarily a site of consumption, young people spend most of their time there chatting, watching and hanging around.

Young people's occupancy of malls is thus not without its detractors, especially as malls have become more sanitised and regulated spaces. Malls mirror some of the methods of control and active enforcement of people, clothing and behaviour through policing found in our earlier focus on the street. Here, young people are watched by close circuit (CCTV) surveillance cameras, subjected to bag checks and move on local bye-laws. At their zenith, Norris and Armstrong (1997) offer a powerful insight into the manifestation of this control, through their ethnographic study of CCTV control rooms in shopping malls. Their research uncovered the way that certain types of young people are labelled 'yobs, 'toe-rags' or 'drug-dealing scrotes' by CCTV operators who impose a normative space-time ecology on the environments they manage, removing all those who they determine to be unwelcome. Such processes reveal the extent to which the quasi-public realm is now privatised and commodified. Despite constant attempts to move young people on, however, they stubbornly remain within the mall asserting a right of presence.

Our third space examined is that of social media. Moving beyond both (in-person) education and leisure spaces, and perhaps cutting across all of them, young people's geographies would not be complete without considering the spaces of Snapchat, Facebook, YouTube, TikTok, Twitter, Instagram and other social media sites that dominate and shape their everyday social encounters. Here our third theme of the social is omnipresent. With the widespread availability of smartphone devices, the seamless interactions that young people adopt, have melded software, data, people and places, in novel ways, and in the process have challenged simplistic online/offline binaries.

These online spaces are often represented and perceived as an arena that is not subject to the same levels of relational policing of young people, as other places considered in this chapter. Indeed, for neurodiverse young people, Leyman (2023) examines a particular play practice, *cosplay* and related companionable fandom activities online that can provide welcoming digital spaces for autistic youth to be social. These fluid relational practices fragment and dissolve many encounters, in turn encouraging geographers to consider the bidirectional links between people's bodies, emotions, devices and locations. Indeed, as Kitchin et al. (2023), contend, the digitisation of young people's (or perhaps all age groups') encounters and media is changing *how geography is done.*

To expand this point, online digital platforms can co-construct different spatialities for young people; for instance, fashion and music have become mediated by digital platforms such as Instagram and Spotify. These have created different spatial affordances for how musicians, fashion designers and young consumers interact (Hracs and Leslie, 2014). In the process, these affordances have fuelled the production of more and more content and collection of more data, fuelling algorithms and bespoke advertising, leading to the datafication of the everyday. Through these practices, the spaces and places of youth are situated within the contexts of globalisation, production, consumerism and neoliberalism. Yet, just as inequality has framed physical encounters seen elsewhere, digital infrastructures have also been uneven, and the sites of production and disposal have led to concentrations of 'e-waste' in the Global South (Lepawsky, 2015).

Geographers have also been interested in investigating the global reach and global spatial imaginaries that young people can navigate through a more explicit crossing of scale through online digital spaces. One interesting example focuses on the experiences of young people with Latin American families in the United Kingdom (Mas Giralt and Bailey, 2010). These young people are depicted as navigating a complex landscape of identity formation, often embracing diverse aspects of their Latin American heritage online while simultaneously adapting to the UK environment. They find and develop their sense of belonging, which may involve local social contexts and transnational ties. Drawing on Bauman's (2000) thesis on liquid modernity, they use the concept of 'liquid life paths' to examine the 'spatial imaginaries' involved in how migration stories impact personal identity. Young people are thus differently positioned in networks of global–local connectivity.

6.4 Conclusions

This chapter has explored the geographies of youth, and the various spaces of encounter young people experience with wider publics, their peers and immediate family members. The chapter serves as our first contribution to Part II of the book, where we begin to see how the foundational concepts in Part I play out including the role of the state (typically underpinned by neoliberalism's guiding force) and the mediating hand of the geographies of production (and consumption) and social reproduction. In particular, we examine how youth is shaped by our three core themes – the social, inequality and social justice. The social is immediately apparent in how youth is socially constructed as a complex period of transition, and as an elongated period that is imbued with shifting and contradictory meanings associated with safety and risk. We also examine how youth is lived and felt in any everyday social sense through intersectional dynamics. Once again, the theme of intersectionality is evident, where the social construction of youth is a delicate balance of rights and responsibilities, shaping varied experiences of dignity, autonomy and independence. These intersectional dynamics shape *inequalities* in how youth can be experienced differently, depending on one's class, gender and so on. Finally, our social justice strand that becomes most apparent is *recognition*. While the recognition of young people has arguably improved over time, the identity politics associated with youth have nonetheless remained complex and multi-layered for young people.

In the second half of the chapter, we explore how these themes play out across the various spaces that young people might use and occupy; the street, the home and spaces of youth (focusing primarily on education, leisure, and social media). Through a variety of measures, these spaces and places are situated within the contexts of globalisation, consumerism and neoliberalism. Here, the chapter offers a critical appraisal of the efforts of older adults (reinforced by state policies) to increasingly expect individual responsibility on the one hand, and to restrict young people's freedoms, on the other. As we saw, this frequently manifested in the geographies of stigma of some young people, marked out by their intersectional characteristics, where ordinary lived citizenship is lived and felt. We continue to build on these core themes in our next chapter on race.

References

Arnett, J. (2004) *Emerging Adulthood: The Winding Road from the Late Teens Through the Twenties*. Oxford: Oxford University Press.

Bauman, Z. (2000) *Liquid Modernity.* Cambridge, UK: Polity Press.

Beck, U. and Beck-Gernsheim, E. (2002) *Individualisation.* London: Routledge.

Clarke, A. (2013) *Childhoods in Context.* Bristol: Policy Press.

Clinton, H. (1995) *It Takes a Village.* New York: Simon & Schuster.

Collins, D.A. and Kearns, R.A. (2001) Under curfew and under siege? Legal geographies of young people. *Geoforum*, 32(3), 389–403.

Collins, R. (2015) Keeping it in the family? Re-focusing household sustainability. *Geoforum*, 60, 22–32.

Dorling, D. (2017) *Do We Need Economic Inequality? (The Future of Capitalism)*. Cambridge: Polity Press.

Dowling, R. (2000) Cultures of mothering and car use in suburban Sydney: A preliminary investigation. *Geoforum*, 31(3), 345–353.

Dwyer, C. (1999) Contradictions of community: Questions of identity for young British Muslim women. *Environment and Planning A*, 31, 53–68.

Dwyer, C. (2008) The geographies of veiling: Muslim women in Britain. *Geography*, 93(3), 140–147. doi: 10.1080/00167487.2008.12094235

Easthope, H. (2004) A place called home. *Housing, Theory and Society*, 21(3), 128–138. doi: 10.1080/14036090410021360

Eriksen, I.M. and Seland, I. (2020) Conceptualizing well-being in youth: The potential of youth clubs. *Young*, 29(2), 175–190. https://doi.org/10.1177/1103308820937571 (Original work published 2021).

Evans, B. (2008) Geographies of youth/young people. *Geography Compass*, 2, 1659–1680. https://doi.org/10.1111/j.1749-8198.2008.00147.x

Florida, R. (2002) *The Rise of the Creative Class: And How It's Transforming Work, Leisure, Community and Everyday Life*. New York: Basic Books.

Hiscock, R., Kearns, A., MacIntyre, S. and Ellaway, A. (2001) Ontological security and psycho-social benefits from the home: Qualitative evidence on issues of tenure. *Housing, Theory and Society*, 18(1–2), 50–66. doi: 10.1080/14036090120617

Holloway, S.L., Hubbard, P., Jöns, H. and Pimlott-Wilson, H. (2010) Geographies of education and the significance of children, youth and families. *Progress in Human Geography*, 34(5), 583–600. https://doi.org/10.1177/0309132510362601

Holt, L. (2010) Young people's embodied social capital and performing disability. *Children's Geographies*, 8(1), 25–37. doi: 10.1080/14733280903500158

Hopkins, P. (2010) *Young People, Place and Identity*. London: Routledge

Horton, J. (2016) Anticipating service withdrawal: Young people in spaces of neoliberalisation, austerity and economic crisis. *Transactions of the Institute of British Geographers*, 41, 349–362. https://doi.org/10.1111/tran.12134

Hracs, B.J. and Leslie, D. (2014) Aesthetic labour in creative industries. *Area*, 46, 66-73. https://doi.org/10.1111/area.12062

Hubbard, P. (2009). Geographies of Studentification and purpose-built student accommodation: Leading separate lives? *Environment and Planning A: Economy and Space*, 41(8), 1903–1923.

James, A. (1986) Learning to belong: The boundaries of adolescence. In A.P. Cohen (Ed.), *Symbolising Boundaries: Identity and Diversity in British Cultures* (p. 15). Manchester, UK: Manchester University Press.

James, M. (2015) Nihilism and urban multiculture in outer East London. *The Sociological Review*, 63(3), 699–719. https://doi.org/10.1111/1467-954X.12219

Jöns, H. and Hoyler, M. (2013) Global geographies of higher education: The perspective of world university rankings. *Geoforum*, 46, 45–59.

Kitchin, R. (2023) *Digital Timescapes: Technology, Temporality and Society*. United Kingdom: Polity Press.

Kumar, P. (2021) *The Kids Are (M)all Right: Youth Culture and Shopping Centers in Late Twentieth Century America*. MSc Thesis in the Department of History, Concordia University Montreal, Quebec, Canada.

Lepawsky, J. (2022) Mapping chemical discardscapes of electronics production. *Geoforum*, 132, 113–124.

Ley, D. and Cybriwsky, R. (1974) 'Urban graffiti as territorial markers. *Annals of Association of American Geographers*, 64, 491–505.

Leyman, A. (2023) Use of digital spaces for cosplay by autistic youth for social interaction, in Lieu of Material Spaces within Urban Contexts. In R. Berman, P. Albanese and X. Chen (Ed.), *Sociological Research and Urban Children and Youth* (Vol. 32, pp. 29–46). Bingley: Emerald Publishing Limited. https://doi.org/10.1108/S1537-466120230000032003

Leyshon, M. (2008) The betweeness of being a rural youth: Inclusive and exclusive lifestyles. *Social & Cultural Geography*, 9(1), 1–26. https://doi.org/10.1080/14649360701789535

Luthar, S.S., Barkin, S.H. and Crossman, E.J. (2013) 'I can, therefore I must': Fragility in the upper-middle classes. *Development and Psychopathology*, 25(4 Pt 2), 1529–1549. doi: 10.1017/S0954579413000758

Malone, K. (2002) Street life: Youth, culture and competing uses of public space. *Environment and Urbanization – ENVIRON URBAN*, 14, 157–168. 10.1630/095624702101286197

Mas Giralt, R. and Bailey, A.J. (2010) Transnational familyhood and the liquid life paths of South Americans in the UK. *Global Networks*, 10(3), 383–400.

McDowell, L. (2003). Masculine identities and low-paid work: Young men in urban labour markets. International Journal of Urban and Regional Research, 27, 828–848. doi: 10.1111/j.0309-1317.2003.00486.x

McDowell, L. (2007) Respect, deference, respectability and place: What is the problem with/for working class boys? *Geoforum*, 38(2), 276–286.

McIlwaine, C. and Datta, K. (2004) 'Endangered youth?': Youth, gender and sexualities in urban. *Botswana Gender, Place and Culture*, 11, 483–511.

Mcilwaine, C. and Datta, K. (2003) From feminising to engendering development. *Gender Place and Culture*, 10, 369–382. doi: 10.1080/0966369032000155564.

McNeish, D. and Roberts, H. (1995) *Playing It Safe: Today's Children at Play*. Barnardo's, Essex.

Mead, M. (1928) *Coming of Age in Samoa*. New York: Morrow.

Norris, C. and Armstrong, G. (1997) *The Unforgiving Eye: CCTV Surveillance in Public Space*. Hull, UK: Mimeo.

Ortega-Alcázar, I. and Wilkinson, E. (2020) 'I felt trapped': Young women's experiences of shared housing in austerity Britain. *Social & Cultural Geography*, 22(9), 1291–1306. https://doi.org/10.1080/14649365.2020.1829688

Punch, S. (2002) Youth transitions and interdependent adultchild relationships in rural. *Bolivia Journal of Rural Studies*, 18, 123–133.

Rhodes, J.E. (2004) The critical ingredient: Caring youth-staff relationships in after-school settings. *New Directions for Youth Development*, 2004(101), 145–161.

Roberts-Holmes, G. (2017) Programming robots for factories: The impacts of baseline assessment measures upon young children. *The Psychology of Education Review*, 41(1), 36–39.

Schissel, B. (2021) *Still Blaming Children, Youth Conduct and the Politics of Child Hating*. 2nd edition. Halifax, Nova Scotia, Canada: Fernwood Publishing, Halifax.

Sear, R. and Coall, D. (2011) How much does family matter? Cooperative breeding and the demographic transition, demographic transition and its consequences. *Population and Development Review*, 37, 81–112.

Seymour, S.C. (2013) 'It takes a village to raise a child': Attachment theory and multiple child care in Alor, Indonesia, and in North India. In N. Quinn and J.M. Mageo (Eds.), *Attachment Reconsidered. Culture, Mind, and Society* (pp. 115–139). New York: Palgrave Macmillan. https://doi.org/10.1057/9781137386724_5

Skelton, T. (2007) Children, young people, UNICEF and participation. *Children's Geographies,* 5(1), 165–181.

Skelton, T. (2013) Young People's urban im/mobilities: Relationality and identity formation. *Urban Studies*, 50(3), 467–483. https://doi.org/10.1177/0042098012468893 (Original work published 2013).

Sullivan, D. and Hickel, J. (2023) Capitalism and extreme poverty: A global analysis of real wages, human height, and mortality since the long 16th century. *World Development*, 161, 106026.

Thomas, M.E. (2005) 'I think it's just natural': The spatiality of racial segregation at a US high school. *Environment and Planning A,* 37, 1233–1248.

Tomanović, S. and Ignjatović, S. (2006) The transition of young people in a transitional society: The case of Serbia. *Journal of Youth Studies*, 9(3), 269–285.

Tyler, I. (2008) Chav Mum Chav Scum. *Feminist Media Studies*, 8(1), 17–34.

UNICEF (2011) *Child Well-Being in the UK, Spain and Sweden*. Paris: UNICEF.

UNICEF Office of Research (2018) An unfair start: Inequality in children's education in rich countries. Innocenti Report Card 15, Florence: UNICEF Office of Research–Innocenti.

Valentine, G. (2003) Boundary crossings: Transitions from childhood to adulthood. *Children's Geographies*, 1(1), 37–52.

Valentine, G. (2004) *Public Space and the Culture of Childhood*. London: Routledge

Valentine, G., Skelton, T. and Chambers, D. (1998) Cool places: An introduction to youth and youth cultures. In T. Skelton and G. Valentine (Eds.), *Cool Places: Geographies of Youth Cultures* (pp. 1–32). London: Routledge.

Vidler, (2002) City transformed: Designing 'Defensible Space'. *Grey Room*, 7, 82–88.

Weller, S. (2006) Situating (young) teenagers in geographies of children and youth. *Children's Geographies*, 4(1), 97–108.

White, N. (2002) 'Not under my roof!': Young people's experience of home. *Youth & Society*, 34, 214–231. doi: 10.1177/004411802237864.

Worth, N. (2015) Youth, relationality, and space: Conceptual resources for youth studies from critical human geography. In J. Wyn and H. Cahill (Eds.), *Handbook of Children and Youth Studies*. Singapore: Springer. https://doi.org/10.1007/978-981-4451-15-4_8

7 SOCIAL GEOGRAPHIES OF RACE

7.1 Race and Social Geography: Definitions, Segregation and Early Interventions

Race is a difficult term to define, given its multiple social meanings. For the purposes of this chapter and for the book, we build on the previous chapter which explored how youth is socially constructed. In that regard, we equally advance a social constructionist perspective on race, with the realisation that race nonetheless has profound social (and spatial) implications. This takes inspiration from earlier work on racial formation theory by sociologists Michael Omi and Howard Winant (1994), in which race was seen as a social construct, not a biological given. To them, different races do not exist biologically – because no human group has ever been isolated long enough or completely enough to allow separate and independent genetic changes. Biological research has in fact shown that there is more diversity within different 'racial' groups than between 'racial' groups.

To understand why this essentialised biological approach, however misguided, has become so popular, we must return to the time when Europeans began exploring and colonising the Americas, Africa, the Middle East, Asia and Australia, from the 15th century onwards. This colonisation was undeniably buttressed by the belief that separate races existed, and that the White race was superior to all other ones. Omi and Winant (1994) refer to this perspective as thinking that race is something *real*, something scientific. But they also signpost (and subsequently oppose) the opposite temptation – to think of race as an *illusion*, something invented by humans and something that could be eliminated if we could all see 'beyond it'. Some anti-racist groups push for looking beyond race because to them, people are all the same in the end. And yet this colour-blind perspective remains problematic, since as we argue in this chapter and throughout the book, *race matters* in profoundly social and geographical ways. To be colour-blind is problematic because it instantly erases all of the *previous* racial discrimination that placed certain populations in inferior positions (and places) for long periods of time, up to and including present-day inequalities. By treating everyone the same, colour-blindness negates past and present experiences of racism, and whose impacts remain potent in terms of truncating life chances, life

DOI: 10.4324/9781003414889-9

courses and even life expectancies, as we saw in Chapter 2. As Ruth Gilmore (2022: 136) succinctly noted, 'if race has no essence, racism does'.

And so, in response to race as 'real' or race as 'illusion', Omi and Winant (1994) propose that we understand race as a social and historical process that has no biological foundation, but nevertheless is fundamental in structuring and representing our social world. They argued that it is difficult to let go of many of our ideas that are central to everyone's identity and understanding of the social world. Race is one of the first things a person notices, and racial stereotypes are built-in to most societies, sometimes implicitly so but usually quite explicitly. We may add that race is also fundamental in terms of structuring the (social) geographies of our social world, which will constitute the main focus of this chapter.

We organise the chapter to focus on key relationalities and encounters between race and social geography, beginning with place-based segregation and then moving onwards to more sophisticated conceptual frameworks that tie directly to the social constructionist view of race and its relationship to various spaces – the body itself, neighbourhoods, cities, colonies and nations. Social geography is a late bloomer when it comes to following the more social constructionist view of race. For many decades, the standard approach, inspired by the ecological model of the Chicago School, was to take racial categories for granted and simply map them across the (usually American) city. This traditional mapping of racial groups across urban space, which largely focused on the theme of spatial distribution and interaction, relied exclusively on essentialised and fixed racial categories derived from the national census. Much of this literature, especially the American one, centred on patterns of segregation in cities and would have an enormous influence on thinking about race within social geography. Segregation is the spatial outcome of particular social forces – especially but not exclusively the drive towards living among similar people while excluding others. Segregation has been a longstanding issue within the social sciences, and relates to the issue of separation among social groupings.

While we will revisit segregation in the next chapter in terms of marginality, we can expand the discussion right here with a brief overview of measuring segregation. Traditionally, two approaches to measuring segregation have held sway since the 1970s in social geography: one focused on pattern and very much in the vein of the Chicago School, the other increasingly focused on process. For pattern, there is a well-established literature on five dimensions of segregation – evenness, exposure, clustering, concentration and centralisation (Massey and Denton, 1988). Evenness measures the distribution of a majority and minority group across geographic areas of a city. The most widely used measure is the Index of Dissimilarity (ID), measuring the percentage of the minority group that would have to move to attain an even distribution across the entire city; a figure of zero indicates no pattern of segregation at all, as the distribution between the two groups is exactly similar across all subareas, while a figure of 1 indicates total segregation, whereby the distribution is totally uneven. Between these two extremes, ID values of over 0.6 are considered high, with 0.7 and above considered very high. For comparison, Belfast is a city divided by religion, and during the peak of the Troubles, the ward of Belfast West had an ID value of 0.807 in 1981 between Catholics and Protestants (Poole and Doherty, 1996).

A key weakness of the ID is that only two groups are considered at a time, and also that the scale is itself aspatial – it does not consider spatial patterns, only the degree of separation. In response, there are indices of exposure that measure the degree to which there may be potential contact between a minority and majority group. The interaction index (P*) captures this interaction, taking into consideration the size of the groups and their spatial distribution. Other measures – clustering, concentration and centralisation – further incorporate the spatial distribution of groups.

Over the past 30 years, there has been an effort to move away from measurement/index-based approaches (Massey and Denton, 1993), and towards a more temporally dynamic approach. This would take into account the lively and synchronic changes in urban social space that would tie into the various processes – including neighbourhood gentrification, regeneration, decline, in- and out-migration and so forth. Moreover, the five dimensions say little about whether segregation is advantageous or disadvantageous. This reflects the sustained interest in the processes that lead up to these patterns, and whether the patterns reflect fundamentally unequal outcomes within broader urban society. There is a certain injustice, of course, if certain groups are able to exclude less desirable groups, who in turn suffer the material deprivation of being separated from mainstream structures and opportunities. For instance, in the Global South, this has led to the emergence of stark divisions in the built environments, between gated communities on the one hand and informal settlements on the other, while in the Global North, the divisions are more subtle, involving the constitution and defence of inequalities via neighbourhood-level and school-level exclusion. Geographers point to how this social polarisation becomes spatially entrenched through place-exclusion by the haves and the separation of the have-nots from the well-off (Davis, 2006). This reduces the social well-being of the very poor, as disadvantages are passed on over the generations in place through spatial and social exclusion, thus ensuring material deprivation and the inability to participate fully in society. Conversely, advantage can be passed on simply by being born in a good place. Spatial segregation therefore has an important role in cumulating, reinforcing and perpetuating inequality and polarisation over time, with direct links to social justice in the form of distribution.

Since the 1990s as well, some social geographers have been less fixated on how to measure segregation, and began to adopt a more theoretically nuanced and sophisticated approach to race and space. Part of this renaissance was inspired by the pioneering work of Kay Anderson (1987), who saw North American Chinatowns as much imagined by the dominant society as actual places. The Chinese were originally brought to North America as labourers on the railways. They were segregated in the worst parts of the city, and were not allowed to work in most areas, which is why they gravitated to what they could do – run restaurants and laundries. Anderson showed how Whites 'imagined' Chinatown in numerous ways, ranging from the exotic to the dangerous, and that 'Chinatown' as a place is a Western construction, embodying processes of cultural imperialism that allowed European settlers the leeway, if not the outright power, to impose upon the district their own images and interests. As such, the Chinese inhabitants became racialised in negative ways. That is to say, their racial characteristics became a proxy for negative feelings among the White population. In this case, racialisation was the process by which 'Chinese-ness' became associated by the White population with negative categorisations (unclean, untrustworthy, incomprehensible) that fused with the spaces of Chinatown itself. Space and race therefore contributed to a specific and combined racialisation of people and place. Studying 100 years of discourses among the dominant White community, from the racist stereotyping of the early 20th century to the shift towards an 'ethnic neighbourhood', Anderson argued that the very existence of Chinatown speaks to the whims of the White power structure, and of cultural imperialism and violence that backed up economic exploitation and powerlessness (but not marginalization), using Young's (1990) five faces of oppression.

7.2 Race and Social Geography Today: Social and Spatial Constructions

Building on this base, race in social geography is now largely understood as a socially constructed category that is contingent and variable over space and time, always unstable. This variability once again counters the timeless and spaceless universality of a biological understanding of race. For instance, certain ethnic groups that came to North America (such as Jews, Italians, Poles) were not even considered 'White' until well after the Second World War. To bring some stability to this variability, dominant groups try to impose their powerful views of race upon space, making them inextricably linked. Mitchell (2000: 258) noted that 'control over the production of space – the ability to create space in particular ways – also lends to powerful groups the ability to actively *create* race. That is, 'race itself' – that social reification, that ideology masquerading as nature – is made "real" on the ground through the maintenance of spatial boundaries'.

Geographical space is imbued with race, ranging from the obviously racialised ghettos of the American city, to which we turn our attention in the next chapter on Poverty and Marginality, to the largely unacknowledged 'Whiteness' of suburbs and rural areas in the Global North more generally. Across these spatial aspects, racialisation is key, the "process by which racialised groups are identified, given stereotypical characteristics, and coerced into specific living conditions, often involving social/spatial segregation and always constituting racialized places' (Peake and Kobayashi, 2000: 393). Since the era of slavery, African-Americans have been separated from the White population in a myriad of ways – and while officially-sanctioned racism is no longer tolerated, there are strong spatial boundaries to the full integration of the former with the latter. The erection of spatial boundaries has also been studied with reference to 'White privilege', which refers to the advantages that Whites enjoy through the largely unquestioned Whiteness that pervades American and other White-dominated societies, including their naturalised ability to occupy prime urban spaces and exclude racialised Others (Peake and Kobayashi, 2000). In this respect, 'space is a resource in the production of White privilege. Indeed, neighbourhoods are not merely groupings of individuals, homes, and commerce, they are *constellations of opportunities* with powerful consequences.... Although Whites must go to ever greater lengths to achieve them, relatively homogenous White spaces for the full exploitation of whiteness' (Pulido, 2000: 30). That is not to say that Whiteness does not also intersect with poverty and marginalisation. In Ellie Jupp's (2022) work in Stoke-on-Trent, a post-industrial city in the British Midlands, low-income, working-class neighbourhoods are predominantly White and at the sharp end of economic hardship and cutbacks to family services and mental health services, and might currently be framed as 'left behind' within contemporary UK policy discourse. We will pick up on this particular UK context in the penultimate section of the chapter.

Race intersects with a variety of core social identifiers, linking back to intersectionality in Chapter 2 and our preceding chapter on youth. Of course, in some nations, poverty is racialised but in others not or it remains rather muddled. In a highly diverse nation such as the United States, racialised populations such as African-Americans and Latinos (although technically not a racial category according to the US Census, based only in national origin from Latin America) suffer disproportionate poverty and marginality, particularly in cities. But at the same time, some segments of this population can avoid poverty and marginality, while the White population is not immune to either as we noted above. In some European, African or Asian nations with little racial diversity, the majority population will suffer from poverty and marginalisation (but not racialisation). Nonetheless, when poverty and race do intersect, they can create intersectional forms of marginality, frontlines that can be incredibly tenacious in society and in certain places. This will be explored in the next section, but also in Chapter 8 on Poverty and Marginality.

7.3 Race and Social Geography: Emerging Topics in the 2020s

In this section we explore the increasingly fertile intersections of social geography and race, tracing its emerging dominance in some parts of the field over the past 15 years. Indeed, to some social geographers, the main focus of social geography *is* race. For instance, the most recent review of social geography literature in the flagship journal *Progress in Human Geography* (Nayak, 2025) places race front and centre in the creation of social geographical scholarship, and promoted as a way to include more non-White thought in the field. In this regard, Nayak (2025) reminds the reader that she is expanding on the longstanding interest in the constructions of race and racism, as well as geographies of exclusion. Moreover, she sees social geography at the forefront of how to study colonisation – that is the long shadows of colonial domination and exclusion, mostly by European powers (but not exclusively – think about the Ottoman and Japanese empires) from the 15th and 19th centuries – and the still-tentative process of decolonisation. This process aims to both recognise that Anglo-American geography is very much built on colonial foundations (e.g. mapping, exploration, exploitation), and that a decolonial strategy would aim to rebuke and replace this exclusionary foundation.

A decolonial strategy does not consign problematic constructions of race to the past, but sees them embedded in present-day social structures outlined in Part 1 of this book, including the state, the market (geographies of production) – and even the university. This can be manifested in terms of a lack of representation in certain key institutions (e.g. the police, directors of companies, the legal sector and university professors) among certain negatively-racialised groups (e.g. Black people, Muslims). Decolonisation is also forward-looking – it does not seek a pre-racial, pre-colonial world but seeks a future-oriented way to build more just worlds (Jazeel, 2014). Decolonisation involves both amends for past abuses – something we will explore in Chapters 13 around quiet forms of social justice, especially in terms of abolition geographies and racially-based reparations, to rethinking how knowledge is produced (Naylor, 2024). Within social geography, decolonisation has involved the latter, in terms not only of race but also empire, slavery and neo-colonialism (Jones, 2025).

Returning to Nayak's (2025) review piece, she names a variety of relationships between race and social geography which, in turn, structure a variety of encounters, and directly tie to the key themes of the social, inequality and social justice. For the idea of the social, we can connect to *White supremacy* as a way to update the now somewhat quaint notion of White privilege. For inequality, we can connect to *racial capitalism* and *White settler colonialism*. And for social justice, we can connect to *environmental racism.* Each of these concepts frame, in overlapping ways, the relationship between race on the one hand and social geography on the other. But they also frame new encounters between race and space in ways that extend well beyond the earlier notions we discussed in sections 7.1 and 7.2.

White supremacy is a more sophisticated and deeper take on White privilege. In terms of the structures and processes that connect societies with space, and how these intersect with issues of personal identity, Whiteness understood through a White supremacist lens is conceptualised as something that:

> describes and locates white racial domination by underscoring the material production and violence of racial structures and the hegemony of whiteness in settler societies. The concept of white supremacy forcefully calls attention to the brutality and dehumanization of racial *exploitation* and *domination* that emerges from settler colonial societies. While white privilege remains an important analytic frame to analyze the taken-for-granted benefits and protections afforded to whites based upon skin color, the concept of privilege emphasizes the social condition of whiteness, rather than the institutions, practices, and processes that produce this condition in the first place.
>
> (Bonds and Inwood, 2016: 715)

In effect, White supremacy shifts the discussion from privilege to domination, leading to more obviously deeper social and spatial exclusions across a variety of scales, from the body to the city to the nation. This directly links to aspects of inequality as discussed in Chapter 2, especially Young's notion of cultural imperialism and violence.

This inequality is entirely unjustified, so it must be continuously upheld by those in power via various mechanisms of social control. One of those that has attracted considerable attention is *racial capitalism* (Simpson and Hugill, 2022). In this relation between race and capitalism, race is central to the logic of capitalism, especially via slavery, colonisation and exploitation. Rooted in Robinson's *Black Marxism* (2000, 26), it contends that capitalism has been a racialised system of exploitation since its beginnings: 'the tendency of European civilization through capitalism was ... not to homogenise but to differentiate – to exaggerate regional, subcultural, and dialectical differences into "racial" ones'. By definition, capitalism exploits – a racial capitalism framework emphasises that racialised groups (and their lands) are disproportionately exploited, not by happenstance, but by design and for profit. Human difference becomes something to exploit but also to naturalise through racialised categories and hierarchies (Pulido, 2017).

Building on the idea of racial capitalism (and colonialism) is a particular form of racism inherent to White settler societies – those places permanently colonised, particularly in the Americas and Oceania (e.g. Australia, Hawaii, and New Zealand) but also South Africa. Settler colonialism, as a form of imperialism rooted in White supremacy, differs from traditional colonialism in its theory, politics and geography. Unlike colonialism, which often centres on military or economic expansion and typically involves the eventual withdrawal of the colonisers, settler colonialism is defined by the permanent takeover of land and the displacement of Indigenous populations to establish a new, ethnically distinct nation (Bonds and Inwood, 2016: 718).

So, unlike the colonisation of most of Africa and Asia, which were superficially decolonised when Europeans went 'home' in the 20th century, places like the US, Canada, New Zealand, Australia, South Africa, Brazil, Chile and Argentina are characterised by a permanent *White settler colonialism*, in which the ancestors of the colonists run the nation to the enduring detriment of the original, Indigenous populations. Building settlements associated with the incoming group necessarily leads to the erasure of the Indigenous population via the violent seizure of land and resources. This creates enormous advantages for the incoming colonisers, who enjoy the fruits of essentially free land without any compensation to the original dwellers. This wide-ranging dispossession is the ultimate expression of racial capitalism, and certainly played a role in the negation of Indigenous culture in the Winnipeg case study from Chapter 4.

Like decolonisation, the point of raising the ills of White settler colonialism is forward-looking and embedded in social justice. It does not expect millions of settlers to repatriate themselves, but rather to seek more equitable ways to share the land and resources with Indigenous populations. The longstanding legacies of White settler colonialism – of replacement and dominance – necessarily bring up issues of apology but also (eventually) repair. While we consider reparations more fully in Chapter 13 under abolition geographies, suffice it to say here that repair usually involves some kind of transfer of land and resources *back* to the original inhabitants, and not just facile recognition of Indigenous culture. An example of the latter is the following: many American, Australian, Canadian and New Zealand universities state that their campuses are built on stolen Indigenous land, and name the original inhabitants. An example of the former would be for universities to give Indigenous people better access to the university (and the land it is built on), acting as a resource to help Indigenous people regain at least part of what was originally lost upon contact with the colonisers. And a further point – White settler colonialism is never 100% complete, despite its logics around elimination and erasure. We will explore the importance of this incompleteness in more detail in Chapters 13 and 14, but suffice it to say that domination is always strewn with resistance, reworking and especially resilience among the Indigenous populations who have survived.

Well before the development of the concept of White settler colonialism, another area of race and social geography was also directly pertained to the theme of social justice: *environmental racism*. A particularly American construct, it focuses on the disproportionate exposure of racialised populations to environmental ills, be they the presence of air/water pollution, the inequitable siting of toxic waste dumps or the general neglect of a racialised population's daily well-being. There does not need to be any particularly racist motivation – an environmental racism approach understands that many inequitable geographical patterns are a result of longstanding structures that were not outwardly 'racist' but end up producing social injustice just the same.

FIGURE 7.1 Recycling and social housing, Brooklyn, 2004 © G DeVerteuil

The image above tries to capture this dynamic. Showing an auto recycling plant in the foreground and various social housing projects in the background in Brownsville (Brooklyn, New York), this image juxtaposes an environmental ill (recycling) with racialised poverty, in that 99% of the tenants in the towers would be African-American. The siting of the recycling plant would not have been random, but connected to the marginal uses that flock to poorly-defended, heavily racialised and renter neighbourhoods such as Brownsville, where the population would have (at least initially) few legal resorts to block such developments and even fewer to expulse them. Racialised minorities must bear this burden while wealthier White communities would have the means to reject it (Pulido, 2017). Racism underwrites these patterns, in the form of racial capitalism seeking places and people to exploit but also a compliant state that does little to rectify these entrenched patterns.

The initial devaluation of people and places is critical in the maintenance of environmental racism specifically, and the larger structures of White supremacy, racial capitalism and White settler colonialism more generally. The example of Hurricane Katrina in New Orleans (2005) brings forth this devaluation before, during and after the disaster (DeVerteuil et al., 2021). Like many American cities, New Orleans before the disaster suffered from deep-seated and racialised urban poverty, high levels of segregation, environmental injustices and racial polarisation (Peck, 2006). Poor infrastructure, including insufficient levees in areas most vulnerable to flooding, was compounded by decades-long decline and neglect of public institutions (police, education, transportation), with a 25% population drop since the

FIGURE 7.2 Lower Ninth Ward, New Orleans, 2015 © G DeVerteuil

1960s. All of these deficiencies produced a dense web of vulnerability that would become obvious during and immediately after the emergency.

The image above shows a typical scene in the African-American neighbourhood of the Lower Ninth Ward, which was entirely submerged by many feet of water from the hurricane once the levees failed. The lack of safeguards before the hurricane was epitomised by the inability of many poor African-American inhabitants to evacuate the city. During and after the disaster, help from the state was disorganised and grudging at best. Even ten years after Katrina when this picture was taken, large parts of the Lower Ninth Ward remained abandoned to the elements and left to fend for itself, a victim of environmental racism that placed disproportionate exposure and burdens on this African-American community. As a disaster, Katrina made only too visible existing and deep-seated injustices, and the long and difficult recovery only confirmed them. As of 2025, a look on Google Earth reveals a Lower Ninth Ward still marked by significant abandonment and neglect that overwhelms the patchwork nature of rebuilding. Environmental racism is thus a subset of environmental justice, becoming in this regard a form of anti-racism. This involves moving beyond not being racist yourself, to calling out racism in all of its forms across society. More on this idea of anti-racism in Chapter 13, as a form of quiet activism.

7.4 Case Study: Immigration in the UK since 2000

In order to showcase some of the relationships between social geography and race, we provide here a case study of immigration and race relations in the UK over the past 25 years, punctuated by some key events including unrest, terrorism, and Brexit but also underpinned by slow-moving demographic shifts. Over this same time period, both Geoff and Andy emigrated to the UK from Canada and Ireland respectively, so in that sense the following narrative is both academic but also felt personally.

Immigration has become a hot-button issue across the Global North, if not a moral panic, to the point where most political parties in places like Canada, the US and the UK actively seek to considerably constrain immigration to only the most affluent and well-educated. Low-skill immigrants, to say nothing about refugees and asylum seekers, are increasingly seen as undesirable, and appear downright poisonous to the far-right. The narrative of loss of control over borders has become dominant in most nations of the Global North, even among those nations with generous welfare states like Denmark and Sweden.

Immigration and race are entwined in many ways. The most straightforward way is that most immigrants by the year 2000 – but certainly not all – were coming (1) from the Global South to other nations in the Global South, or (2) to the Global North. This is a byproduct of inequality at a global scale, between greater poverty and instability in the Global South and greater opportunities and stability in the Global North. The large gaps between these two realities fuel migration northwards. We focus on this migration to the Global North, despite the fact that most immigrants go from the Global South to adjoining Global South nations, especially in Africa, Latin America and the Middle East. For instance, while the large-scale forced migration of Syrian migrants to Europe attracted lots of attention in 2015, most Syrians remained nearby, in Lebanon or Türkiye. This connects to an important distinction frequently lost in the current anti-migrant narrative – that forced migration is increasing because of war and environmental issues, rather than presenting all migrants as motivated by economics. The spectre of mass migration of non-White people to the Global North must be seen as racialised, part of a longer fear of 'colonisation in reverse', amplified by ideas of a 'Great Replacement', a debunked online White nationalist far-right conspiracy theory that, White European populations at large are being demographically and culturally replaced by non-White peoples with the complicity or cooperation of 'replacist' elites. More widely, this is shaped by spatial and social constructions of many Global North nations, as decidedly White spaces, even when the nations were settler-based. These narratives have travelled from place to place, abetted by social media, right-wing rhetoric and over-heated identity politics.

The UK was a long-time colonial power (arguably, *the* colonial power in the 19th century), and perhaps not surprisingly a large proportion of immigration in the 20th and 21st centuries were from former colonies, including Bangladesh, Ghana, India, Nigeria, Pakistan, Sri Lanka and Ireland. But this list has expanded in the past 40 years to include places such as Poland and Romania, as part of their accession to the EU. Just the same, until the

1970s, the UK was a remarkably White nation and remains to this day in many peripheral rural areas and left-behind formerly industrial small towns. It was not until 1991 that the UK census even asked about ethnic origin and race. Racially, Whites still constituted 81% of the total population according to the 2021 census, with Asians at 9% and Blacks at 4%. These numbers, however, obscure a very uneven racial landscape, with many cities featuring very diverse populations and others not. The same goes for racial groups – Asians are found across a wider set of cities compared to Blacks (including Caribbean and African populations), who are far more concentrated in just a few cities, London especially.

We begin the UK story with the early 2000s, when it was segregation of (previous) immigrants, rather than mass migration, which was drawing the most attention. After all, segregation is more than just an academic exercise – it fires up the popular imagination and becomes part of very public discussions around how cities and societies ought to be structured. If one follows the public debates and media discourse in the UK since the early 2000s, there has been a distinct fear of US-style hyper-segregation in British cities, stemming from unrest in Bradford, Burnley, and Oldham 2001, and compounded by the bombings in London in July 2005. A moral panic emerged around fears of immigrant groups leading 'parallel lives', and that UK society was 'sleepwalking to segregation'; in these discourses, certain minority groups (especially, but not exclusively, Muslims) were blamed for self-segregating and operating independently of British society and norms.

To illustrate this with an example, the former chief executive of Nottingham Council (a British metropolitan council) and chair of the Community Cohesion Review Team, Ted Cantle, was charged with analysing the causes of disturbances across the north of England in 2001. He visited the scenes of riots in Oldham, Burnley and Bradford, as well as touring Southall, West London, Birmingham, Leicester and Sheffield. The riots injured more than 100 people and caused millions of pounds worth of damage, the spark of which was police brutality against the Muslim South Asian population. In response, Cantle called for a national debate on a shared British identity. His report highlighted separate education, communal and voluntary organisations, employment, places of worship and cultural activities. It stated that 'Whilst the physical segregation of housing estates and inner-city areas came as no surprise, the team was particularly struck by the depth of polarisation in our towns and cities. Many communities operate on the basis of a series of parallel lives. These lives often do not seem to touch at any point, let alone overlap and promote any meaningful interchanges'. To him, the lack of spaces of encounter between the Muslim and non-Muslim communities was a product of self-segregation on the part of the former, a point that will be refuted below.

In the wake of the July 7 2005 bombings in London, Trevor Phillips, Commission for Racial Equality Chair, warned again of 'sleepwalking towards segregation'. More specifically, he said that:

> the fact is that we are a society which, almost without noticing it, is becoming more divided by race and religion. We are becoming more unequal by ethnicity. This is not only or even principally about Muslims. But the aftermath of 7/7 forces us to assess where we are. And here is where I think we are: we are sleepwalking our way to segregation. We are becoming strangers to each other, and we are leaving communities to be marooned outside the mainstream.

Drawing on the example of New Orleans and Katrina (a city which was well over 50% Black, with no equivalent in the UK), Mr Phillips warned against complacency in the face of divisions, of the dangers of leaving this segregation unaddressed:

> When the hurricane hits – and it could be recession rather than a natural disaster, for example – those communities are set up for destruction. Even if there is no calamity, these marooned communities will steadily drift away from the rest of us, evolving their own lifestyles, playing by their own rules and increasingly regarding the codes that the rest of us take for granted as outdated behaviour that no longer applies to them.

One of the main assumptions made by both Cantle and Phillips is that immigrant groups actually have the ability to truly self-segregate, when in fact their spatial patterns are far more constrained, yielding to a largely involuntary pattern of settlement into the areas of least resistance. A further shortcoming was the neglect of seeing defensive clustering as something positive for immigrants. As Phillips, Davis and Ratcliffe (2007: 219) aptly put it, 'the more positive aspects of minority clustering, namely, the social and cultural capital inherent in supportive social networks, and the better health and well-being associated with community support in otherwise deprived inner urban areas, tend to receive less attention'. But at an even more basic statistical level, Ceri Peach (2006) argued that British levels of segregation are much lower than those found in the US and, for the Black Caribbean population, they are falling. South Asian levels of segregation are higher than for the Caribbean population, but still show considerable internal variation. Bangladeshis, the most recently arrived of the groups, show the highest levels of encapsulation, followed by the Pakistanis, while Indian rates are relatively modest. Unlike the US, there were only three major cities in which ethnic minorities constituted a majority using the 2021 census (Birmingham, Leicester and Manchester), alongside 10 of London's 33 boroughs. Compared to hyper-segregated cities such as Chicago, Detroit, Memphis, New Orleans and Philadelphia where African-Americans live primarily in areas that are 100% co-ethnic, London and other major UK cities feature relatively scattered minority populations. In this respect, we invite the readers to examine images from George Georgiou (*The Guardian*, 2015) on is website (Last Stop – George Georgiou), which shows a level of racial mixing in London that would be unlikely in the United States.

In the same year that Andy came to teach at Southampton, In August 2011, unrest broke out in most major UK cities – including London, Birmingham and Manchester – in the aftermath of the shooting death of an unarmed Black man by the police in Tottenham. The unrest itself was not seen to be especially racially based, although the initial spark certainly was. Indeed, of the 2,000 accused brought to court in London by the autumn of 2011, 90% were male, and 74% were 24 and under, with 46% Black, 42% White and 7% Asian (Lewis et al., 2011). There was a link between deprivation and rioting, with 70% of those arrested living in the 30% most deprived areas. Around three-quarters had previous convictions. There were relatively few racially-tinged attacks or targeting once the unrest got underway, confirming that despite the dire warnings of Ted Cantle and Trevor Phillips, the levels of UK segregation did not lead to nor inflame the 2011 unrest. Rather, it was class-based antagonism that played the major role, in the midst of government cuts and welfare state austerity (Hall, 2019).

In June 2016, a bare majority of voters in the UK passed Brexit, which would eventually enable the exit of the UK from the European Union a few years later. The lead-up to the vote was fixated on issues around migration, especially the sense that the UK borders were being overwhelmed by EU migrants from places such as Poland, the Baltic States, Romania and Bulgaria (and unfounded fears about Turkey, which was not even an EU member back then and still not to this present day). A fundamental principle of the EU has been the freedom to migrate across the borders of the various member states. The pro-Brexit vote was complex, and in some ways was a protest vote against the current government, austerity and gross regional imbalances between the prosperous Southeast and the rest of the nation. But, it was also certainly to do with 'securing' borders and stopping seemingly runaway migration, even though most EU migrants were White and contribute to key components of the economy, including construction, hospitality and the National Health Service. Interestingly in the vote, the more diverse a place was the more it was likely to vote against Brexit, while the opposite was certainly true: non-diverse places in the UK were more likely to vote for leaving the EU. Interestingly, the ire against EU migrants from Eastern Europe in the run-up to Brexit is a reminder of how race is quite malleable, and that White skin does not always translate into acceptance by White society. The aftermath of the Brexit vote was still being felt when the UK was hit was several terrorist attacks in May–June 2017 (Manchester Arena bombing, London Bridge attacks) as well the Grenfell Tower fire (see Chapter 4). This again raised racial tensions around immigration, but more to do with Muslims than with EU migrants.

In the aftermath of Brexit, EU in-migration plummeted but was more than matched by a rising tide of visa holders coming from certain war-torn or politically-vulnerable nations. These included Hong Kong (a former colony), Afghanistan (after the 2021 fall of Kabul), and Ukraine (invaded by Russia in 2022). The welcome list is quite short however, and did not include other war-torn nations or conflict zones (e.g. Sudan, Syria, Myanmar, Gaza) or places with ongoing strife (e.g. Iraq, Pakistan, Venezuela). The surge from the selective welcome list once again ignited concerns about too many immigrants entering the nation, particular during the pandemic.

Finally, there was significant unrest in many Northern English cities in August 2024, in the aftermath of a shocking knife attack on young girls in Southport, in which three were killed. Within a few hours, right-wing social media was spreading rumours around that the killer was a migrant (when, in fact, he was a UK citizen). This prompted coordinated rioting in various public spaces, and against perceived immigrant-related infrastructure such as hotels housing asylum seekers. Unlike the multi-racial nature of the 2011 unrest, however, in this case it was overwhelmingly White males who were out in the streets, and it was the police who were trying to protect infrastructure and public order. Much of the unrest occurred in depressed towns in Northern England, where anti-migrant sentiments had been readily stoked by 15 years of rhetoric from the Conservative government. But these same areas had also been systematically ignored and even harmed by 15 years of government austerity and welfare cutbacks, and featured relatively few migrants when compared to cities like London. The mix of these two feelings – of being resentful and of being abandoned – proved quite powerful, and it is a mix to which we will return in Chapter 14, which puts into conversation urban unrest and broader government neglect of 'bread and butter issues'. This is *not*, however, a conversation that can justify anti-immigrant violence, only to contextualise it within larger societal, spatial and justice issues.

This wide-ranging review of the past 25 years shows, at least in the UK, that race and immigration are entangled in complex and contingent ways. The examples used touch on issues of what constitutes the social, in terms of who is 'British' or 'English' and who is not. They also touch on issues of inequality, in that cities and societies that tended to be more unequal also featured a more combustible mix of pro- and anti-immigrant sentiments, as well as complicated patterns of segregation that cut both ways. The British case also touches on issues of social justice. In 2001 and 2011, the unrest was clearly linked, at least initially, to issues of exclusion along racial lines. Conversely, the 2024 unrest was clearly nativist and populist – that is, against those perceived as foreign, among a socially- and economically-disconnected population encouraged to blame immigrants for all of their problems. In Chapter 11 on housing, and 14 on social justice, we will spend more time connecting social unrest with social justice.

As a way to conclude this section on a more positive note, we turn to a UK-based framework to understand race, immigration and nativism in the UK, using Paul Gilroy's (2004) sense of *conviviality*. Conviviality is the boisterous interplay of different races in the city, with specific reference to certain aspects of British cities and parts of British life. An example of Gilroy's (2004) concept of conviviality in British cities can be seen in the everyday multicultural interactions that occur in neighbourhoods like Hackney in London or Moss Side in Manchester. In these urban spaces, people from diverse ethnic and cultural backgrounds live in close proximity and share public spaces, schools, workplaces and transport systems. Using a symbiotic framing from Chapter 4, conviviality could be seen as a mutually beneficial relationship, in that both sides benefit. In particular, Gilroy's celebration of the ways in which race has become, in his view, almost irrelevant for many young and urban Britons in recent years is shadowed by another, gloomier insight. For the rest of the population, race has become more important: panics about asylum seekers; hostility to new eastern European immigrants; fear of 'Black' gun and knife crime; fear of infiltration by foreign or Muslim terrorists – in many ways, he suggests, race relations in Britain in 2004 were going as much backward as forward. This divide was certainly captured in 2016 with Brexit, and again in the 2024 unrest – that the UK population is increasingly divided between those open to immigration in well-off cities, and those closed to immigration, in left-behind cities.

In other words, conviviality only thrives in cities that fully embrace existing diversity, and shrivels in the very cities that lack the diversity upon which conviviality depends. Why? To Gilroy, it is the lack of meaningful spaces of encounter, that in turn leads to suspicion, racism and disdain – and likely Brexit in 2016, if he could have foreseen it back in 2004, where the least diverse places were the most likely to worry about immigration, while the most diverse places were the least worried. Further into the future, the right-wing anti-migrant unrest in many British cities in August 2024 (and the counter-protests that followed) very much underlines the contested nature of who belongs and who does not. To some, belonging remains narrowly defined to native-born inhabitants, to others it is far more ecumenical and globe-spanning, even dare we say convivial. We explore more on belonging in Chapter 10, when we consider the relationships between society and mental health. As such, conviviality is an open sense of belonging, not a narrow or defensive one, more utopian than reality. More on the idea of the utopian in Chapter 13, but we can signpost here that to be utopian means imagining alternative possibilities, so as to then create an alternative world. Conviviality is utopian given the sustained hysteria around immigration in many Global North nations that goes beyond what happened in British cities in August 2024. These nations include traditional immigrant-receiving nations such as the United States and the UK, to those relatively unaccustomed to immigrants, such as Ireland and Japan, which for a long time featured high levels of social coherency but low levels of diversity. As a final note, the new Labour Prime Minister Keir Starmer gave a speech in May 2025 that was tinged with fear that immigration to the UK remained too high, that Britain risked becoming a 'nation of strangers'. This speech was very much targeted to those British voters who felt the same way.

Overall then, the UK case study relates back to some of the key terms we have presented in this chapter. These first included segregation, of how each round of unrest brought up issues of self-segregation by certain racialised groups, even if overblown. The second involved racial capitalism, in that the UK's wide-ranging colonisation and exploitation has subsequently created certain immigrant streams (especially from South Asia) that have proven durable. Finally, and connected to more recent debates about belonging, is the sense of Whiteness in terms of British identity, but also how this identity may be surmounted, including with the use of conviviality.

7.5 Conclusions

Race has increasingly become a dominant social identifier as societies such as Australia, Canada, the UK and the US become more diverse. This arguably reflects both growing fears but also growing acceptance of racial diversity and immigration. Social geographers have acknowledged the need to engage with these trends. From its very basic beginnings in the 1990s to the wealth of conceptual material by the 2020s, social geographers have taken up the task of conceptualising the growing importance of race in everyday social life. But while race is a highly versatile and potent social category, it cannot hope to cover or explain everything in a world as complex as the one in which we are writing. In some places, race is a non-starter because there is no actual racial diversity at all – for instance, most of the Global East. In other places, any discussion of racial difference is suppressed as it would constitute an attack on social coherency, such as France or the current Trump administration, which seeks to deny racial diversity in all of its forms for the sake of White-based social coherency. Still in other places, there is a backlash against discussing race on its own, as part of the grievance game that sidesteps other key social identifiers as well as larger class inequalities. So we are certainly aware of the pitfalls of treating race as all-embracing and all-encompassing.

In the next chapter, we seek to further take race out of its silo by putting it in conversation with class and marginality, as a way to show the interactions from a relational if not intersectional approach. But crucially, we concentrate on the various spaces of encounter where race and class intersect negatively, whether the stigmatised spaces of the African-American ghetto, and the service hub or the invisible spaces of the immigrant hub, but also intersect more positively, in terms of places of hustle and survival.

Further Reading

Atkinson, R. and McGarrigle, J. (2009) Segregation, urban. In R. Kitchin and N. Thrift (Eds.), *International Encyclopedia of Human Geography* (Vol. 1, pp. 76–80). Amsterdam: Elsevier Press.

Bledsoe, A. (2020) The primacy of anti-Blackness. *Area*, 52(3), 472–479.

Coates, T.-N. (2014) *The Case for Reparations. The Atlantic*, June.

Deener, A. (2012) *Venice: A Contested Bohemia*. Chicago: University of Chicago Press.

Duneier, M. (1999) *Sidewalk.* New York: Farrar, Straus and Giroux.

Evans, J. (2011) Exploring the (bio) political dimensions of voluntarism and care in the city: The case of a 'low barrier' emergency shelter. *Health & Place*, 17(1), 24–32.

Fairbanks, R. (2009) *How it Works*. Chicago: University of Chicago Press.

Harvey, D. (2012) *Rebel Cities.* New York: Verso Books.

Iveson, K. (2013) Cities within the city: Do-it-yourself urbanism and the right to the city. *International Journal of Urban and Regional Research*, 37(3), 941–956.

Klinenberg, E. (2002) *Heat Wave: A Social Autopsy of Disaster in Chicago*. Chicago, IL: University of Chicago Press.

Knowles, C. and Harper, D. (2009) *Hong Kong: Migrant Lives, Landscapes, and Journeys*. Chicago: University of Chicago Press.

Lancione, M. (2016) The assemblage of life at the margins. In M. Lancione (Ed.), *Rethinking Life at the Margins* (pp. 3–25). London: Routledge.

Lancione, M. and McFarlane, C. (2016) Life at the urban margins: Sanitation infra-making and the potential of experimental comparison. *Environment and Planning A: Economy and Space*, 48(12), 2402–2421.

Ley, D. (1974) *The Black Inner City as Frontier Outpost: Images and Behaviour of a Philadelphia Neighbourhood*. Washington, DC: Association of American Geographers.

Leyshon, A. (1995) Missing worlds: Whatever happened to the geography of poverty? *Environment and Planning A*, 27, 1021–1028.

Marcuse, P. (1997) The enclave, the citadel, and the ghetto. *Urban Affairs Review*, 33, 228–264.

McKittrick, K. (2011) On plantations, prisons, and a black sense of place. *Social & Cultural Geography*, 12(8), 947–963.

Mitchell, D. (2001) Postmodern geographical praxis? The postmodern impulse and the war against homeless people in the 'post-justice' city. In C. Minca (Ed.), *Postmodern Geography: Theory and Praxis* (pp. 57–92). Oxford: Blackwell.

Park, R. and Burgess, E. (1925) *The City*. Chicago: University of Chicago Press.

Philo, C. et al., editors (1995) *Off the Map: The Social Geography of Poverty in the UK*. London: Child Poverty Action Group.

Pickerill, J. and Chatterton, P. (2006) Notes towards autonomous geographies: Creation, resistance and self-management as survival tactics. *Progress in Human Geography*, 30(6),730–746.

Reese, E., DeVerteuil, G. and Thach, L. (2010) The contradictions of containment: Deconcentrating poverty in Downtown Los Angeles. *International Journal of Urban and Regional Research*, 34(2), 310–327.

Sheppard, E., Sparks, T. and Leitner, H. (2020) World class aspirations, urban informality, and poverty politics: A North-South comparison. *Antipode*, 52(2), 393–407.

Simone, A.M. (2004) People as infrastructure: Intersecting fragments in Johannesburg. *Public Culture*, 16(3), 407–429.

Smith, N. (1996) *The New Urban Frontier: Gentrification and the Revanchist City*. London: Routledge.

Stuart, F. (2016) *Down, Out, and Under Arrest: Policing and Everyday Life in Skid Row*. Chicago: University of Chicago Press.

Takahashi, L. (1997) The socio-spatial stigmatization of homelessness and HIV/AIDS: Toward an explanation of the NIMBY syndrome. *Social Science and Medicine*, 45, 903–914.

Thieme, T. (2018) The hustle economy: Informality, uncertainty and the geographies of getting by. *Progress in Human Geography*, 42(4), 529–548.

Vergara, C. (1995) *The New American Ghetto*. New Brunswick: Rutgers University Press.

Wacquant, L. (2008) *Urban Outcasts: A Comparative Sociology of Advanced Marginality.* Cambridge, UK: Polity Press.

Wilton, R. (1998) The constitution of difference: Space and psyche in landscapes of exclusion. *Geoforum*, 29, 173–185.

Wilton, R. (2000) Grounding hierarchies of acceptance: The social construction of disability in NIMBY conflicts. *Urban Geography*, 21, 586–608.

References

Anderson, E. (1990) *Streetwise: Race, Class and Change in an Urban Community*. Chicago: University of Chicago Press.

Anderson, K. (1987) The idea of Chinatown: The power of place and institutional practice in the making of a racial category. *Annals of the Association of American Geographers*, 77(4), 580–598.

Bonds, A. and Inwood, J. (2016) Beyond white privilege: Geographies of white supremacy and settler colonialism. *Progress in Human Geography*, 40(6), 715–733. https://doi.org/10.1177/0309132515613166

Davis, M. (2006) *Planet of Slums*. New York: Verso.

DeVerteuil, G. (2015) *Resilience in the Post-Welfare Inner City: Voluntary Sector Geographies in London, Los Angeles and Sydney*. Bristol: Policy Press.

DeVerteuil, G., Golubchikov, O. and Sheridan, Z. (2021) Disaster and the lived politics of the resilient city. *Geoforum*, 125, 78–86.

Gilroy, P. (2004) *After Empire: Melancholia or Convivial Culture?* London: Routledge.

Gilmore, R.W. (2022) *Abolition Geography: Essays Toward Liberation*. New York: Verso.

Hall, S.M. (2019) A very personal crisis: Family fragilities and everyday conjunctures within lived experiences of austerity. *Transactions of the Institute of British Geographers*, 44, 479–492.

Jazeel, T. (2014) Subaltern geographies. *Singapore Journal of Tropical Geography*, 35, 88–103. https://doi.org/10.1111/sjtg.12053

Johnston, R. and Gregory, D. (1994) *The Dictionary of Human Geography* (1st edition). London: Blackwell Publishers.

Jones, C. (2025) Geographies of war and violence I: Decolonising war. *Progress in Human Geography*, 49(2), 194–214. https://doi.org/10.1177/03091325251319019

Jupp, E. (2022) *Care, Crisis and Activism: The Politics of Everyday Life*. Bristol: Policy Press.

Lewis, P., Newburn, T., Taylor, M., Mcgillivray, C., Greenhill, A., Frayman, H. and Proctor, R. (2011) *Reading the Riots: Investigating England's Summer of Disorder*. London, UK: The London School of Economics and Political Science and The Guardian.

Massey, D. and Denton, N. (1988) The dimensions of residential segregation. *Social Forces*, 67(2), 281–315.

Massey, D. and N. Denton (1993) *American Apartheid: Segregation and the Making of the Underclass*. Cambridge MA: Harvard University Press.

Mitchell, D. (2000) *Cultural Geography (Critical Introductions to Geography)*. London: Wiley.

Naylor L. (2024) Political geographies of the pluriverse in practice: Towards a decolonial feminist approach. In F. Menga, C. Nagel, K. Grove et al. (Eds.), *Political Geography in Practice: Theories, Approaches, Methodologies* (pp. 69–86). Cham: Springer International Publishing.

Nayak, A. (2025) Social geography I: Anti-racism, implacable whiteness and decolonizing Anglo-American geography. *Progress in Human Geography*, 49(1), 109–118. https://doi-org.cardiff.idm.oclc.org/10.1177/03091325241296067

Office of National Statistics (2021) UK Census of Population.

Omi, M. and Winant, H. (1994) *Racial Formation in the United States: From the 1960s to the 1990s*. New York: Psychology Press.

Peake, L. and Kobayashi, A. (2000) Racism out of place: Thoughts on Whiteness and an antiracist geography in the new millennium. *Annals of the Association of American Geographers*, 90(2), 392–403.

Peck, J. (2006). Liberating the City: Between New York and New Orleans. *Urban Geography*, 27(8), 681–713.

Phillips, D., Davis, C. and Ratcliffe, P. (2007) British Asian narratives of urban space. *Transactions of the Institute of British Geographers*, 32, 217–234.

Poole, M.A. and Doherty, P. (1996) *Ethnic Residential Segregation in Northern Ireland.* Coleraine: University of Ulster.

Pulido, L. (2017) Geographies of race and ethnicity II: Environmental racism, racial capitalism and state-sanctioned violence. *Progress in Human Geography*, 41(4), 524–533. https://doi-org.cardiff.idm.oclc.org/10.1177/0309132516646495

Pulido, L. (2000). Rethinking environmental racism: White privilege and urban development in Southern California. *Annals of the Association of American Geographers*, 90(1), 12–40.

Robinson, C. (2000) *Black Marxism: The Making of the Black Radical Tradition*. Chapel Hill, NC: University of North Carolina Press.

Simpson, M. and Hugill, D.W. (2022) The settler colonial city in three movements. *Progress in Human Geography*, 46(6), 1311–1330. https://doi.org/10.1177/03091325221114115

Statistics Canada (2022) *Canadian Census of Population 2021*. Ottawa.

The Guardian (2015) A view from the bus: Double-decker visions of London life – in pictures I Art and design I The Guardian.

8 POVERTY AND MARGINALITY

8.1 Social Geography and Poverty

This chapter first asks you to interrogate the geographies of poverty, framed by the larger concern for the hard edge of inequality introduced in Chapters 1, 2 and 4, as well as the social identifier of class and economic status introduced in Chapter 1 and developed in Chapter 2. In terms of social justice, the focus is on distribution, of who gets what and where – in this case, we focus on deprivation, the opposite of 'getting'. As we saw in Chapter 2 and the focus on inequality, poverty is caused by many factors. In this chapter we are especially interested in various labour market exclusions, whether completely separated from employment or suffering from precarious and/or low-paying employment. But this labour market focus is tempered by a concern with the representation of poor people and poor places, and how this affects the procedures for managing poverty (i.e. all three strands of social justice are relevant). All these factors shape the experience of poverty and class within the city.

Poverty has long captured the academic and political imagination, ranging from 19th-century social reformer investigations into the impoverished inner city, to the Chicago School of Urban Sociology's fixation with the 'zone of transition' (Park and Burgess, 1925), to explorations into the vengeful politics of the post-justice city (Smith, 1996; Mitchell, 2001), to be developed in more depth in Chapter 11 with regards to homelessness. Chronicling the academic (and, critically, geographical) fascination with poverty – especially but not exclusively urban – requires a robust definition of the term itself, both in isolation from and in relation to other key components of marginality. Poverty is usually defined in *absolute* and *relative* terms. Absolute poverty usually involves a fixed benchmark, recognising the fact that there are people who make insufficient money to afford minimum necessities for survival, such as food, shelter, running and clean water, safety and medical care. While useful in the Global South where upwards of half the population lack these fundamental requirements, absolute measures are less relevant in the Global North, where the vast majority of people currently have access to the basic necessities of life (Philo, 1995). Further, it can be exceedingly challenging to choose an objectively-defined and stable set

DOI: 10.4324/9781003414889-10

of basic necessities. For example, indoor plumbing and electricity would now be regarded as givens in the Global North, but this was not the case 100 years ago, and nor is it necessarily and currently the case in parts of the Global South.

Conversely, a relative conceptualisation of poverty describes individuals with significantly less of something than others around them, usually (but not always) in terms of income. Poor individuals may be defined as having less than some percentage of median income. In many European countries and the UK, *relative poverty* refers to people living in households with income below 60% of the median in that year. A major advantage of this approach is its simplicity and transparency, as it requires no decisions about what constitutes the barest necessities. For example, in the Canadian context, those 'worse off' are defined by federally determined low-income cutoffs: income levels at which families or unattached individuals spend 20% more than average on food, shelter and clothing (i.e. 70% of their income on food, shelter, and clothing), based on the consumer index, adjusted for different family sizes and the size of the city (Statistics Canada, 2022). A general consensus among poverty scholars is that the relative approach is the most straightforward to use in the context of measuring poverty in the Global North. One caveat is that relative poverty measures are unduly sensitive to the median income. All gains made by the poorest on this measure will not show up if the median income also rises, while in times of recession it can seem like fewer people are in poverty because the median drops. Consequently, this measure is typically used alongside a measure of *absolute poverty*, which is adjusted for inflation, by measuring current household income relative to 60% of median income in a base year.

Like inequality, social geography's interest in issues of poverty has waxed and waned over the past 40 years, and overall appears downright meagre when cast against the long and sustained tradition of poverty-ethnography in urban sociology and anthropology (Fairbanks, 2009). Indeed, there are generally few equivalents in social geography (with perhaps the exception of Ley's work in Philadelphia, 1974) to the intensive, single-site ethnographic works of Anderson (1990), Klinenberg (2002), Stuart (2016), Deener (2012), Duneier (1999) and Wacquant (2008) that draw inspiration from to the fine-grained methods of the Chicago School of Urban Sociology. These works explored the life-worlds of impoverished people in urban settings, attempting to interpret the inner workings of everyday life and survival among the poor and marginalised.

Arguably, interest in the geography of poverty peaked in the 1970s with the emergence of an energetic Marxist and radical project, linking back to Harvey's work (1973) on social injustices in the city. It may well have reached its nadir in 1995 when Andrew Leyshon bemoaned the omission of 'poverty' from the *Dictionary of Human Geography* (Johnston et al., 1994). Leyshon (1995: 1023) was particularly concerned that 'a broad, socially orientated political orientation is being replaced by a more inward-looking "identity politics", which is practiced in the main by and for middle-class academics, a development which augers badly for emancipatory political projects'. At this time, there were debates around the importance of distributional social justice when compared to more post-structuralist readings where identity and recognition were understood as more prominent, linking back to works by Nancy Frasier and Iris Marion Young in Chapter 2. As mentioned previously, his fears now appear somewhat unfounded, as there has been renewed interest in the geographies of poverty from a variety of geographical and non-geographical quarters since 1995. One example is the recent interest in food banks in the UK. Austerity measures imposed since 2010 fuelled a dramatic rise in individuals using food banks. This speaks to a rise of absolute poverty in the Global North, that it is not just an issue for the Global South. Cloke et al. (2017) saw food banks as not just sites of dire need, but also as spaces of care and voluntarism that we saw in Chapter 5.

Social geographers also began linking ideas of poverty to the larger issues of *exclusion*, which we further explore in the next chapter, on disability and social/spatial exclusion. Exclusion refers to the relegation of groups and individuals to less desirable status and places, in this case impoverished groups. Persons who are part of the non-stigmatised, 'included' group wish to homogenise, purify and protect their own status by erecting boundaries that exclude those that do not belong or who are defiling. In the Global South, exclusion has become more blatant and drastic, in terms of certain areas being completely disconnected from infrastructure and basic resources. This will be connected to the idea of 'let go' areas and the 'hustle' in Section 8.6, as well as larger discussions around the meanings of 'informality'.

8.2 Stigmatisation, Invisibility and Marginalisation

In this subsection, we try to understand the relationship between poverty and the conditions of stigmatisation and marginalisation that produce specific outcomes of exclusion, both social and especially spatial. In effect, the focus around stigma and marginality implies that some people occupy a societal position and spatial location that is somehow less than most others. This means expanding our treatment of 'status', that is the relative social positioning of people and places based on class distinctions.

The first area of concern is the labelling and stigmatisation of the poor, something Takahashi (1997: 903–904) defined as the 'processes of social relations that lead to and reproduce definitions of "'outsider" and "other"; such processes serve to define and label specific groups as being undesirable and dangerous'. Linking back to Goffman in Chapter 2, she pointed to three dimensions of stigma that devalues impoverished individuals and groups: (1) constructions of non-productivity, which relates to functionality (i.e. the more functional individuals being more acceptable than less functional individuals) and participation in the labour market, consumption and stable membership in social and spatial communities; (2) constructions of dangerousness relates to unpredictability, criminality, contagion, risky behaviour that lie outside of social norms; and (3) personal culpability that relates to the degree to which the individual can be held responsible (and thus blamed) for their impoverishment. Her focus was on homeless populations that are seen to be non-productive, dangerous and blameworthy, and how the areas they inhabit inherit the stigma of their presence – what she calls the 'socio-spatial stigmatization' of poor places (Takahashi, 1997).

The second area of concern is marginality, which usually follows stigmatisation. To Lancione (2016), the urban margins are both a position in social space and a subject, usually framed as a condition of sustained territorial stigma. Here we incorporate but also go beyond Young's (1990) version of it from Chapter 2 ('marginalization'), which is essentially and narrowly about being excluded from the labour market. Rather, we use *all* of Young's five faces of oppression (Chapter 2), including the economic ones that are tied to poverty – marginality, powerlessness and exploitation – but also moving beyond poverty to considering issues of cultural imperialism (being rejected by mainstream society) and violence (being oppressed in violent ways). Marginality is a complex set of relationships and encounters, but usually involves the sense of being subordinate to the powerful – socially but also spatially – and that the powerful seek to exclude and undermine marginal people and places. The margins can be understood as sites on the periphery of society both real and imagined, excluded but also subordinated and delineated by the centre.

This delineation is enacted by a process of bordering, which can include physical segregation, making the marginalised group *invisible*. This sense of invisibility from mainstream society was articulated by Jean O'Leary, an American gay and lesbian activist, to characterise the root of gay people's oppression: 'Our invisibility is the essence of our oppression. And until we eliminate that invisibility, people are going to be able to perpetuate the lies and myths about gay people'. Consequently, life becomes a political strategy for groups at the margins, to make visible their presence and right to occupy public life. Places can also suffer the same fate. Rob Wilton (1998; 2000), for instance, explored the visibility of an AIDS hospice in an urban neighbourhood and found that even with local residents self-identifying as 'accepting' people, they characterised the users of the hospice as 'outsiders'. This place was seen as a threat to social order and inferior to the imagined norm. In a quote from a local police officer:

> We got a couple of calls from this woman who said that she couldn't walk her dog down the street, because she can't breathe the 'AIDs air'... We had an open patio by the hospice ... we started getting complaints from people calling by; they didn't want to see these people with AIDS, these sick people.
>
> (cited in Wilton, 1998: 180)

As a result, the residents sought to make the users and purpose of the facility invisible through constructing the meaning of their neighbourhood as belonging to certain people or symbolising a particular way of life.

Although both these examples (gay people and people with AIDS) are not about poverty, they serve as potent examples of how invisibility can operate as a form of segregation. Similar accounts of poor people being made invisible – both physically separated from and made to feel out of place – in mainstream society is captured in Lisa McKenzie's book, *Getting By* (2015). She documents her personal experiences of living in a working-class estate in Nottingham, UK for more than 20 years to reveal how these communities are often overlooked, misrepresented and stereotyped in both academic discourse and public policy, rendering their voices and realities largely invisible in mainstream narratives about class and poverty. Invisibility here is not just cultural but political – they are excluded from decisions that affect their lives, despite being directly impacted. McKenzie reveals the daily acts of care, solidarity and aspiration that go unseen by broader society. These are not the kinds of behaviours that are captured in statistics or political speeches, but they form the social glue of these communities. Reflecting later, McKenzie argues that this process of making invisible can operate at a macro level and affect whole communities that have suffered badly from deindustrialisation. Here, major infrastructural projects like HS2 (high-speed rail development), the Channel Tunnel and London's Crossrail have all moved meticulously around these communities, never stopping, moving the financially mobile around the country and into Europe, while their local buses and services are stripped back to almost nothing. In short, McKenzie uses the concept of invisibility to highlight how structural inequality, cultural misrecognition and exclusion from research all contribute to the marginalisation of working-class communities, despite their very real presence and agency in urban Britain.

Invisibility can also be felt in relation to one's position to the state, as we saw in Chapter 3 on citizenship. If persons do not fit within the system, a 'specific form of sovereign power operates in the practice of banishing the non-compliant, disruptive or unruly' (Evans, 2011: 29). Here, a form of institutional abandonment is in operation, encompassing various suspensions, absences, lapses and blind spots in the protection of certain lives (Agamben, 1998, cited in Evans, 2011). Yet marginal spaces in particular are crucial for marginalised people's everyday survival, of those 'forced to the economic, cultural, and political edges of urban society, located there because of the inequalities of the urban world they live in, not because of their own actions' (Lancione and McFarlane 2016: 2405). Although covered in more detail in Chapter 13, suffice it to say that marginal space can also be liberating and contrarian, a sort of autonomous zone that captures 'those spaces where people desire to constitute non-capitalist, egalitarian, and solidaristic forms of political, social, and economic organisation through a combination of resistance and creation' (Pickerill and Chatterton, 2006: 730).

As explained in the previous chapter, marginal groups in society can also seek to self-segregate for several reasons. One is for defence. This is most prominent when group discrimination is widespread and intense, and groups seek to occupy a territorial heartland to withdraw from the hostility of the wider society. Second, groups congregate for mutual support. This was seen in the miners' communities in South Wales during the decline of the coal industry in the 1980s, particularly around the time of the 1984–1985 miners' strike. In towns like Merthyr Tydfil and Ebbw Vale, communities came together in powerful ways to support each other through severe economic hardship. A third reason groups congregate in an area is for cultural preservation. This was the case for Harlem in New York in the 1920s, when it became a site for writers and artists, many of whom lived in Harlem, to produce a wide variety of fine and highly original works dealing with African-American life.

For bell hooks, occupying a position at the margins is much more than inhabiting a site of deprivation. Connecting to our points made previously about people in poverty living in the margins, hooks says the opposite: that it is also the site of radical possibility, a space of resistance. It offers the possibility of radical perspectives from which to see and create, to imagine alternatives, new worlds:

> To be in the margin is to be part of the whole but outside the main body. As black Americans living in a small Kentucky town, the railroad tracks were a daily reminder of our marginality ... Living as we did – on the edge – we developed a particular way of seeing reality. We looked both from the outside in and from the inside out. We focused our attention on the centre as well as on the margin. We understood both. This sense of wholeness ... provided us an oppositional world view – a mode of seeing unknown to most of our oppressors that sustained us, aided us in our struggle to transcend poverty and despair, strengthened our sense of self and our solidarity.
>
> (hooks, 1989: 341)

8.3 Marginality and the Racialised American Ghetto

The American ghetto is an extreme yet insightful way to illustrate these insights around poverty, exclusion, stigmatisation, marginalisation and invisibility. The example of the American ghetto is further galvanised by how it strongly intersects with race, racialisation and racism, and how these tendencies show up in the very materiality of certain American cities. This is why we take a primarily visual approach to the racialised American ghetto in this section. Following both Vergara (1995) and Bledsoe (2020), we use the term 'American' ghetto, rather than African-American, since they had no part in its construction. We trace how the powerful have conspired to create the racialised ghetto as, following Marcuse (1997), an imposed segregation of people involuntarily denigrated by their race and class, and having little economic attachment to the rest of the city.

Building on the discussion of segregation in Chapter 7, Marcuse made some key differences between self-segregation and imposed segregation. He defined enclaves(1997: 231) as 'a spatially concentrated area in which members of a particular population group, self-defined by ethnicity ... congregate as a means of enhancing their economic, social, political and/or cultural development'. In this sense, enclaves are voluntary and self-segregated spaces, defensive clusters designed to ensure daily survival and, potentially, long-term social mobility. With enclaves, self-segregation is seen as positive. Conversely, ghettos are also seen as 'spatial concentrations used to separate and to limit a particular involuntarily defined population group (usually by race) held to be, and treated as, inferior by the dominant society' (Marcuse, 1997: 231). In other words, creating environments where segregation is imposed, thus involuntary and necessarily negative. More to the point, imposed segregation, like that of African-Americans in US inner cities, creates certain *place effects* – the idea that living in disadvantaged neighbourhoods has a negative effect on residents' life chances over and above the effect of their individual characteristics. Simply put, the spatially contiguous and temporally stable nature of high-poverty neighbourhoods lead to *social isolation* from mainstream (job) opportunities and norms, which in turn produced high unemployment, and a tangle of social pathologies (crime, dropping out of school, out-of-wedlock births, welfare dependency, intergenerational poverty, oppositional street culture, etc.). The spatial separation and concentration of the ghettoised poor magnified their problems in a myriad of ways. Children growing up in high-poverty neighbourhoods encounter a hostile and generally negative environment, attend low-performing schools that have few resources, lack networks for durable employment and are exposed to the temptations of street life including Class-A drugs without strong social buffers.

The material realities of the ghetto clearly reinforce its subordinate and largely outcast nature, yet crucially point an accusing finger at larger economic structures. Since we do not wish to further stigmatise these particular places, we accuse White supremacy and neoliberal ideas of blaming individuals for their perceived failures to be productive, self-sufficient citizens. Images can be used to accuse the powers-that-be. Indeed, it requires an enormous prosperity to lay low so many useful everyday places, just to recreate them elsewhere from scratch. Racialised American ghettos are the victims of slow violence and deliberate urbicide (Vergara, 1995) connected to specific racialised bodies in the American city. Yet the outcomes of urbicide, classified as 'imperilled and dangerous, or spaces "without"/spaces of exclusion' (McKittrick, 2011: 951), can also produce dissidence through everyday practices of survival, resilient Black life and identity-making through everyday fabrics, moving beyond accusations and the sense, taken from the TV show

FIGURE 8.1 West Baltimore, 2019 © G DeVerteuil

The Wire, that the (Baltimore) ghetto is nothing but a gigantic game of chess that cannot be won by those who live there.

These images (Figures 8.1– 8.3) capture the two-sided nature of marginality – abandonment on the one hand, and dissidence on the other, in terms of bringing up past heroes (Martin Luther King (MLK), Muhammad Ali). There are many sites of former African-American settlement and presence that have been left to rot, urbicide by invisibility and neglect. The image that follows (Figure 8.4) shows the Mount Zion Cemetery in Washington DC, which has fallen into disrepair but contains the graves of runaway slaves and earlier settlers to the capital (Fandos, 2016), and contrasts sharply with nearby cemeteries of the powerful and well-known.

The take away of these images is to show the extra burdens imposed when race and class combine, in the form of the materiality of the racialised American ghetto. The effects of living in such environments are largely negative – the very dilapidation of the racialised American ghetto tells those who live there they count for little in society, that they are marginalised and outside of the mainstream. This position of being 'outside' will be unpacked in Chapter 10, in terms of the mental health effects of living in particularly excluded places and social positions.

8.4 Marginality, Waste and Excess

Another way to illustrate the bordering, cultural imperialism and violence inherent in marginality is through the notions of waste and excess. Waste and excess have volume and weight and must go somewhere and sometimes

FIGURE 8.2 Atlantic City, 2019 © G DeVerteuil

cannot be recycled or re-used. Marginal areas invariably endure the most unwanted land uses and activities, and not just the processing of waste and excess. Figure 8.5 shows the 40 Expressway that bisects West Baltimore, a staunchly African-American poorer neighbourhood we have just seen. To add insult to injury, the 40 literally goes nowhere and connects nothing. Below this image is the Mahul colony of Mumbai, in the shadow of massive petrochemical plants and the frequent recipient of families displaced from demolished informal settlements elsewhere in the city. Figure 8.7 shows informal settlements that dot the land along the Ring Railway in Delhi, another hardly-used transportation infrastructure.

These images give a sense of disproportionate exposure to unwanted and negative land uses, a central point to the concept of environmental racism (see Chapter 7). What follows are more Global North examples of land uses and activities beset and shunned by society, including environmental toxic sites mixed in with residential areas (Figures 8.8 and 8.9).

In more detail, Figure 8.8 illustrates the toxic intermingling of industrial and residential uses in Southeast Los Angeles, an area of Latino settlement.

FIGURE 8.3 Newark, 2012 © G DeVerteuil

FIGURE 8.4 Washington DC, 2016 (photo taken by copyright Stephanie DeVerteuil)

However, even the milder version of 'badlands' in Canadian cities still contain unwanted mixes (industry and residential) as in the case of Hochelaga in East End Montréal, but also in places such as the North End of Winnipeg or Vancouver's Downtown Eastside (safe injection sites, which are life-saving but place an unfair burden on the locality; see Chapter 10 on harm reduction).

FIGURE 8.5 Baltimore, 2019 © G DeVerteuil

FIGURE 8.6 Mumbai, 2019 © G DeVerteuil

FIGURE 8.7 New Delhi, 2017 © G DeVerteuil

The images we have shown in Sections 8.3 and 8.4 certainly raise the troubling issue of double subjugation – places already poor and marginal (and sometimes racialised) that are then further subjugated by the (supposedly intrepid) photographer. And certainly some of the places pictured – including the Downtown Eastside and West Baltimore – could be considered over-studied places by academics, although remaining largely invisible in the eyes of society, in the sense that they are invisible enough to be ignored. As noted in the pedagogic approach in Chapter 1, our visual production therefore must be in the service of exposing structural inequalities found in the original subjugation, of pointing an accusatory finger at the poverty and marginalisations in the city, for if we do not, all we will end up with is an 'Instagram' version of the city. Rather than problematic representations of the poor and marginal, in that version we get no representation of them at all, no possibility for accusation, only a whitewashed version of the city where inequality and structural deprivation no longer are shown.

FIGURE 8.8 Southeast Los Angeles, 2003 © G DeVerteuil

FIGURE 8.9 Montréal, 2021 © G DeVerteuil

8.5 Case Study of the Poor, Racialised and Marginalised: Photo Essay of Homeless Service Hubs in Los Angeles

Building on the extensive use of images in sections 8.3 and 8.4, we want to present a photo essay of how poverty, race and marginality combine in a particularly stigmatised, and marginalised part of Los Angeles – Skid Row.

FIGURE 8.10 Downtown Eastside Vancouver, 2022 © G DeVerteuil

FIGURE 8.11 Skid Row, Los Angeles, 2015 © G DeVerteuil

FIGURE 8.12 Skid Row, Los Angeles, 2017 © G DeVerteuil

A photo essay is driven by images, rather than words. It seeks to 'explore specific facets of city life where even small observations are woven into a sophisticated cultural and political critique' (Gandy, 2011: 5). Yet by relying on quickly-taken images, there is the danger of exploiting the place, of a 'shoot-and-go' mentality. One way to avoid such an exploitative approach is to draw upon our longstanding connections to certain poor and marginal places. As McNeill (2019: 155) argued, 'to understand the depth of a city involves staying in one place for a long time and getting to know its nuance, its path dependencies, its durable institutions'.

One such place is the *service hub*, an intensely spatialised social infrastructure, which is the co-location of service-dependent populations and the services (e.g. housing, health care, rehab) upon which they rely. Service hubs are contact zones between two populations, i.e. between the residents and the services that cater to them. Geoff wants to present the example of Skid Row Los Angeles as a service hub, but also to show how to research poor and marginal places in a connected and sympathetic manner. He has been visiting Skid Row on and off for 25 years (1995–2023): alone, on guided tours, as part of interviews for his research (DeVerteuil, 2015), or with students on fieldtrips (2016, 2017, 2018, 2023). It is a place apart, marked by a hyper-concentration of services and those that depend on them, a microcosm of the welfare state, the voluntary sector, and care.

The first two images show the strong place attachment – to the point where even the fire trucks have a 'Skid Row' insignia on them. But Skid Row is also a real community, not just Hell's half-acre. There exists endlessly repetitive and lurid YouTube exposés by self-proclaimed experts on poverty, but who inevitably miss the forest for the trees, the complexity of the place and the structural pressures, focusing instead on damaged people and the squalid, makeshift materiality. By contrast, Geoff's long-term approach to Skid Row enabled him to see changes in the way extreme poverty is managed over time, from laissez-faire in the 1990s to lockdown in the 2000s, and then back again in the 2010s. Skid Row is definitely one of the more challenging places to photograph, given that Geoff stuck out like a sore thumb, or as Stuart (2016: 193) noted, 'White is Black' with its preponderance of African-Americans. Not blending in at all led to inevitable accusations of trespassing, or of just passing through. Still, Geoff's personal connections to some of the service providers allowed him crucial access and vantage points.

Figures 8.13 and 8.14 show that the mid- to late 1990s, where Skid Row was a place of last resort for the extremely poor, but also subject to occasional forays to clean the streets using bulldozers. By the early 2000s, a more concerted effort to move visibly homeless people into services and off the streets, part of a zero-tolerance campaign initiated by the LAPD, created a state of siege (Reese et al., 2010; DeVerteuil, 2015; Stuart, 2016), with sidewalks cleared out and an exodus to sites nearby (Figures 8.15 and 8.16).

FIGURE 8.13 Skid Row, Los Angeles, 1999 © G DeVerteuil

FIGURE 8.14 Skid Row, Los Angeles, 1998 © G DeVerteuil

Finally, after numerous challenges to the legality of the Safer Streets campaign in the early 2010s, Skid Row reverted back to a more laissez-faire node of survival. But with one key change – the growing tide of homelessness and desperation, the product of a housing crisis familiar to many global cities in the 2010s (see Chapter 11) was crowding the sidewalks with

FIGURE 8.15 Skid Row, Los Angeles, 1998 © G DeVerteuil

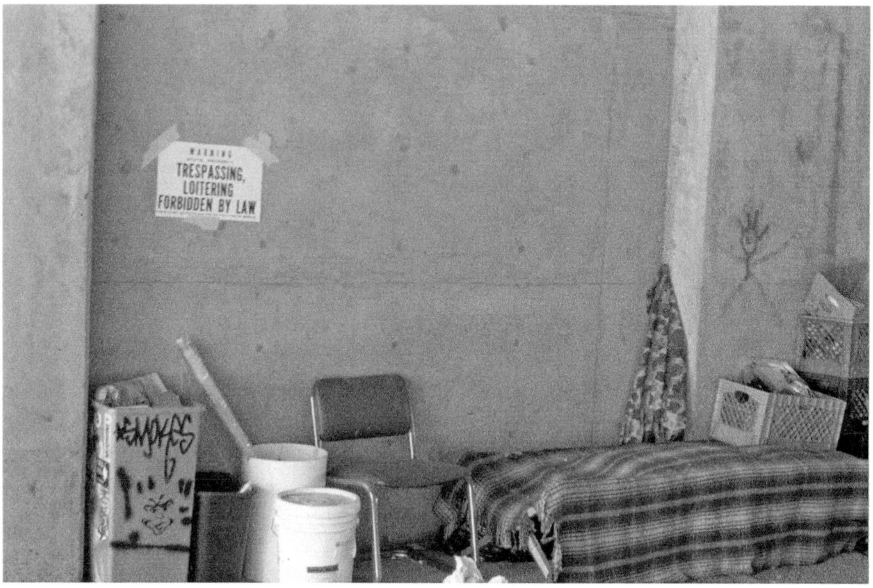

FIGURE 8.16 Skid Row, Los Angeles, 1999 © G DeVerteuil

ever more people, to the point where most were impassable. During the pandemic, this impassibility was 'discovered' by a variety of YouTube gurus, ranting about the sad state of affairs in Los Angeles while driving through Skid Row with their dashboard cameras. This was hardly news to anyone

FIGURE 8.17 Skid Row, Los Angeles, 2015 © G DeVerteuil

FIGURE 8.18 Skid Row, Los Angeles, 2020 © G DeVerteuil

who had been coming down there for 25 years, or more pointedly for those who must live there. In March 2022, Geoff revisited Skid Row for the first time since the pandemic – and found it much the same as in 2020, with large-scale encampments and crowded sidewalks (Figure 8.19).

Finally, in March 2023, during a fieldtrip to Los Angeles, Geoff revisited Skid Row twice – the first alone, the second with students. The first revisit was about re-taking some of the photographs taken back in the late 1990s or 2000s, and they showed for the most part a dramatic increase in the number of tents and encampments. The second revisit was with 40 students and staff, and stayed to the western margins of Skid Row. Needless to say, such a large and clumsy group attracted attention, hackles and a mix of incomprehension and disdain. But also some insight – several inhabitants made the point to the group that not everyone is damaged on Skid Row, that for some people it is the only place they can afford to stay. This suggests a spatial aspect to stigma, that anyone living in a stigmatised place will themselves be stigmatised. And the students brought up a key aspect of Skid Row, once they were through it – that it is a male place, a place of failed masculinity, rather than toxic masculinity. It is composed of men who have been discarded by society for many different reasons, including mental ill-health, substance abuse and long stints in prison; more on social problems and masculinity in Chapter 12. And finally, it is a deeply racialised (African-American) place in a city with a dwindling African-American population – an

FIGURE 8.19 Skid Row, Los Angeles, 2022 © G DeVerteuil

all-too durable hub that combines some of the ghetto model we saw earlier in the chapter with a dense mix of voluntarism and care from Chapter 5. So gender can complicate the class/race approach we have used in the photo essay.

Overall then, the key takeaway from the Skid Row photo essay is the importance of an intersectional (race/class) and relational understanding of particular places that suffer from marginalisation. While largely invisible to the rest of society, Skid Row offers a useful space of encounter to understand the combined weight of poverty and pernicious stigmatisation and racialisation. Skid Row makes visible the outcome of larger relationships-not just race and class but also the welfare state, voluntary sector and care.

8.6 Precarity and the Perpetual Hustle

While the previous sections on marginality were rather downbeat, in this section we underline the more hopeful and energetic means through which to endure, if not rise above. In many ways, this penultimate section is about survival and making-do in a world that does not want you. The relationalities of poverty, race and marginality produce many conditions – exclusion, but also precarity, that is the lack of stability and marked vulnerability across various social arenas, especially housing and employment. While housing will be the focus in Chapter 11, here we want to focus on employment: those experiencing low-wage work, informalisation, casualisation, unemployment, underemployment, and for global cities, a workforce that is increasingly comprised of exploited migrant workers. At its essence, employment precarity is marked by in-work poverty and powerlessness, which in turn creates social marginality, of being (and feeling) peripheral to mainstream society and places. The rise of the so-called Gig Economy fits into this schema, connecting back to Chapter 6 and its focus on youth.

While employment precarity necessarily produces poverty and marginality, and therefore despair, victimisation and entrapment, it also engenders a variety of more positive and agentic responses, hinted at by bell hooks, which we quoted earlier in the chapter. One of those responses is surely the notion of *hustle* (Thieme, 2018: 530), which to Tatiana Thieme is the 'collective condition of individual insecurity disproportionately distributed among young people navigating uncertainty in irregular employment through prolonged states of "waithood"'. In other words, responding to a lack of regular work opportunities by being shrewd, moving off the grid, living beyond and around the various barriers that block societal expectations of adulthood. To her, hustling is not just a negative pathway, a last resort means of survival, a trap – it can also be a first resort, a rational response that no longer sees hustling as 'something extra on the side' to the norm. Hustling comes out of need, not greed. The hustle also comes out of informality, that is the various practices that are outside of the legal structures of society, a mode of self-help in the face of abandonment and insecurity (Tonkiss, 2013). The informal is not always associated with marginality, however, since in many countries most economic activities are technically outside of legal structures, for both the precarious groups who seek survival and the powerful ones who seek separation from the former. But for the purposes of this chapter, we mostly focus on the precarious groups and their everyday survival.

Crucially for building a global social geography (as noted in Chapter 1), hustle shows remarkable parallels in the Global North and South: of stunted opportunities in the informal sector of the makeshift Global South city, or across the austere, incomplete post-welfare city of the Global North. Hustling is therefore globally experienced and globally understood, increasingly so among the younger generations. The memorial images of the rappers the Notorious BIG, Tupac, Eazy-E and Coolio – passing away in 1997, 1996, 1995, and 2022 respectively – show how the notion of the street hustle can transcend the parochial surroundings of Brooklyn and South Los Angeles and reach a global audience as far as London, Paris and Bogotá (Figures 8.20 to 8.25).

Hustle can also be seen in terms of migrant survival spaces in both Global North cities (usually international migrants) and Global South cities (usually internal migrants from the countryside). Here we can connect to immigrant enclaves, which Marcuse (1997: 242) defined as 'a spatially concentrated area in which members of a particular population group, self-defined by ethnicity ... congregate as a means of enhancing their economic, social, political and/or cultural development'. In this sense, immigrant enclaves are

FIGURE 8.20 Brooklyn, New York, 2016 © G DeVerteuil

FIGURE 8.21 Brixton, London, 2017 © G DeVerteuil

more voluntary than racialised American ghettos, which are seen to be involuntary, and more connected to congregation, defence, mutual support, and economic opportunity, tendencies that were raised in Sections 8.2 and 7.4.

The spaces of encounters now shift, from the racialised American ghetto or service hub to immigrant enclaves, areas that are even less connected to the state and more conditioned by DIY (Do-It-Yourself) urbanism and care. This connects to the sense that certain parts of the city are, in Abdoulmaliq Simone's (2004) words, 'let go' to hustle as they see fit, with maximum informality (and invisibility), and minimal state oversight. Usually found in marginal inner-city areas, these spaces 'underwrite the means of subsistence for a monetarily poor population largely excluded from formal capitalist labour and housing markets. These spaces provide housing often in the absence of clear title, and informal income generating activities such as street tradition, under the table employment'(Sheppard et al., 2020: 397). This links to DIY urbanism, where amateur designers create and implement small-scale

FIGURE 8.22 South Los Angeles, 2016 © G DeVerteuil

FIGURE 8.23 Brixton, London 2017 © G DeVerteuil

interventions in urban public spaces. While tactics vary widely and may have some benefits for certain users, they nonetheless have the potential to complicate careful and considered long-term planning and urban design strategies (Iveson, 2013). This further links to Lefebvre's right to the city, of not just appropriating spaces but also producing them, literally.

FIGURE 8.24 Bogotá, 2016 © G DeVerteuil

FIGURE 8.25 Paris, 2024 © G DeVerteuil

The crowded yet let-go nature of Old Delhi enables a plethora of social geographies to emerge around precarious work (Figure 8.26).

In the Middle East, migrants from the Global South recreate some of these 'let go' geographies, such as in the example from Dubai with a population from South Asia (Figure 8.27).

FIGURE 8.26 Old Delhi, 2017 © G DeVerteuil

The vertical marketplace of Chungking Mansions in Hong Kong has become a key immigrant node through remittances, cheap stalls for electronics and street food, emerging as a first stop (and one-stop) for new immigrants from other parts of the Global East. Similarly, the now-defunct Elephant and Castle Mall in South London played such a role for Latin American migrants, as does the high street of Southall in West London for South Asian migrants. These places all initially had different purposes, all are currently shabby and threatened with demolition, yet all work as global pathways, an 'entranceway to a global matrix' (Knowles and Harper, 2009: 116). They become conspicuous nodes of survival, part of a two-way tolerance between the city that exploits them but also provides employment. With reference to Hong Kong, Knowles and Harper (2009: 76) deemed the city as a 'node of income generation, tolerated rather than enjoyed, in which a small sustainable life is carved out'. Figure 8.28 from 2015 shows the brisk interactions of Filipinos sending their goods from Hong Kong back to the Philippines.

Another example of a let go area is the Tel Aviv Central Bus Station, which ostensibly was originally designed for travellers in mind, not migrants. And yet, its labyrinthine and Brutalist architecture performs many duties – bomb shelter, subterranean art galleries and of course immigrant commerce. It is considered an eyesore and disorienting, but it is also too big and solid to readily demolish, and so its intransigent materiality sustains and protects everyday uses.

Already outdated when it opened in 1993, the Central Bus Station is a warren of remittance shops, dentists, clothing stores and so forth for those migrants who have come to Israel over the past 30 years – Eastern

FIGURE 8.27 Bollywood Theatre, Dubai, 2005 © G DeVerteuil

FIGURE 8.28 Hong Kong, 2015 © G DeVerteuil

Europeans, Ethiopians, Filipinos. The neighbourhood in which it lies is far removed from the glamorous edge cities and beachfront redevelopment. A brooding and inflexible presence, the Central Bus Station may not be pretty, but its multipurpose interior serves its immigrants communities in an organic, unplanned yet tolerated way (Figures 8.29 to 8.31).

FIGURE 8.29 Central Bus Station, Tel Aviv, 2019 © G DeVerteuil

FIGURE 8.30 Central Bus Station, Tel Aviv, 2019 © G DeVerteuil

Finally, we can briefly highlight the La Chapelle area of northern Paris, which has long been a traditional reception point for immigrants. Over the past ten years, a large contingent of (mostly male) Afghan refugees and immigrants have appropriated the commercial artery and nearby Métro station as a place of ethnically-linked commerce. Figure 8.32 shows a barber shop named after Kabul, the capital of Afghanistan:

FIGURE 8.31 Central Bus Station, Tel Aviv, 2019 © G DeVerteuil

FIGURE 8.32 Paris, 2024 © G DeVerteuil

All of these places (and the activities therein) are largely invisible to the wider city and to society more generally, be they racialised American ghettos, service hubs or immigrant enclaves, even the hustle itself. The distance, both socially and spatially, between these let-go areas and wider society means that spaces of encounter become more tenuous, and thus more subject to mis-understandings, mis-representation, stigma and stereotyping.

Yet in some nations of the Global South, and perhaps contradicting the notion that 'let go' areas are themselves marginal, hustle is the primary mode of everyday survival, rather than a marginal activity in marginal spaces rendered invisible. In cities like Addis Ababa and Dakar, both of where Geoff has visited, the vast majority of citizens are engaged in informal commerce, recycling, peddling, transporting, cleaning, salvaging, begging and so forth, in a bid to survive where formal jobs are few and the state remains, as Simone (2004) noted, profoundly indifferent. Hustle and precarity become mainstream, a majority activity.

8.7 Conclusions

The relationalities, intersectionalities and encounters in this chapter have built on those from Chapter 7. In effect, they revolved around poverty, precarity and marginality, largely from a class and racial/migrant perspective. Spaces of encounter in this chapter were fundamental to the survival and hustle of those living in these neighbourhoods, but rather elusive for the rest of society, in that they are often actively avoided by most city-dwellers (or ignored). Few residents of Los Angeles, Tel Aviv or Hong Kong have actually visited the places that were photographed in this chapter – and the state has also largely left these places to their own devices, in a condition of marginality and precarity. So while constituting zones of encounter, relatively few meaningful interactions (e.g. spaces of encounter) across society occur there, and if they do occur, they are tinged with stigma and inequality.

As for social justice, the chapter was especially interested in the mal-distribution of resources that create poverty. Poverty has multiple causes, but for the purposes of this chapter have a lot to do with being excluded from the (formal) labour market or suffering from precarious and low-paying employment. This 'marginalization', in Young's (1990) words, is among the worst oppression to suffer from, because it brings with it exclusion from mainstream society. Even with labour market participation, many who live in marginality are exploited and left powerless (e.g. immigrant enclaves). These exclusions are backed up the cultural imperialisms and violence of previous and current rounds of racism and place-based stigma and discrimination. Recognising these mal-distributions is a first step but not the last step in terms of seeking justice – something has to change in terms of the way resources are distributed, including the procedures of who gets what and where. We would expect at least two changes to procedures and outcomes (Fincher and Iveson, 2008). The first is a greater focus on equity in the allocation of expenditure on different local areas and populations. We might expect to see efforts made by government to make resources available to local agencies in such a way that some specified measure of equity will have been achieved, i.e. spend more in the neediest areas on schools, libraries, public health, etc. Second, we might expect a greater recognition of social mix, whereby the state would seek to ensure that concentration of disadvantaged residents in a locality is no greater than a certain percentage of the population, so that this group of residents is placed in a socially mixed setting in which a range of income groups or households live in proximity to each other. This is based on the fear of poverty concentration as a generator of social problems, something we will return to in Chapter 11 with regards to social housing in the Global North.

References

Anderson, E. (1990) *Streetwise: Race, Class and Change in an Urban Community*. Chicago: University of Chicago Press.

Bledsoe, A. (2020) The primacy of anti-Blackness. *Area*, 52(3), 472–479.

Cloke, P., May, J. and Williams, A. (2016). The geographies of food banks in the meantime. *Progress in Human Geography*, 41(6), 703–726. https://doi.org/10.1177/0309132516655881 (Original work published 2017).

Deener, A. (2012) *Venice: A Contested Bohemia.* Chicago: University of Chicago Press.

DeVerteuil, G. (2015) *Resilience in the Post-Welfare Inner City: Voluntary Sector Geographies in London, Los Angeles and Sydney*. Bristol: Policy Press.

Duneier, M. (1999) *Sidewalk.* New York: Farrar, Straus and Giroux.

Evans, J. (2011) Exploring the (bio) political dimensions of voluntarism and care in the city: The case of a 'low barrier' emergency shelter. *Health & Place,* 17(1), 24–32.

Fairbanks, R. (2009) *How it Works*. Chicago: University of Chicago Press.

Fandos, N. (2016) At 2 Georgetown Cemeteries, History in Black and White, New York Times. 20 October 2016.

Fincher, R. and Iveson, K. (2008) *Planning and Diversity in the City: Redistribution, Recognition and Encounter*. London: Palgrave Macmillan.

Gandy, M. (2011) *Urban Constellations*. London: Jovis.

Harvey, D. (1973) *Social Justice and the City*. Baltimore, MD: Johns Hopkins University Press.

hooks, b. (1989) Choosing the margin as a space of radical openness. *Framework: The Journal of Cinema and Media*, 36, 15–23. www.jstor.org/stable/44111660. Accessed 14 August 2025.

Iveson, K. (2013) Cities within the city: Do-it-yourself urbanism and the right to the city. *International Journal of Urban and Regional Research*, 37(3), 941–956.

Johnston, R. and Gregory, D. (1994) *The Dictionary of Human Geography* (1st edition). London: Blackwell Publishers.

Klinenberg, E. (2002) *Heat Wave: A Social Autopsy of Disaster in Chicago*. Chicago, IL: University of Chicago Press.

Knowles, C. and Harper, D. (2009) *Hong Kong: Migrant Lives, Landscapes, and Journeys*. Chicago: University of Chicago Press.

Lancione, M. (2016) The assemblage of life at the margins. In M. Lancione (Ed.), *Rethinking Life at the Margins* (pp. 3–25). London: Routledge.

Lancione, M. and McFarlane, C. (2016) Life at the urban margins: Sanitation infra-making and the potential of experimental comparison. *Environment and Planning A: Economy and Space*, 48(12), 2402–2421.

Ley, D. (1974) *The Black Inner City as Frontier Outpost: Images and Behaviour of a Philadelphia Neighbourhood*. Washington, DC: Association of American Geographers.

Leyshon, A. (1995) Missing worlds: Whatever happened to the geography of poverty? *Environment and Planning A*, 27, 1021–1028.

Marcuse, P. (1997) The enclave, the citadel, and the ghetto. *Urban Affairs Review*, 33, 228–264.

McKenzie, L. (2015) *Getting By: Estates, Class and Culture in Austerity Britain.* Bristol: Policy Press.

McKittrick, K. (2011) On plantations, prisons, and a black sense of place. *Social & Cultural Geography*, 12(8), 947–963.

McNeill, D. (2017) World-making. In *World-Making* (pp. 124–148). London: Sage Publications Ltd. https://doi.org/10.4135/9781473921870.n7

Mitchell, D. (2001) Postmodern geographical praxis? The postmodern impulse and the war against homeless people in the 'post-justice' city. In C. Minca (Ed.), *Postmodern Geography: Theory and Praxis* (pp. 57–92). Oxford: Blackwell.

Park, R. and Burgess, E. (1925) *The City*. Chicago: University of Chicago Press.

Pickerill, J. and Chatterton, P. (2006) Notes towards autonomous geographies: Creation, resistance and self-management as survival tactics. *Progress in Human Geography*, 30(6), 730–746.

Reese, E., DeVerteuil, G. and Thach, L. (2010) The contradictions of containment: Deconcentrating poverty in Downtown Los Angeles. *International Journal of Urban and Regional Research,* 34(2), 310–327.

Sheppard, E., Sparks, T. and Leitner, H. (2020) World class aspirations, urban informality, and poverty politics: A North-South comparison. *Antipode*, 52(2), 393–407.

Simone, A.M. (2004) People as infrastructure: Intersecting fragments in Johannesburg. *Public Culture*, 16(3), 407–429.

Smith, N. (1996) *The New Urban Frontier: Gentrification and the Revanchist City*. London: Routledge.

Statistics Canada (2022) *Canadian Census of Population 2021*. Ottawa.

Stuart, F. (2016) *Down, Out, and Under Arrest: Policing and Everyday Life in Skid Row*. Chicago: University of Chicago Press.

Takahashi, L. (1997) The socio-spatial stigmatization of homelessness and HIV/AIDS: Toward an explanation of the NIMBY syndrome. *Social Science and Medicine*, 45, 903–914.

Thieme, T. (2018) The hustle economy: Informality, uncertainty and the geographies of getting by. *Progress in Human Geography*, 42(4), 529–548.

Tonkiss, F. (2013) *Cities by Design: The Social Life of Urban Form*. Cambridge: Polity Press.

Vergara, C. (1995) *The New American Ghetto*. New Brunswick: Rutgers University Press.

Wacquant, L. (2008) *Urban Outcasts: A Comparative Sociology of Advanced Marginality*. Cambridge, UK: Polity Press.

Wilton, R. (1998) The constitution of difference: Space and psyche in landscapes of exclusion. *Geoforum*, 29, 173–185.

Wilton, R. (2000) Grounding hierarchies of acceptance: The social construction of disability in NIMBY conflicts. *Urban Geography*, 21, 586–608.

Young, I.M. (1990) *Justice and the Politics of Difference*. New York Princeton University Press.

9 DISABILITY AND CAREGIVING

9.1 Disability and the Social Geographies of Care

Our decision to include the geographies of disabled people and family caregivers together in this chapter is justified by the opportunity to explore the obligatorily relational experiences of both groups – and to highlight some of the difficult tensions that can create a divide between the two. This also serves as an opportunity to revisit and build on the earlier themes of social infrastructure and care (Chapter 5), to get a deeper understanding and explore individuals' everyday experiences of space within this landscape of care. However, we do acknowledge that not all disabled people either wish for or require family members to become caregivers. In many cases, disabled people live relatively independent lives – although we unpack some of the assumptions of this idea later. Nonetheless, in those cases where family caregivers are involved in disabled people's lives, complex relational dynamics can occur, and serve as a compelling account of the contact zones between disability and care.

The three strands of social justice are also relevant in this chapter, in terms of how disabled people's or carers lives have been affected by a lack of recognition, including the positive representation, of disability and care, the distributive allocation of statutory entitlements (e.g. disability or carer benefits) and the procedural provision of rights and services. Disabled people's lives can be shaped on the one hand by experiences of applying for, waiting for, paying for, gaining independence from, and relying on support. On the other hand, life can be shaped by being denied access to, being disempowered by or accepting low expectations in support. Thus, gaining access to appropriate support can be controlled and mediated through rules of eligibility and testing of functioning and financial resources for government benefits. Access to informal support can also be at the discretion and willingness of family members. The experience of disability, when support is required (not always), is fundamentally one of *encounters*, of unequal spaces of encounter, of how certain individuals are slotted into society. The experience of disability is also fundamentally about *embodiment*, although as we explore, the body has been rejected by disabled people as a sole cause of their disablement. Each of these strands has shaped disabled people's

DOI: 10.4324/9781003414889-11

or carer's experience of ordinary citizenship, through everyday encounters with various shifting divides between 'them' and 'us', impacting their sense of belonging and inclusion. These encounters often reinforce a sense of difference and exclusion.

We are mindful of our commitment to hope though, and so we explore how disability activism, underpinned by disability scholarship, has rejected the dominance of medical definitions of disability (where a person's impairment or bodily difference is understood to define a person's experience of disability). Instead, there has been a movement towards highlighting the role that social barriers play in *disabling* a person. In recognising this shift, we explore the concept of social exclusion drawing on material in Chapters 1 and 6, illustrating with disabled people's experiences of public encounters in everyday life. We also seek to examine its antonym, 'social inclusion' and consider what it may look and feel like.

Finally, in this chapter we focus mainly but not exclusively on physical and intellectual disability (see definitions in Section 9.2), while in the next chapter we focus on mental health. Suffice it to say here that while we feel that this division is sometimes arbitrary given the overlaps between the body and the mind, we do think that issues of mental health merit their own chapter, given the enormous attention paid to mental health during and since the pandemic, its crucial importance to students' everyday lives, and its specific connections to society, inequality and social justice. More on this division and transition will be raised in the conclusions to this chapter (Section 9.5).

9.2 Disability and the Historical Geography of Care

Disability is not a homogenous group and can include differences in functioning at the level of the body (including physical, sensory, neurodiversity and cognitive differences as well as chronic ill-health) or through participation restrictions that differently abled people may experience. This acknowledgement of participation recognises that the functioning of a person is determined by their social interactions with society and experiences of ordinary citizenship. A person's differences can be experienced and judged as limitations or merely as different modes of thinking, acting or moving, that are deemed to be non-typical and stigmatised as a result. For example, neurodiversity describes the idea that people experience and interact with the world around them in different ways. For some people with autism, differences may not cause any distress to the person themself. Instead, if distress is felt, it is often the result of barriers imposed by societal norms, which set narrowly defined behaviours that are deemed 'acceptable' by the local population, causing social exclusion. Thus, we reject binary thinking of disability/ability, as this distinction is often quite porous, inconsistent and not fixed. We continue with this insight in Chapter 10, which is more about mental health and ill-health as relational. As Chapter 2 illustrated, people have varying characteristics and capabilities that can change over time and space. In this case, these differences can determine how groups can become more or less independent than others. In many cases, environmental factors can affect these experiences and whether these factors are facilitators or barriers.

Disabled people have faced a long history of 'care' in institutional settings, initially starting with private hospitals in the 17th century, and expanded to include large public institutions in the 19th and 20th centuries. The historical context of institutions is covered in more depth in Chapter 10 on mental health. This long history of institutions evolved across the world, in Ango-European, (Post-) Soviet and Asian countries. Power, Lord and DeFranco, (2013) provide an account of their history in Ireland, the UK, France, Sweden, the US and Canada. In Ireland, where Andy was born, the rate of institutionalisation was the highest in the world per capita throughout this earlier history, as the country was used as a colonial 'testing ground' for the treatment of disability (and mental health). These large public institutional settings absorbed more and more people with impairments, chronic conditions and mental ill-health, who were deemed ill-equipped to participate fully in public life with the emergence of industrialisation and rapid urbanisation, especially during its zenith in the 19th century. These public provincial 'asylums' as they became known, evolved into a complex system for warehousing disabled people, in often restrictive, semi-rural, over crowded settings where they lived often for extended, sometimes lifelong, periods. For those with intellectual/learning disabilities, specific settings were developed known as idiot asylums, reflecting the language of the time (which we do not endorse). Home care by default was portrayed as a poor site of care among the emerging medical profession, although it did continue to persist outside the walls of the asylum (Bartlett and Wright, 1999).

While institutional segregation was a primary way disabled people were treated, they also experienced periods of acute hostility within wider society. An example of this was the *Ugly Laws* which were prevalent across several American cities from the late 1860s, including San Francisco, California; Chicago, Illinois and Portland, Oregon (for a full history, see Schweik, 2009 from which we draw upon here). These were public ordinances that sought to remove 'unsightly beggars' from urban areas. It was illegal for persons who were deemed 'unsightly' or had what was understood as 'disgusting' disabilities to appear in public. The San Francisco law of 1867 deemed it illegal for 'any person, who is diseased, maimed, mutilated or deformed in any way, so as to be an unsightly or disgusting object, to expose himself or herself to public view'. To help explain the context behind these laws, during this period, there was an influx of new residents in settler areas, which 'locals' felt had placed strain on their existing communities. People were sometimes impoverished by the harsh conditions and more susceptible to acquiring disability and poor health due to malnutrition and punishing work. Second, people born with impairment were more susceptible to economic insecurity, given the lack of health care. A third group constituted post-war veterans who were disabled through conflict, following the late American Indian Wars and the American Civil War (1861–1865). This created antagonism with people of Puritan Christian faith who felt that pauperism was a disease upon the community and disabled beggars as a 'pitiable nuisance'. The Ugly Laws were, therefore, a method by which lawmakers attempted to remove the poor from sight, subtly removing responsibility away from the government and criminalising disabled paupers in the process. The term Ugly Laws was retrospectively applied in the mid-1970s to describe these punitive laws by proponents of reform at the time (and which led to their removal).

The mention of conflict as a cause of disability and related exclusion in the above example echoes more recent accounts from countries grappling with the legacies of war in the Global South. In Cambodia, following the US-Vietnam war in which the country became a target for heavy bombing, a communist-led dissident party called the Khmer Rouge took control of the government from 1975 to 1979. An account by Owens (2024) reveals how educated citizens such as healthcare workers were specifically targeted by the Khmer Rouge, resulting in the health system being almost completely dismantled. Millions of lives were lost due to starvation, poor living and working conditions and mass executions. Poliomyelitis (polio) was left untreated. During this period, most of the population had to rely on traditional healers and medicine, with access to modern medicines reserved for leaders of the Khmer Rouge. Also, due to the conflict, a large amount of unexploded ordnance such as mines and explosive remnants of war remained scattered throughout the landscape. This meant impairment became prevalent and was often left untreated. Despite this shared collective history, disability is still misunderstood, and disabled people still experience stigmatisation and exclusion, which can have a significant impact on their livelihoods.

Returning to our earlier history, one final early strand of historical exclusion of disabled people was rooted in a newly emerging scientific area of eugenics which emerged between the late 19th century and the early 20th century. This involved an emphasis on selective breeding and breeding out of those deemed as 'carriers' of defective genes, as well as compulsory and 'voluntary' sterilisation of people labelled 'defectives' (typically people with intellectual disabilities). These procedures were largely recorded as voluntary, with individuals made to feel that by consenting to sterilisation, they were behaving in a responsible and socially useful manner, working for the best of society. These punitive social policies inflicted harm on disabled people and drove widespread negative attitudes towards them. Eugenics thus serves as an example of the darker edge of modernity briefly noted in Chapter 3. Eugenics extended across America, Sweden, Germany and the US and, to a lesser extent, the UK. To illustrate the role of the state in endorsing eugenics, in the UK, the Report of the Mental Deficiency Committee (Wood Report) 1929 used a new category, the 'Social Problem Group' to describe what they deemed the 'subnormal tenth of the nation'. Large numbers were regarded as 'carriers' of defective genes. This prompted a growing movement to campaign for sterilisation legislation throughout the 1930s. Home Secretary Winston Churchill was a noted advocate, and his successor Reginald McKenna introduced a bill that included forced sterilisation. However, sterilisation only gained partial legitimacy within the scientific community, and in the UK's case. The fall-out from the atrocities of the Holocaust in Nazi Germany, which was heavily informed by eugenics thinking, also played a role in diminishing its influence.

From the 1960s, various high-profile allies of disabled people began to challenge the use of institutions and sterilisation (and the Ugly Laws in the US) and sought to empower disabled people to become more included in society. In the US, Senator Robert Kennedy (John F. Kennedy's brother) toured the Willowbrook institution in New York in 1965. Calling Willowbrook a 'snake pit', Kennedy stated that individuals in the overcrowded facility were 'living in filth and dirt, their clothing in rags, in rooms less comfortable and cheerful than the cages in which we put animals in a zoo'. Shortly after, Geraldo Rivera, a reporter in the US released a TV Exposé, 1968 called *Willowbrook: The Last Great Disgrace*. In the UK, Ely Hospital in Cardiff built in 1862 as a Poor Law institution and converted to a long-stay NHS hospital for people with intellectual disabilities in 1948, was hit with similar allegations of endemic maltreatment of its patients, including cruelty, verbal abuse, beatings, stealing of food, clothes and other items, lack of medical care and the use of medication to sedate patients. The subsequent Ely inquiry in 1969 exposed a hospital cut off from the mainstream with little or no staff training and overcrowded wards. The surfacing of various scandals from institutions was pivotal in securing reform.

Within this context, de-institutionalisation movements took place around the world, including the US, UK, Canada, New Zealand, Sweden and many European nations. The process took several decades and was not without its challenges. One lesser-known challenge relates to the fact that family members in some nations lobbied against their governments to prevent institutional closure. In Ontario, Canada, a class action lawsuit was brought forward by parents against the provincial government to prevent institutional closure (Power et al., 2013). While we can somewhat justify feeling dismissive of this, we need to be careful about not criticising the asylum period too much, given the many deficiencies of care in the community – at least the patients in the asylum period had relatively stable (enforced) geographies and were not wandering the streets and dying alone. These tensions will be explored in more depth in Chapter 10 with regards to the management of mental ill-health. Moreover, systematic violence and negative attitudes had for a long time framed many (urban) social policies which were antagonistic to disabled people.

The above accounts serve as formidable (and relatively recent) historical examples of the extreme end of social and spatial exclusion that disabled people have faced – and continue to do so in some parts of the world. As we explore later in this chapter (and return to in more detail in the next), the transformation to community care has not been without its challenges. Indeed, some governments that advocated care in the community held underlying assumptions that this should mean care *by* the community. Indeed, this was explicitly stated in UK policy (DHSS, 1981). While disabled people have deservedly come to expect more dignified lives, it should also be noted that institutions do still exist in many parts of the world (e.g. Russia, Hungary, China). Moreover, institutional practices have also been found to persist in community settings where disabled people require support, and through low expectations that frame everyday life, even with family members within the family home. Consequently, we must remain vigilant to call out such practices and remain attuned to the complex relational dynamics that have unfolded since, which we turn to next.

9.3 Challenging Social and Spatial Exclusion of Disabled People

Thankfully, disabled people have a long and proud history of engaging in overhauling the harmful and inequitable treatments explored above. The roots of the disability movement emerged from the 1970s, after disabled people became emboldened by the various struggles for civil rights, gay rights, and women's rights. Disability advocates saw the opportunity to join forces alongside other minority groups to demand equal treatment, equal access and equal opportunity, including the closure of institutions. In doing so, they sought to challenge negative attitudes and stereotypes, rally for political and institutional change and lobby for the self-determination of their minoritised status. The movement was not only focused upon engaging in protest and direct action but also on theorising disability. They sought to explore what it meant to be disabled.

Box 9.1 Engagement Questions

- How would you define disability?
- What is ableism and disablism?
- What barriers do you think disabled people may face today where you live?

One of the foremost ideas that emerged from this movement was a challenge to the *medical model* of disability. The medical model had dominated thinking within the various professions involved in the treatment of disability (or exercising control over disabled people), including psychiatry, rehabilitation, medical general practice and nursing, education and architectural design, for hundreds of years. The medical model of disability understood that people are disabled by their impairments or differences, with the aim that these should be 'fixed' or changed by medical or other treatments – or overlooked if not. This understanding also shaped how disability was considered more widely by society, and informed the recognition of disabled people as dependent and disempowered by their impairments.

In response, the disability movement rejected the medical approaches to disability. Following an initial emergence of a disability rights movement in the US (explored further in Chapter 14 on direct action), two activists in the UK, Paul Hunt and Vic Finkelstein, took up the mantle by establishing the Union of the Physically Impaired Against Segregation (UPIAS) in 1972. UPIAS directed attention towards the social and structural barriers that *disable* people with impairments. In other words, it was society, built and run by non-disabled people, which was preventing disabled people from actively participating in a fuller working and social life. As outlined in their founding statement:

> In our view it is society which disables physically impaired people. Disability is something imposed on top of our impairments by the way we are unnecessarily isolated and excluded from full participation in society.
>
> (Vic Finkelstein & Paul Hunt, UPIAS, 1975)

In 1983, the disabled academic Mike Oliver who was part of the movement coined the phrase the *social model* of disability to refer to this emerging idea. This shift in thinking and its impact on understandings of disability more generally cannot be overstated. It challenged several hundred years of perceived wisdom. It also allows us to expose the debilitating nature of neoliberal economies and precarious post-welfare states explored in Part 1 of this book as a *cause* or *agitator* of disability (Macpherson et al., 2021).

The social model exposed several barriers disabled people experience, including environmental (inaccessible buildings, services, language and communication), attitudinal (prejudice, stereotyping and discrimination), and organisational (inflexible procedures and practices). These barriers serve as reminders of the three strands of social justice which inform this book – recognition, distribution and procedure. These can manifest either through *ableism*, which Chouinard (1997: 380) defines as the 'ideas, practices, institutions, and social relations that presume able-bodiedness, and by so doing, construct persons with disabilities as marginalised, oppressed, and largely invisible "others".' In a sense, this is a more subtle practice of overlooking disability or assuming that all people are non-disabled. As citizenship has ostensibly become intertwined with the ideologies and practices of neoliberalism, ableism has become a foundational feature of the neoliberal state, in terms of individualised expectations on disabled people to overcome their adversity and earn their way out of exclusion (Goodley, 2014). On the flip side is *disablism* which refers to more direct acts of discrimination against disabled people (Scope, 2024). These are frequently related, where the ableism of a neglectful administration connects with the harms of disablism.

While the social model is still routinely advocated within disability rights groups and remains a powerful tool for change, the conceptual underpinnings have more subtly evolved to accommodate a more relational approach, consistent with the core thread running through our book. Some disability activist-scholars, i.e. disabled academics speaking from personal experience and their non-disabled allies (referenced below), have argued that the social model overlooks the difficult embodied experiences of pain, discomfort, and tiredness that impairment can sometimes produce – irrespective of a more positive representation or the removal of barriers. Social geographers of disability have been part of this conceptual journey and have sought to combine some elements of the medical and social models, by using the relational accounts of *embodiment* that was already signposted in Chapter 2, the notion that disability as a lived experience is, by its very nature, experienced through the body. This concept is used to move beyond the impairment/disability binary of early social model formulations (Wilton and Evans, 2009: 205). Research by Moss and Dyck (1996) used fine-grained qualitative methods to investigate the micro-geographies and embodied nature of disability and chronic illness at both the workplace and home, thereby combining the material conditions of the (disabled) body with the discourse, identity and representations of it. This effectively extended the social model beyond the focus on societal oppression. Similarly, Longhurst's (2005) research on obese bodies has used these more cultural and bio-social approaches to understand different experiences of disability.

As a last point, there is a movement afoot in architectural and policy circles to undo some of the embedded exclusions of disabled people, particularly through the concept of 'universal design'. The idea here is to produce buildings, products and environments that are accessible to people with and without disabilities (Gleeson, 1999), as opposed to the first group, as seen in Figure 9.1. Elements of universal design have surfaced in a wide variety of national disability legislation, including the UK, US, Australia and New Zealand. This idea has certainly helped challenge the ableist assumptions within architecture and design and given them the tools to advance new modes of accessible design. This movement is not without its detractors though, as it has been critically re-appraised through a relational lens, that acknowledges that the experience of using 'accessible' design still relies on relational encounters (e.g. local business owners, bus drivers, members of the public) which can be disabling and must not be overlooked – a point we return to below. Through theorising disability in new ways, individuals were able to create a new politics of identity with disabled people being recognised as a class of people, enabling them and others to view their disabilities in a more positive light and to lay claim over their identities. A more complete account of disability activism is explored in Chapter 14.

Despite the advances made by the disability rights movement, disabled people in more contemporary times still experience social exclusion in their

FIGURE 9.1 Allotment, Southampton UK, 2025 © Andrew Power

everyday lives, and the work of disability activism has had to keep pace. Social exclusion within wider society refers to broadly similar practices to marginalisation (and is often used interchangeably with this term), which was introduced in both Chapter 8 and earlier in this chapter. Revisiting these different terms, each has a different lineage. The following is a useful starting point: *marginalisation* refers to the process by which individuals or groups are pushed to the edges of society – socially, economically, politically or culturally. They are still present within the system (e.g. they can work) but are denied full participation or influence (their voices may be ignored in policymaking or media representation). *Social exclusion* goes a step further – it means complete denial or removal from access to resources, rights or spaces within society. Social exclusion is seen as an end point or product of marginality, where people can feel alienated and disenfranchised with more enforced boundaries. Social exclusion thus involves the processes whereby individuals or entire communities are systematically blocked from rights, opportunities and resources which are normally available to members of society i.e. in housing, employment, education, healthcare and democratic participation. Consequently, social exclusion has evolved more in disability studies from scholars and activists seeking to make sense of disabled people's experiences.

Social exclusion can thus emerge as an outcome from various forms of marginalisation, which relate broadly to the recognition of disabled people, including attitudinal barriers, ableist assumptions, spatial marginalisation, stigma and oppression (Gleeson, 1999). Social exclusion impacts disabled people's de facto citizenship based on who is believed to have the rights to fully participate in society. As we learnt in Chapter 3, citizenship reflects dominant ideologies of social membership. What it means to be a citizen, or an accepted member of a society will vary depending on space and time, consequently affecting people's exercise of power, economic engagement through the labour market and access to resources, including economic, social and cultural. Disability also intersects with other societal characteristics which may further impact on their citizenship, i.e. class, educational status, living standards (i.e. poverty), sexuality, gender identity, ability, ethnicity, religion and age.

Without a doubt, the disabled citizen is always the subject of special governmental scrutiny – in terms of being defined by it, assisted by it or disparaged and abandoned by it (Imrie, 2014). This marginalisation heightens vulnerability in an increasingly post-welfare world, one 'shaped ... by the human worthiness of disabled being...discredited in a context whereby welfare policy reform is placing onus on self-active and self-starting individuals as the basis of a good society' (Imrie, 2014: 13). In the United States, it has always been burdensome to secure disabled assistance – Supplementary Security Income (SSI) – while in the UK, recent rounds of austerity have brought increased scrutiny to current benefit holders, to ensure that they are indeed 'disabled', which can be challenging.

One of the most visible manifestations of social exclusion is *spatial* exclusion. Geographers have explored the myriad ways that disabled people encounter, occupy, move through and shape place within a broader context of spatial exclusion. The contemporary spatialities of disability and disabled people's lives has continued to be barriered and bounded, however in more subtle ways. Similar, to what we discussed in regard to the social model of disability, Rob Imrie (1998) proposes the term *Design Apartheid* to characterise how building form and design in Western culture is inscribed with the values of an able-bodied society (a form of ableism), which serves to segregate and separate disabled people from the mainstream in the process, thus alienating and constraining their movement.

Shortly after Imrie's important work, Rob Kitchin (1998) expands upon this by explaining how this design apartheid can create clear demarcations, whereby disabled people are made to know when they are either in or out of place, impacting upon their sense of citizenship and self-worth. Kitchin seeks to highlight the attitudinal barriers that disabled people face, drawing on cultural geographies' literature on sense of place. In his paper, he argues that space is instrumental in keeping disabled people 'in their place' and 'written' to convey to disabled people that they are 'out of place'. While Imrie and Kitchin's work is close to 30 years old, the durable nature of the built environment has meant many of these obstacles still remain.

As noted, disability activism has sought to keep pace, and further gains have been made in some nations. Over recent years, we have seen an increased *recognition* of disabled people in the media from disabled musicians such as Stevie Wonder, actors including Michael J. Fox (*Back to the Future*), Liz Carr ('Silent Witness'), Roy Frank 'RJ' Mitte III (Walter White's son, 'Silent Witness'), and comedians such as Lee Ridley and Alex Brooker ('The Last Leg'). Geoff himself has a visual impairment (non-arteritic anterior ischemic optic neuropathy) in his right eye which, while relatively minor, is permanent and thus life-changing, and certainly invisible as a disability. We have also seen an increased presence of disabled people across all spheres of social life including politics, law, business, education, media, sports, entertainment and the arts. Might the increased presence of disabled people in the media suggest they experience more *social inclusion*? As the phrase suggests, inclusion implies an absence of social exclusion. However, as the Engagement Questions below ask, does it mean more than this?

Box 9.2 Engagement Questions

Does social inclusion merely mean the absence of social exclusion?
How would you define social inclusion?
What other key words would you use to help explain the concept?

To help answer these questions, the relational focus of this book serves us well here, as social inclusion is an inherently relational concept. As van de Ven et al. (2005) remind us in their aptly titled paper 'It takes two to Tango', it must rely on the mutual responsibility of disabled people and wider society (although Ryan, 2006 reminds us that both people have to be able to metaphorically dance and, moreover, want to). Rather than continuing to see disabled people as a minority group that should accept the norms and values and, therefore, the regulations of the majority in society, society needs to adapt and adjust its narrow accounts of people who are perceived not to live up to these 'common' norms and values. This serves as a useful critique of some of the narrower understandings of inclusion that have been articulated. In the UK, for example, the government heralded 'inclusion' as a key pillar of their intellectual disability policy at the turn of the last century, *Valuing People* (2001), alongside rights, choice and independence.

Hall (2005) explains how social inclusion is commonly conceived too narrowly in policy as relating solely to employment and independent living. His analysis traces how disabled people are in a double bind, from being excluded *from* and *within* the labour market – a timely reminder of the barriers that exist within the geographies of production. This obscures many other factors which may be needed to ensure the social inclusion of disabled people, including better transport links, health services, leisure facilities, and the need for greater choice and control over their lives. It also ignores the greatest barrier to social inclusion – the slow or complete lack of change in attitudes from people in society towards disabled people. Simply being in work or living in the community therefore does not equate with social inclusion. Equally, the real difficulties in providing supports to enable inclusion are made to appear of little significance and thus likely to incur little further expenditure (Burton and Kagan, 2006).

Several geographers have since sought to enrich our understanding of *spatial inclusion* by foregrounding place-making as an everyday relational practice, comprising various performances, feelings, negotiations, narratives and tactics in encounters between disabled people and wider publics, and the built environment. Kubentz (2022) examines disabled people's interactions with strangers while using accessible parking spaces. While being an accessible space for disabled people, her research reveals that the use of these spaces is sometimes met with difficult encounters with members of the public over contestations over 'invisible' impairments. She explores what strategies disabled people employ to navigate (or avoid) encounters and their effect on disabled people's emotions and self-identity. Her study reveals the emotional labour that goes into 'performing' disability, while facing prejudicial attitudes over who is perceived as a 'legitimate' disabled person.

Andy's own contributions to this field have centred on the social inclusion of adults with intellectual disabilities (Power and Bartlett, 2015). This work builds upon earlier critiques of disability services by disability scholars and activists who have sought to challenge the block-funded, inflexible nature of disability services, which have stymied social inclusion. These critiques have contributed to a landscape of social care provision that has sought to become more person centred, flexible and respecting the choice and control of disabled people in their lives. With these changes, responsibility for daytime social care and support has increasingly been handed to more people with intellectual disabilities. While some evidence suggests that more people with intellectual disabilities are taking up new community-based opportunities, Andy's research with colleagues (Kaley et al., 2022) explored their sense of belonging (see Chapter 10) which provides a richer understanding of social inclusion within this wider context. In this paper, they demonstrate how belonging involves a complex configuration of actors, places, times, relationships and institutional roles (i.e. opportunities for people to 'give something back' to their community – for example through volunteering, supporting others or campaigning). However, while belonging is often presented to people as a desirable and realisable outcome of social inclusion policies, their findings echo those from Burton and Kagan above that cuts in funding and a lack of appropriate support frustrate people's desires to meaningfully belong with other people in their local community. Moreover, neoliberal government rhetoric casting disabled people as 'work-shy' 'spongers' (i.e. akin to a sponge soaking up money from government) (Cameron, 2006) can in turn fuel abusive encounters within the immediate neighbourhood, as shared by a research participant:

> people picks on people who are disabled because they think they are bums … they said I shouldn't be living there because I am a sponger.
>
> (Billy, quoted in Power and Bartlett, 2018: 571)

This quote reveals how spatial exclusion is still pervasive. But the findings from this study are not entirely bleak. There were some positive accounts. The image below is of an allotment, taken by a disabled man who talked about learning to grow vegetables from his father. He recounted how he was able to share this wisdom with fellow allotment holders. The site was therefore a democratic space, where he could feel included and welcome.

One account was from a participant who regularly went to football matches with his neighbours. Being a football fan and wearing the jersey of the local football team was a way that the participant could share a positive collective identity – which comprises other key factors in social inclusion; feeling valued and reciprocating with wider publics. There was also evidence of participants self-building 'safe havens' in the community (Power and Bartlett, 2018). These were in bingo halls, fish and chip shops, pubs and peer-support groups. In each of these spaces, disabled people were able to find a sense of belonging. These sites have been similarly characterised by Hall and Bates (2019) as moorings, which recognise that rather than changing all of society, disabled people have created spaces of inclusion and belonging *within* a broader context of exclusion. Perhaps this is a more accurate reflection of the experiences of ordinary lived citizenship that many other groups that we have covered in other chapters feel.

9.4 Caregivers In/By the Community

From here, we turn the focus around to the other side of disabled people's social relations, to consider the experiences of one group commonly involved in the support of older, sick or disabled people: family carers. As noted at the start of the chapter and building on the version of 'care' brought up in Chapter 5, this does not mean that all disabled people require or wish for their family members to become 'carers'. In some cases, the recognition of 'carers' is contested, which gives rise to a range of ethical concerns that emerge when attending to disability justice and circumstances of dependency. Some of these issues were briefly raised in Chapter 5; in this section, we dive deeper into the issues at ground level and on an everyday basis.

Family carers or caregivers are terms that have evolved to define any relative, partner, friend or neighbour who has a significant personal relationship with, and provides a broad range of assistance for, an older person or those with a chronic or disabling condition. These individuals may be primary or secondary caregivers and live with, or separately from, the person receiving care (Family Caregiver Alliance, 2024). While a lot of family care is thought to take place naturally in the community, an emerging political agenda has sought to recognise the work of family carers as distinct, and for their identity and work to be recognised and supported. As we shall see, this has sat uncomfortably alongside some disability agendas, as it has been perceived as undermining the focus on the primary client (i.e. disabled people). An immediate connection opens up to the social justice strand of recognition, of how disabled people are represented in debates around care and autonomy. Consequently, this space of encounter has been contentious and involves delicate boundary making, where family carers have gotten caught up in debates over legitimacy and ownership of rights and control.

In terms of sources of support, the informal sector of the family remains the largest and most fundamental in most countries today, in terms of the number of people involved (most of us will be a carer or require care at some point in our lives). This role for family members is somewhat shaped by the political orientation of governments. In China and East Asian countries, a Confucian ethos prevails, where the responsibility to care for family members is enshrined in law. In countries with strong religious affiliations (e.g. Catholic or Muslim faith), there are strong moral obligations to do so. In the UK, where Geoff and Andy are located, there is an estimated 7 million carers at any one time (out of a national population of 67 million) and the estimated saving that carers make to the economy (and especially the welfare state) every year is £119bn per year (NHS, 2024). This figure is based on calculations by carer rights organisations to present an estimated 'value' to society, but hinges on a questionable assumption that if they did not provide the care, the state would step in to do so.

The number of carers is also expected to rise due to an aging population globally, better survival rates from medical conditions and a reduction in public funding for social care in many nations. In countries with ageing populations the proportion of the population undertaking family caregiving is much higher. Japan's 'super-aged' society is the oldest in the world: 28.7% of the population are 65 or older, with women forming the majority. The country is also home to a record 80,000 centenarians. This places a heavy obligation on women, due to a traditional gender division of caregiving tasks in the country, to take care both of children, older relatives and their household.

There is also a prevalent assumption that caregivers occupy only one side of the dis/ability binary. In reality, many carers are themselves older and may have frailty, chronic ill-health and disability themselves. Beyond adult or older women (and occasionally men), it is also worth highlighting that young people can also become young carers, taking on a range of roles when supporting a parent, guardian, relative or sibling who is chronically ill, disabled or otherwise requiring care for medical reasons. Their positions as non-normative or hidden caregivers are often erased from research, policy and public awareness. Because young caregivers are comparatively absent from popular imaginaries and contemporary policy debates about the care needs of families, they navigate caregiving with limited recognition despite concerns surrounding mental and physical health consequences (Lacey et al., 2022).

Another justification for looking separately at family carers is that they themselves constitute a group that are situated at the margins and are frequently affected by social injustice – across its three strands: distribution, recognition and procedure. In terms of recognition, carers occupy an ambiguous position within the social care system and face complex identity politics of their own. The representation in turn affects the distribution of resources and the procedures which they must face (e.g. waiting lists for services). To unpack this further, the initial wave of family care literature from the mid-1970s to 1980s was rooted in an appraisal of how 'community care' was unfolding in a way which *presumed* family members, typically mothers, would shoulder the responsibilities involved. This sentiment is echoed in government rhetoric at the time: 'We hear a lot today about social work – perhaps the most important social work is motherhood' (1977 Patrick Jenkin, future Secretary of State for Social Services, UK).

These gendered assumptions point to the way in which kinship obligations have been reinforced in government rhetoric and serves as an example of how citizenship status is shaped by political norms. Some of this ingrained assumption resorted to explanations of women's perceived altruism. Two feminist geographers, Land and Rose (1985), critiqued this assumption, arguing that whatever the source of the motivation, altruism can hardly explain why people provide care beyond levels that are good for their own health and well-being. They use the term 'compulsory altruism' to describe how this altruism is enforced. Recognition of carers is also shaped by the wider care landscape in which they are situated. Abrams (1978) here made the distinction between informal and formal care, with the former meaning unpaid care based on personal relationships, thought to exist 'naturally' in the community. Twigg (1989) argues that many of the problems that arise in thinking about carer issues relate to confusion over the way the relationship between social care agencies and informal carers should be perceived. The paper outlines three models that provide frames of reference for this relationship: carers as resources; carers as co-workers; and carers as co-clients.

In the first framing, family members are not perceived to be subject to the formal laws of supply and demand; instead, their 'resource' is shaped by cultural and societal kinship norms. The pattern of the availability of informal carers thus remains in any extensive sense beyond the influence of agencies. This has implications for distributive justice: Informal is *prior to* formal and is presumed, so government support is typically only minimally extended to ensure that this resource is not exhausted. The second framing where carers are seen as co-workers has evolved in response to calls by carer groups to more formally recognise and *value* their roles. The social construction of the term 'carer', and the rapid growth of its use within social service departments and other social care agencies is itself part of this process whereby kin and friendship relations are semi-professionalised and brought within the orbit of the formal system. Carers thus became co-workers in the care enterprise. The third frame relates to carers as co-clients. Here, carers are seen as legitimate cases for welfare intervention. Carers become fully integrated into the concerns of agencies and government. As Twigg notes, these three accounts are variable, and the degree to which carers become recognised as co-clients depends on broader cultural and political factors. Here, it is worth a reminder of Drewett's (1999) assertion that social rights are not 'rights' at all but akin to party manifestos, explored in Chapter 3. Where recognition is given, this is at best partial. Carers are often on the margins of the social care system. They are in some senses within its remit, part of the subject of its concern and responsibility, and yet are at the same time beyond its remit, part of the taken-for-granted background to provision, the 'out-there' against which agencies act. Again, this recognition affects wider debates about distribution of resources and the procedures which carers must navigate.

Beyond the role of government and support agencies in shaping the recognition of carers, they are also shaped by *who* they are caring for. For carers of disabled people, they are subject to the perspectives among the disability movement that are very sceptical about notions of carers being prioritised. Some disabled activists have argued that a focus on carers is 'ideological dynamite' (Oliver and Barnes, 1998: 8):

> Constructing the category and developing policies in respect of this group has neither enabled disabled people to become independent nor freed other family members from their 'caring' duties.

We acknowledge the logic underpinning this quote and accept that the barriers in contemporary support services make disabled people 'derivatively dependent' upon the care relationship (Kittay, 2021: 293). The social and political undervaluing of care can determine the conditions of dependency. The disability movement argument is also rooted in a belief that carer groups often represent disabled people as being dependent and as 'burdens' on the family. This burden has been described in numerous ways, including feelings of burnout, depression, isolation, sadness, fear, anxiety, grief and loneliness (Murphy et al., 2007). Many carers are themselves older people providing care to spouses. Other accounts have emphasised how the care work may involve around the clock care.

The term 'care' is thus one that still evokes deep feelings of distrust, which arguably heightens tensions and raises the spectre of the contact zone between 'them' and 'us'. These very tensions informed Andy's earlier research with family caregivers and adults with intellectual disabilities. As he explores in a 2008 paper, carers often feel caught in the middle, promoting independence while grappling with overprotection (Power, 2008).

This highlights the ambiguity between providing needed care and perpetuating dependency.

To fully capture the tasks that carers undertake, Abrams (1978) makes a helpful distinction between caring *about* and *caring for* someone (echoing our opening discussion in chapter 5). We ask you to pause and consider what these two tasks may involve.

Box 9.3 Engagement Questions:

- Can you think what are the differences between caring about and caring for people?
- What personal resources and characteristics may a person have or need to draw on for each?

To help explain Abrams' (1978) classification, 'caring about' conveys feelings of (altruistic!) concern for one's state of being, whereas 'caring for' is an active engagement in doing something to positively affect it. Beyond this, geographers have expanded our understanding of both these spheres in the various spaces that carers use and occupy in their lives.

9.5 A Geography of Carers

In the public sphere, Ryan (2005) explores how going out in public spaces for mothers of children with intellectual difficulties can involve considerable layers of negotiation, mediation and management. This can involve repairing disruptions to social order:

> There was this lady in Tesco [supermarket]. She was massive, about 6'2" and Charlotte grabbed her. She turned around and she said 'What an awful child'. Like that. She was like really snotty. It was like a red rag to a bull. I just went mad at her and I said 'How dare you say that to my daughter? She's got special needs. She didn't mean you any harm'.
>
> (Lucy, individual interview)

These difficult encounters reflect the disabled barriers addressed in the earlier part of the chapter. Though the oppression associated with ableism is not equally experienced by disabled individuals and their carers, they do both experience and navigate ableism. In caring for someone, this stigma can extend to their carers who are confronted with dealing with the representation of the disabled body in society and the experiences of embodiment, and frequently mediate these public attitudes. They in effect, become 'honorary members' of the stigmatised group. This is compounded by the proximity of the care recipient to the carer throughout life. Given that these experiences are repeatedly mediated by carers over time, the participants in Sara Ryan's (2008: 736) study had found ways to cope:

> I used to worry about what other people thought but now I just think, well you get passed the point where you think well I don't care. Sophie is just Sophie and so I'll sing along with her. I just don't think about it anymore. I used to. I used to think, God, you know. I used to not want to take her out anywhere, not anywhere, not even to the local shops.
>
> (Tessa, interview)

The response above echoes similar accounts in a study by Power (2009), where carers had adopted both practical and emotional coping strategies. In some cases, staying home was adopted as a strategy; many of the participants spent a significant amount of their time in their home directly affecting how they and their care recipients can feel tied to the home (Power, 2009).

Consequently, the home space can change from being relatively carefree to a careful monitoring of movement and traffic and noise. At the scale of the body, Twigg (2000) also explores the bodywork in caring such as intimate tasks like bathing of those with significant support needs (Twigg, 2000). One of the other impacts, according to Wiles (2003), is that the home becomes no longer a place for social entertaining. Some carers felt uncomfortable inviting neighbours or friends. As Milligan (2000) argues, domestic space can be seen to have merged with public services to become a site of caring. She argues that the restructuring of care into the home may be more aptly described as 'an institutionalisation of the carer's private space' (p. 55).

While we have some sympathy for this view, it is also important to challenge normative ideas of the home. As Andy and Sophie Bowlby seek to explain in their latest research (in press), it is important to reject idealised conceptions of the home, particularly those that inadvertently reinforce the idea that care disrupts 'normalised' ideas of everyday life. Equally, for disabled people, Imrie (2004) highlights the paradoxes and problems of idealised conceptions of the home that hinge on no recognition of impaired corporeality of disabled people as a potential (dare we say 'normal') part of home life. This challenge to normative assumptions serves as a useful endpoint to this chapter, as it reminds us of how experiences across the contact zone are socially constructed and inexplicably linked, yet we must remain conscious of how each side of the line affects the other. We will revisit the home (and housing more generally in Chapter 11, building on the issues and insights from this chapter.

9.6 Conclusions

This chapter has focused primarily on people with physical, sensory and cognitive (intellectual) disabilities and family caregivers. We have extended notions of care from Chapter 5 in terms of various relationships but especially how they translate into discrete spaces of encounter, the physical spaces of embodiment. The role of the state looms large in this chapter, as the conditions of gaining full citizenship – working and living independently – often poses barriers to disabled people. The neoliberal welfare state is also always conditional, and forces disabled people into difficult and uncertain positions. The theme of embodiment is woven through, as the scale of the body is where these wider structural constraints can be seen most vividly, and where disabled people are socially excluded in a more complete sense. As we sought to show, disabled people have nonetheless been vocal agitators for greater recognition and for fairer distribution of resources. They have also resolutely navigated this terrain and carved out safe havens for them and their communities.

Our additional focus on family carers serves to extend our relational engagement in the chapter. Although we acknowledge at the outset that family members do not necessarily *care for* their disabled relatives, when they do, these family members often must thread complex identity politics. Their position is often ambivalent, which means that their position is not easily defined – by the state, by health and support services and by disabled people themselves. This ambivalence as we saw means that family carers typically do not get recognised as a discrete group requiring support in their own right and become neglected by the welfare state.

In our next chapter, we examine mental health and ill-health, where some parallel issues are evident around the various neglects of the welfare state, the materiality of the city and the impacts of larger societal pressures and problems. The next chapter continues and expands on some of the themes raised in this chapter, but in this case more around mental health as a catch-all for how individuals and society relate to each other in mental spaces, rather than strictly in terms of disability and physical/material spaces. In this respect, we return to issues of social inclusion/exclusion and the prickly issue of *social problems* (Chapter 1, Section 1.4), of how society deals with and manages unmet social needs, particularly in cities. We also build on the insight that disability is more than just physical and social – it is also mental. We further extend the disability discussion by focusing equally on what makes for good mental health, and relate it to mental ill-health, just as in this chapter we focused on both inclusion and exclusion.

References

Abrams, P. (1978) *Neighbourhood Care and Social Policy*. Berkhamsted: Volunteer Centre.

Bartlett, P. and Wright, D. (1999) *Outside the Walls of the Asylum: on Care in the Community in Modern Britain and Ireland*. Linton: Athlone Press.

Burton, M. and Kagan, C. (2006) Decoding valuing people. *Disability & Society*, 21(4), 299–313.

Cameron, A. (2006) Geographies of welfare and exclusion: Social inclusion and exception. *Progress in Human Geography*, 30(3), 396–404.

Chouinard, V. (1997) Making space for disabling differences: Challenging ableist geographies: Introduction: Situating Disabling Differences. *Environment and Planning D: Society and Space*, 15(4), 379–387.

Department of Health (2001) *Valuing People.* London: Stationary Office.

DHSS (1981) *Care in the Community: A Consultative Document on Moving Resources for Care in England.* London: DHSS.

Drewett, A.Y. (1999) Social rights and disability: The language of 'rights' in community care policies. *Disability & Society*, 14(1), 115–128. https://doi.org/10.1080/09687599926415

Finkelstein, V. and Hunt, P. (1975) Fundamental Principles of Disability: summary of the discussion held on 22nd November, 1975. The union of the physically impaired against segregation and the disability alliance, London.

Gleeson, B. (1999) *Geographies of Disability.* London: Routledge.

Goodley, D. (2014) Dis/ability studies: Theorising disablism and ableism. https://doi.org/10.4324/9780203366974.

Hall, E. (2005) The entangled geographies of social exclusion/inclusion for people with learning disabilities. *Health and Place*, 11, 107–115. doi: 10.1016/j.healthplace.2004.10.007

Hall, E. and Bates, E. (2019) Hatescape? A relational geography of disability hate crime, exclusion and belonging in the city. *Geoforum*, 101, 100–110.

Imrie, R. (1996) *Disability & the City*. London: Sage.

Imrie, R. (2004). Disability, embodiment and the meaning of the home. Housing Studies, *Housing Studies*, 19, 745–763. doi: 10.1080/0267303042000249189.

Imrie, R. (2014). Space, place and policy regimes: The changing contours of disability and citizenship. In Soldatic, K., Roulstone, A. and Morgan, H. (Eds.), *Disability, Spaces and Places of Policy Exclusion* (pp. 13–30). London: Routledge ISBN 9780415854801

Kaley, A., Donnelly, J.P., Donnelly, L., Humphrey, S., Reilly, S., Macpherson, H., Hall, E. and Power, A. (2022) Researching belonging with people with learning disabilities: self-building active community lives in the context of personalisation. *British Journal of Learning Disabilities*, 50(3), 307–320. https://doi.org/10.1111/bld.12394

Kitchin, R. (1998) 'Out of place', 'knowing one's place': Space, power and the exclusion of disabled people. *Disability & Society*, 13(3), 343–356.

Kittay, E.F. (2021) Precarity, precariousness, and disability. *Journal Society of Philosophy*, 52, 292–309. https://doi.org/10.1111/josp.12391

Kubenz, V.I. (2022) The (re)construction of disability through encounters in accessible parking spaces. RGS-IBG Annual International Conference, Newcastle, UK, 30 August 2022.

Lacey et al. (2022) *Japan's Ageing Society: Briefing to European Parliament.* Brussels: European Parliament.

Land, H. and Rose, H. (1985) Compulsory altruism for some or an altruistic society for all? In P. Bean, J. Ferris and D. Whynes (Eds.), *In Defense of Welfare* (pp. 74–96). London, England: Tavistock.

Longhurst, R. (2005) Fat bodies: Developing geographical research agendas. *Progress in Human Geography*, 29(3), 247–259. https://doi.org/10.1191/0309132505ph545oa

Macpherson, H., Hall, E., Power, A. and Kaley, A. (2021) Debilitating landscapes of care and support: Envisaging alternative futures. *Social & Cultural Geography*, 24(1), 140–156. https://doi.org/10.1080/14649365.2021.1922736

Moss, P. and Dyck, I. (1996) Inquiry into environment and body: Women, work, and chronic illness. *Environment and Planning D: Society and Space*, 14(6), 737–753. https://doi.org/10.1068/d140737

Murphy, N.A., Christian, B., Caplin, D.A. and Young, P.C. (2007) The health of caregivers for children with disabilities: Caregiver perspectives. *Child Care and Health Development*, 33(2), 180–187. doi: 10.1111/j.1365-2214.2006.00644.x. PMID: 17291322.

Oliver, M. and Barnes, C. (1998) *Disabled People and Social Policy: From Exclusion to Inclusion*. London: Longman.

Owen, C. (2024) *Access to healthcare and assistive technologies for people with physical disabilities in Cambodia*, PhD Thesis, University of Southampton.

Power, A. (2008) Caring for independent lives: Geographies of caring for young adults with intellectual disabilities [in special issue: Ethics and the ethnography of medical research in Africa]. *Social Science & Medicine*, 67(5), 834–843. doi:10.1016/j.socscimed.2008.05.023. PMID: 18573581.

Power, A. (2009) 'It's the system working for the system': Carers' experiences of community care services in Ireland. *Health & Social Care in the Community*, 17(1), 92–98. doi: 10.1111/j.1365-2524.2008.00807.x

Power, A. and Bartlett, R. (2015) Self-building safe havens in a post-service landscape. *Social & Cultural Geography*, 19(3), 303–313.

Power, A. and Bartlett, R. (2015). Self-building safe havens in a post-service landscape: How adults with learning disabilities are reclaiming the welcoming communities agenda. *Social & Cultural Geography*, *19*(3), 336–356. https://doi.org/10.1080/14649365.2015.1031686

Power, A. and Bartlett, R. (2018) 'I shouldn't be living there because I'm a scrounger': Negotiating everyday geographies by people with learning disabilities. *Disability & Society*, 33(4), 562–578.

Power, A., Lord, J. and DeFranco, A. (2013) *Active Citizenship and Disability: Implementing the Personalisation of Support*. Cambridge, NY: Cambridge University Press.

Ryan, S. (2005) 'People don't do odd, do they?' Mothers making sense of the reactions of others towards their learning disabled children in public places. *Children's Geographies*, 3(3), 291–305. https://doi.org/10.1080/14733280500352920

Ryan, S. (2006) It takes two to tango ... but what if one can't dance and the other doesn't want to: A response to van de Ven et al., *Disability & Society*, 21(1), 91–92, DOI: 10.1080/09687590500375721

Ryan, S. (2008) '"I used to worry about what other people thought but now I just think ... well I don't care": Shifting accounts of learning difficulties in public places.' *Health & Place*, 14(4), 730–739.

Schweik, S.M. (2009) *The Ugly Laws: Disability in Public*. New York: New York University Press.

Scope (2024) Disablism and Ableism. London: Scope www.scope.org.uk/about-us/disablism (Accessed on 18 September, 2024).

Twigg, J. (1989) Models of carers: How do social care agencies conceptualise their relationship with informal carers? *Journal of Society Policy,* 18(1), 53–66. doi: 10.1017/s0047279400017207. PMID: 10292254.

Twigg, J. (2000) *Bathing – the Body and Community Care*. London: Routledge.

van de Ven, L., Post, M., de Witte, L. and van den Heuvel, W. (2005) It takes two to tango: the integration of people with disabilities into society. *Disability & Society*, 20(3), 311–329. https://doi.org/10.1080/09687590500060778

Wiles, J. (2003) Informal caregivers' experiences of formal support in a changing context. *Health Society Care Community*, 11(3), 189–207. doi: 10.1046/j.1365-2524.2003.00419.x. PMID: 12823424.

Wilton, R. and Evans, J. (2009) Disability and Chronic Illness chronic illness. In R. Kitchin and N. Thrift (Eds.), *International Encyclopedia of Human Geography* (pp. 205–210). Elsevier.

Wood Report (1929) *Report of the Mental Deficiency Committee*. London: Mental Deficiency Committee.

10 MENTAL HEALTH AND ADDICTION

10.1 Introduction

This chapter asks you to engage with the long-neglected yet increasingly crucial component of the social geographies of mental health. Globally, the World Health Organization (2022) defines mental health as a:

> state of mental well-being that enables people to cope with the stresses of life, realize their abilities, learn well and work well, and contribute to their community. It is an integral component of health and well-being that underpins our individual and collective abilities to make decisions, build relationships and shape the world we live in. Mental health is a basic human right. And it is crucial to personal, community and socio-economic development.

Mental health exists on a complex continuum, which is experienced differently from one person to the next, with varying degrees of difficulty and distress and potentially very different social and clinical outcomes. Mental ill health include mental disorders and psychosocial disabilities, as well as other mental states associated with significant distress, impairment in functioning or risk of self-harm. People with mental health conditions are more likely to experience lower levels of mental well-being, but this is not always or necessarily the case.

More specifically in the UK, the National Health Service defines mental health as a 'a positive state of mind and body, feeling safe and able to cope, with a sense of connection with people, communities and the wider environment' (NHS, 2024). This definition also underlines an important social aspect to mental health. Mental health is not simply the absence of mental ill-health, which is usually inherited, caused by trauma, adverse life events or bereavement, the treatment of which is usually subsumed within a biomedical approach (e.g. medication to treat outward physical manifestations; see also Section 10.2). Rather, mental health is also a product of one's social environment, connections and expectations, yielding a multiple set of causes and effects. Yet despite the relative neglect, the issue of mental health is especially resonant in our post-pandemic world, and for the current

DOI: 10.4324/9781003414889-12

generation of students; 75% of mental health problems are established by the age of 24 (Mental Health Foundation, 2015).

This chapter considers both mental good health and mental ill-health, and for the latter, between severe and stigmatising mental ill-health, and more benign issues of anxiety and societal malaise, each setting with their own relationalities and encounters. In between, we also consider the ambiguous social spaces of addiction, especially to intoxicants, as well as so-called 'deaths of despair'. In particular, the focus on long-term, stigmatising ill-health illustrates the entanglements of power between medical professions, the judiciary and the welfare state, and is very much an issue between those in need and those that serve. We connect back to the discussion of stigma (citing Goffman) in Chapter 2 and Chapter 8 (poverty, marginality). But, and moving beyond the previous chapter, we also seek to understand the wider malaise that has gripped many societies during and since the pandemic, around anxieties about an uncertain future articulated by a variety of current social and environmental crises, and the social justice issues it has raised with regards to distribution, recognition and perhaps procedure. Overall then, mental health acts as an uneven (perhaps even unfair) space of encounter between the individual and society, always within a specific spatial context.

10.2 Mental Health and Social Geography

What is mental health, or should we say *good* mental health? In other words, what are its determinants, both at the individual and societal scales? Good mental health can be defined as mental well-being, if understood from a health geography perspective that values a socio-ecological model of health, rather than a biomedical model with its focus on disease ecology, spatial analysis of communicable diseases and health care provision. We can now proceed to three concepts that have helped to frame mental health in recent decades: happiness; social capital; and belonging (which brings up sense of place, from Chapter 1). The first is very much related to a larger agenda around well-being, an increasingly commodified presentation of what makes happiness. In *The World Happiness Report 2023*, the following points were made:

- The natural way to measure a nation's happiness is to ask a nationally-representative sample of people how satisfied they are with their lives these days.
- A population will only experience high levels of overall life satisfaction if its people are also pro-social, healthy and prosperous. In other words, its people must have high levels of what Aristotle called 'eudaimonia' (e.g. happiness, flourishing). So at the level of society, life satisfaction and eudaimonia go hand-in-hand.

- At the individual level, however, they can diverge. As the evidence shows, virtuous behaviour generally raises the happiness of the virtuous actor (as well as the beneficiary). But there are substantial numbers of virtuous people, including some carers, who are not that satisfied with their lives.
- When we assess a society, a situation or a policy, we should not look only at the average happiness it brings (including for future generations). We should look especially at the scale of misery (i.e., low life satisfaction) that results. To prevent misery, governments and international organisations should establish rights such as those in the United Nations' Universal Declaration of Human Rights (UDHR). They should also broaden the Sustainable Development Goals (SDGs) to consider well-being and environmental policy dimensions jointly in order to ensure the happiness of future generations. These rights and goals are essential tools for increasing human happiness and reducing misery now and into the future.
- Once happiness is accepted as the goal of government, this has other profound effects on institutional practices. Health, especially mental health, assumes even more priority, as does the quality of work, family life, and community.

In these respects, happiness can be compartmentalised enough to be measured, and that happiness figures strongly with mental health. Certain societies are understood to be 'happier' than others, a spatial variation that will be explored in more detail throughout the chapter, including directly below in terms of social capital, ontological security and belonging.

Related to happiness but more academically inclined is the notion of *social capital*. This concerns the relationship among people that help society function properly. It involves social groups getting along through functioning interpersonal relationships, shared identity, understandings and norms (echoing our social understanding of the self in Chapter 2), thereby creating feelings of trust, cooperation and common purpose. The French sociologist Pierre Bourdieu (1986) distinguished four kinds of capital: economic, cultural, symbolic and social. For him, social capital bundles actual or potential knowledge and recognition that can be shared within social groupings. Bourdieu's version of social capital is not enough to explain social hierarchy in itself. Rather, its principal role is to sustain the other kinds of capital, such as creating networks to gain economic capital or cultural capital. In his book *Bowling Alone* (2000), Robert Putnam argued that social capital is accrued through social networking. As a result, social capital can bind together groups, or it can also act to bridge different groups via reciprocity and cooperation. To both authors, social capital is essential for democracy, and at the individual level brings advantages in terms of resources, peace of mind and presumably good mental health – and of course happiness. Finally, social capital is gendered, skewed towards women who bear the greatest burden in terms of organising groups, networking and reciprocal activities. This parallels the skewed nature of caregiving that ends up being mostly female (Chapter 9, but also Chapter 12). Social capital varies by place – some have very high levels, and others not so much, suffering from social isolation and a lack of coherency (such as the American ghetto featured in Chapter 8), themes we pick up in the next section.

Belonging is the most obviously spatial understanding of the three framing concepts. It is about how people come to belong to certain places and groups, and with what implications for identity and the politics of identity/place. Allied to belonging is the sense of ontological security, or the extent to which a place of residence offers a level of security beyond merely providing a roof over one's head (hugely important though that is). Ontological security can be discerned by the following: permanence and stability; privacy and control; a base for the practice of daily routines and activities; and scope for personal development. A place of residence thus comprises both 'hard' (i.e. material) and 'soft' (i.e. imaginative) aspects) and it should be viewed as being a verb as much as a noun, representing the outcome of embodied practices that occur within and around this specific place. It is therefore also best understood at an emotional level, and the ways in which it is created, used, experienced and felt. More on this in the next chapter around housing.

In 2025, there is a definite unease around the degree to which 'outsiders' are accepted or even catered to, and who belongs where. The issue of belonging becomes a divided space of encounter, first in the minds of individuals, and eventually at the scale of the city and the nation. While more on this will be brought up in Chapter 11 with regards to the housing market, populist ideas around who belongs and who does not (see Chapter 7 on race, especially in the context of the UK and immigration) further relate to geographically-inclined concepts of 'sense of place' and topophilia. Sense of place connects to how people feel about certain places – positive, negative or even neutral. A positive sense of place bolsters happiness, social capital and especially, belonging. Commonly referred to as a 'happy place', this positive sense of place is related to topophilia, which is a strong sense of place often mixed with positive understandings of personal identity. Yet, a positive sense of place for some can be excluding to others, leading to xenophobia (i.e. fear of others), stigma and removal. One person's mental well-being can be at the expense of another's sense of place and belonging.

And so good mental health can be captured, to a certain extent, by happiness, social capital and belonging, each creating specific relationalities between the individual and society. The forces arrayed against good mental health are many, some societal and others individual. In the next section, we explore when mental ill-health dominates a person's identity and everyday reality, particularly in how society stigmatises those suffering from it. This discussion connects back to the key insights from Takahashi's (1997) stigma in Chapter 8; mental ill-health has elements of non-productivity (many cannot find work, facing exclusion from the labour market), dangerousness (certainly not contagious, but through the societal expectation of erratic or anti-social behaviour), but rarely culpability (can someone actually be personally blamed for being mentally unwell? See Section 10.3 below).

10.3 The Social Geographies of Stigmatising Mental Ill-Health

Long-term stigmatising mental ill-health has been a sustained focus of social geography, which parallels some of the discussion in Chapter 9 around (physical and intellectual) disability. Here we want to underline the overlap between the medicalised issues around such ill-health, and the more socially-conditioned responses to such ill-health which largely structures how certain conditions (schizophrenia, bipolar, depression, etc.) are managed. The medical approach to mental ill-health is encapsulated in the *Diagnostic and Statistical Manual of Mental Disorders* (DSM), published by the American Psychiatric Association every few years. This manual presents mental ill-health as something that is tangible, visible and understandable to medical professionals (e.g. psychiatrists, pharmacists), but increasingly recognising that mental ill-health is somewhat subjective and prone to exclusions based on various social identifiers, e.g. gender, race, sexuality and other co-disabilities.

In effect, the manual cannot escape its societal and historical context – for many decades, being LGBTQ+ was considered a mental illness, and still is considered the case in some places in the world. For many decades as well, activists argued that being racist – that is, discriminating on the basis that race is a biologically-based difference – should be considered a mental illness, given how it formulates opinions about others not based on their individual merits, rather on their membership in a group with assumed negative characteristics (see Chapter 7). Interestingly, the most recent manual (5th edition with revised text, 2022) now includes a comprehensive review of the impact of racism and discrimination on the diagnosis and manifestations of mental disorders, a major step forward. Of course, criticisms of the DSM remain substantial, including a tendency for over-diagnosis of ADHD (attention deficit hyperactivity disorder) and autism among exuberant young boys especially, pressure from the pharmaceutical industry around 'anti-social' and erratic behaviour that is behaviour that runs against societal norms (if they even exist) of individuals in public, cultural insensitivity and the generally amorphous nature of mental distress in a currently over-medicated post-pandemic society, at least in the Global North.

As such, societal context matters, and not just in terms of how mental ill-health is defined. Some have argued that societal structures may themselves *cause* certain mental illnesses among genetically-predisposed people, an important point to which we return in the concluding section on the pandemic. Here we can point first to groundbreaking work by a variety of early 20th-century sociologists (especially Georg Simmel) who considered the relationship between mental health and society, and particularly urban society. Simmel recognised that the sheer volume of stimuli in cities threatened to overwhelm those who lived in them. This can be linked to Chapter 2 and industrial urbanism, of how society became that much more regimented with the onset of mass production and mass consumerism. This became a preoccupation of health geographer, John Giggs (1993) in Nottingham, who sought to understand whether higher levels of schizophrenia in the inner-city area was due to the conditions of living there (the 'breeder' model) or due to inward migration by people to utilise mental health services (the 'drifter' model). The answer to this question is that both factors are at play. More generally though, city-dwellers could be regarded as becoming more blasé, increasingly selective in what they respond to within the much larger assault on their senses. This dulled sensibility may manifest itself in terms of not seeing, for instance, desperate need and gaping inequality in the city.

Perhaps a more drastic example, at the scale of society itself, is that of *anomie*. A cornerstone of sociological inquiry, anomie refers to the breakdown of standards and values in society, leading to a vacuum and lack of purpose among the population. Introduced by French sociologist Émile Durkheim over one hundred years ago, it was linked to suicide rates. He argued that suicide rates (and other evidence of social breakdown) went up when there were a lack of social standards that regulate behaviour – common values have eroded, and no new ones have emerged. A sense of futility and despair emerges when the rules of society shift dramatically. An example of this was the replacement of socialist systems in Eastern Europe after the fall of Communism in the early 1990s. Those who had benefited from the previous system were marooned in a new system that did not value their experience or expertise, causing alienation and outward resentment. This could lead to 'anti-social' behaviour, and constitute a clear component to the idea of 'social problems', which we touch on in this chapter but were introduced in Chapter 1.

One way to draw together these various social forces bearing down on mental health, as well as controversies surrounding the DSM, is through the emerging focus on *outsider status*. By this we mean how those deemed outsiders in society – racially-marked minorities, or immigrants whose social capital is low – are consistently diagnosed at a higher rate of mental illness than the population deemed to be 'mainstream' or majority (Anglin et al., 2008). In the US and UK, this outsider status is occupied by Black and African immigrants; in Australia and Canada, by those of an Indigenous identity; and in many European nations, as minority immigrants from Muslim backgrounds. For these populations, suffering from stigmatising mental ill-health is not so much the exclusive domain of genetics, but rather how genetic predisposition to mental ill-health meshes with societal structures that often let them know they are not part of the mainstream, that people look down on them or are afraid of them, and that they never quite belong (see

Section 10.2). Moving beyond the individualisation of mental ill-health, this notion of outsider status and its burdens links to the slow shift – signposted already with regards to the DSM – in understanding mental ill-health collectively, that the social environment has an increasingly important role to play (echoing the social model of disability explored in the preceding chapter).

10.4 A Societal View on Managing Stigmatising Mental Ill-Health

If society can be a causal agent in the development of mental ill-health, it also inevitably influences how mental ill-health is managed, presented and socially constructed – as well as spatially anchored *to particular places*. As such, mental illness cannot be properly understood without recourse to the additional layers of stigma, exclusion and inequality that place extra burdens on those who suffer from them. Individuals with mental illness are more likely to be homeless, incarcerated and unemployed than the so-called mainstream population: 'as a result of these over-arching social and economic factors, many individuals … are systematically excluded from full participation in civic and social life, and are constrained to live lives that are shaped by stigma, isolation, homelessness and denial of rights' (Kelly, 2005: 721).

One way to frame societal responses is through the concept of *structural violence*. This is a social way to approach violence, not psychological. Rather than seeing violence as something physical and interpersonal, like visibly hitting someone in a fit of rage, structural violence is understood as part of collective everyday life, the 'often catastrophic consequences of the smooth functioning or our economic and political systems' (Zizek, 2008: 1). Structural violence is the role that institutions and social practices play in negating the survival of whom we deem marginal. This invisible structural violence among those with stigmatising mental ill-health occurs given the pervasive exclusions in the labour market and the associated, disproportionate reliance on welfare, government assistance and benefits. This has worked in tandem with greater responsibilities placed on those receiving welfare, a trend signposted in detail in Chapter 4 in terms of workfare.

How is geography implicated in the managing of individuals with enduring and severe disease such as schizophrenia and bipolar disorder? It would seem worthwhile, then, to consider how we arrived at the current situation, with a particular emphasis on the shifting relationalities and spaces of encounter that have served to constrain those people with mental health problems and ultimately have led to their social and spatial marginalisation. In effect, the spaces of mental ill-health have shifted appreciably in the past 200 years, and so we trace the historical geography of society's management of people with mental health problems, largely from a Global North perspective (e.g. Philo, 1995; Dear and Wolch, 1987). In the late 18th and early 19th century in Western nations, a fundamental spatial shift occurred in how certain populations deemed disruptive (e.g. the poor, criminal, sick, insane) were treated by the state. Previously, low population densities, the predilection for outdoor punishment, and a weak central state combined with strong local attachments ensured that such populations were treated locally in a non-systematic fashion (Rothman, 1971; Foucault, 1977).

From the late 1700s, private hospitals, known as asylums, first began to emerge. These were envisaged as a therapeutic setting and were typically situated in semi-rural, parkland estates. As noted in the last chapter, this infrastructure of care expanded considerably with the emergence of industrialisation and rapid urbanisation from the mid-1800s. A system of public provincial asylums began to emerge (like the one pictured in Figure 10.1) to deal with the casualties of urban poverty, housing squalor, poor health that emerged. This prompted a long history of institutionalising those with mental health (and disabilities) within many nations across the Global North and East (Power, Lord and DeFranco, 2013). So, by contrast to the earlier non-systematic fashion of managing mental ill-health and difference, by the mid-19th century, these same populations were now confined in a constellation of spatially isolated institutions, such as prisons, asylums, poorhouses and juvenile halls. The locations of these institutions were based on therapeutic aims of removing those deemed 'anti-social' to more bucolic settings. However, while the therapeutic motivation informed the initial emergence of asylums, these public institutions quickly became over crowded, understaffed and under-resourced.

FIGURE 10.1 St. Loman's Hospital in Ireland (formerly Mullingar District Lunatic Asylum), built in 1847. Hospital closed on 5 December 2013. 2010 © Andrew Power

The sociologist Erving Goffman, whom we already heard about in Chapter 2, coined the term 'total asylum' to convey how these settings became all-serving settings, with chapels, water towers, farms, laundries and so forth, all on site. The 'caring' profession saw it fit to remove the patient from 'all the influences and all the associations' of the home:

> It seems to have been too much forgotten that in every case of insanity the first object should be, as in other maladies, the recovery of the patient: and that is often impossible *without a removal of the patient from home*. All the influences and all the associations of home become perverted, in the large majority of cases, in this unhappy malady.
>
> (cited in Philo, 1987: 402 our emphasis)

An obvious shift had occurred in terms of the prevailing correctional principle. Foucault's (1977) *Discipline and Punish* is particularly useful in accounting for this change. In effect, Foucault explicitly links the birth of the 'lunatic asylum' (again, reflecting the terminology of its time) to a decisive transformation in the internalised exercise of power. Punishment was once a horrifying public display, an expression of the sovereign's power, a principle of spectacular punishment, but by the 19th century, people regarded as 'deviants' were spatially confined within the walls of a host of new institutions, most obviously the asylum and prison but also extending throughout society in the form of hospitals, reformatories, schools and the like, creating a wider 'carceral archipelago'.

Distance from mainstream society, accentuated by physical separation, thus becomes a technique in the hands of the state to transform those deemed problematic. In terms of institutional location, specific sentiments about certain places acted to both justify mass segregation as well as to offer the potential for individual transformation and rehabilitation. Rothman's (1971) *The Discovery of the Asylum* is particularly revealing of this shift. Using the US as example, he traces the rise of total institutions to the dramatic transformation of American society between the colonial (1780s) and Jacksonian periods (1830s). He contrasts the localised and generally haphazard treatment of the those deemed 'insane' in colonial America with the segregating policies of Jacksonian America. The catalyst for change was not so much the shift from an agricultural to an anonymous, increasingly mobile urban society, but rather 'the way it was organized into an alarmist interpretation of disorder and dislocation by philanthropic reformers' (Ignatieff, 1981: 164). In response, traditional mechanisms were jettisoned for new institutions devised to isolate disruptive elements and protect the social order from contagion. As a result, social control shifted from local to state responsibility, outside relief to indoor confinement, and laissez-faire treatment to professionalised and bureaucratic routine.

This alarmist interpretation was grounded in contrasting ideas about two different places: the city and the countryside. Indeed, worrisome images of increasingly anonymous and immigrant cities acted as 'potent catalyst[s] for social speculation and social action' (Boyer, 1978: vii). An intimate connection between health and place was envisioned – rural therapy versus urban disease and overstimulation, anticipating Simmel by a century. In this sense, the environment was both cause and cure for insanity (Rothman, 1971: 68–71): the basic thrust was to remove the insane person from his 'corrupt' environment and isolate him in a 'better' one, in this case the asylum, in order to transform (rehabilitate) him and protect society as a whole. This improved environment represented a preventive measure, a re-creation and repository of the supposedly placid, moral and rural surroundings of colonial America. Asylums thus shared a common goal (individual transformation through isolation), and morality (prevention of social disorder through insulation).

Despite successive attempts at reform throughout the 19th, and particularly during the Progressive era in the early 20th century, the basic locational dynamics of asylums in America (and throughout the Global North and Global East) remained remarkably stable, a combination of inertia and pragmatism (Rothman, 1980). Throughout the 1960s and 1970s, however, the majority of Western nations experimented with the deinstitutionalisation of former mental patients (Scull, 1984). The ostensible goal was therapeutic: to reintegrate formerly isolated clients into the community through a system of decentralised clinics, thus initiating a fundamentally new set of locational dynamics in service provision. As Dear and Wolch (1987: 62) indicated, 'instead of spatial isolation of deviant, dangerous and dependent classes, the modern prescription for client rehabilitation called for spatial integration, engagement in community activities and interaction with the non-dependent population in the course of everyday life'.

This new principle emphasised a different route to individual transformation and by extension a new, and more prominent, image of urban places. Proponents of deinstitutionalisation, including health care professionals, bureaucrats and defenders of clients' rights envisioned transformation through a *reintegration* with their former (urban) community. This alternative space developed not only in response to demands for reform, but also a means to defray the costs of renovating an outdated system. However, the deinstitutionalisation experiment – one of the most crucial welfare state restructurings of the 20th century in the Western World – went awry for a variety of reasons, including lack of fiscal resources, and of particular interest to social geography, an intense set of locational conflicts. These locational conflicts arose due to the siting of new (and necessary) mental health care facilities in urban communities (compared to the semi-rural locations of institutions at the urban-rural hinterlands), which was feared to lower property values and lead to a concentration of potentially unpredictable individuals. Community attitudes to mental health care facilities were subject to community opposition and stigmatisation, thereby ensuring that deinstitutionalisation would be limited to the poorest, most socially disorganised neighbourhoods in cities.

Building on the discussion of stigma in Chapter 8, NIMBY is a widely used acronym for 'Not in My Back Yard' which ultimately summarises the oppositional rhetoric of those residents concerned about the potential impacts of new developments on their neighbourhood or locality. NIMBY can manifest itself in many ways, but it is usually and narrowly applied to: (1) opposition to potentially hazardous facilities (landfill sites, power plants, incinerators, and mobile phone masts), and (2) opposition to 'undesirable' human service facilities (such as alcohol rehabilitation centres, AIDS hospices, prisons, and homeless shelters). These Locally Unwanted Land Uses (LULUs), however, can also include new housing developments that ostensibly threaten the character of existing neighbourhoods. These NIMBY campaigns usually cast new housing and/or higher densities in a negative light in terms of putting pressure on local amenities and resources. Opposition is usually a function of both distance (e.g. the closer to the impact, the more opposition) and saturation (e.g. the more of something negative, the more opposition). The latter suggests that sometimes already marginal communities will fight against yet more unwanted facilities and development, although this blurs the line between NIMBY (usually seen as pejorative) and justice/environmental justice movements (largely seen as positive). At the very least, geographic distance has a determining role to play in the production of community opposition and locational conflict (Dear and Taylor, 1982: 1). Faced with the possibility of having a mental health facility located in their neighbourhood, residents are typically sympathetic in principle toward the policy of treating the mentally ill in community rather than in institutional settings, but resistant in practice to a facility in the immediate neighbourhood.

In their 1987 book *Landscapes of Despair*, Michael Dear and Jennifer Wolch chronicled the unintended and negative consequences of deinstitutionalisation in the US, namely the clustering of facilities and clients in the least resistant neighbourhoods, giving rise to what they called service-dependent ghettos (or what we deemed, perhaps in less stigmatising language, the service hub in Chapter 8). Although acting as crucial coping mechanisms, these ghettos also constituted spaces of neglect, marginalisation and confinement. Worse still, some of these service-dependent ghettos were targeted for dismantlement through gentrification and neighbourhood redevelopment, thereby leading to homelessness given the lack of a strong safety net via the welfare state. Conversely, rural-based asylums suffered a variety of fates. Some were redeveloped as residences, usually quite upscale (Moon et al., 2006). In effect, the authors explored the revalorisation of old asylum spaces, not for the re-institutionalisation of patients, but rather for respite and therapy. These spaces capitalise on the private and bucolic nature of the old asylums, but do so for private profit among well-heeled clients. Interestingly, some nations never abandoned the asylum model. In Russia and China, asylums are part of a larger criminal justice camp system for those deemed anti-social (or anti-government) (Yin and Cohen, 2020). These asylums are basically psychiatric outposts of the larger state system of control. Since the 2022 war in Ukraine, some patients in the Russian system have been offered generous payouts to join the Russian army (BBC, 2024). In these places, the carceral nature of the asylums therefore remain firmly entrenched, and the space of encounter between mental health and ill-health has not budged an inch.

More recently, social geographers have investigated the increased residential cycling and re-institutionalisation of people with mental ill-health. There is increasing evidence of misassignment and circulation of clients to inadvertent, informal or inappropriate settings, including jails, prison, hostels and shelters and the street; a 'revolving door' approach to care (Knowles, 2000; DeVerteuil, 2003). Caroline Knowles' (2000) work shows how informal services can be used to deliver mental health care – in her case study of Montréal, this was sometimes fast-food joints like Burger King, but libraries also were recruited. The resulting displacement has led to a very complex and dispersed geography of mental health care, homelessness and service dependency that extends to settings far beyond any kind of hospital-based or community-based care model. Without access to basic services, as well as even a spatial base of operations (e.g. the service-dependent ghetto), the homeless mentally ill are increasingly churned across a bewildering array of residential and medical settings, especially between the street and the criminal justice system, the latter of which has replaced psychiatric hospitals as institutions of first resort.

This revolving door is partly a product of the welfare system's long and checkered connection to the well-being of those with mental ill-health. Large-scale governmental social policies have long bounded the individual and collective geographies of people with mental ill-health, particularly the transition from asylums to community-based health to re-institutionalisation. The current welfare system acts both as a way to support people with mental ill-health in their everyday lives but also limits their lived experiences by paying only poverty-level benefits that severely restricts everyday circumstances and geographies, as well as imposing multiple conditions. The spaces of encounter may have shifted from the asylum to the community, yet the spaces remain socially unjust, marred by a lack of resources and understanding, and persistent stigma. Recognition remains an unresolved issue but perhaps clarified in the wake of the pandemic (see the final section on mundane mental health).

Overall then, the spaces for mental health care have gradually shifted, from the rural asylum to community-based treatment to a hodgepodge of urban-based sites that barely address the many individual and societal issues at play and make a mockery of the promise of individual autonomy that lay at the heart of the deinstitutionalisation model. But deinstitutionalisation did not necessarily bring de-stigmatisation, and some nations never gave up their older institutional model (e.g. Russia, China). In the next section, we explore the ambiguous relationalities and encounters that lie between good mental health and ill-health through the case of addiction to intoxicants.

10.5 In-Between Encounters: The Social Geographies of Addiction

We now enter a space of encounter between mental health and ill-health: the social geographies of addiction, an ambiguous space that sits alongside stigmatising mental ill-health, independently of it, as a precursor to or as a direct outcome of it (i.e. self-medicating as a way to deal with untreated mental ill-health). Indeed, the relationships between mental health and addiction is fraught. Some behaviour is genetic, but as social geographers we are most interested in how genetic predisposition to addiction is shaped by the social and spatial environment. Having outsider status, for instance, could create the added stress that pushes someone into seeking a way to numb the pain and rejection – but so could a life-changing trauma (e.g. family tragedy, accident) or an immediate social environment saturated with addiction. We will try to develop some kind of coherent view on the relationships among addiction, society, and space, bearing in mind that there is much that remains to be studied. As such, and while a massive topic on its own, the social geographies of addiction remain relatively underdeveloped conceptually, and the specific zones of encounter can be difficult to perceive.

While addictions come in many guises, ranging from gaming to food to social media, we are narrowly interested in addiction to intoxicants. By intoxicants, we consider substances that have psychoactive, consciousness-altering effects (Tracy and Acker, 2004). Not surprisingly, intoxicants range widely, from commonly (mis)used substances such as alcohol, illicit drugs (e.g. crack cocaine, crystal meth, heroin) and prescription drugs (e.g. painkillers). The degree to which these intoxicants are framed is quite varied, from abolition to moralisation, control and legalisation. What constitutes an abuse of intoxicants is also complex. The World Health Organization (2008) defined substance abuse as:

> [T]he harmful or hazardous use of psychoactive substances, including alcohol and illicit drugs. Psychoactive substance use can lead to dependence syndrome – a cluster of behavioural, cognitive, and physiological phenomena that develop after repeated substance use and that typically include a strong desire to take the drug, difficulties in controlling its use, persisting in its use despite harmful consequences, a higher priority given to drug use than to other activities and obligations, increased tolerance, and sometimes a physical withdrawal state.

Despite this very clinical definition, there remains no consensus on whether addiction is a medical disease, when addiction is actually present and whether its causes are individual, social or both, and the most effective ways to treat addicts. But as social geographers, we are drawn to the more social and environmental causes for the abuse of intoxicants, rather than the genetic or moral failings at the individual level. In Geoff's classroom, when discussing community opposition, students are challenged to see addiction as something at least as much environmental as moral, the product of chance happenings such as meeting the wrong people, becoming addicted to painkillers after a work (or car) accident, being born into an addicted family setting or using intoxicants to cope with trauma (which is also a common cause behind smoking cigarettes), and stunted opportunities. In these ways, addiction can become an issue of public health concern, rather than involving solely the individual interacting with the criminal justice system.

One way to frame the shift from moralising and criminality to public health is the recent interest in so-called *deaths of despair* (Case and Deaton, 2020; DeVerteuil, 2022). As economists, Anne Case and Angus Deaton located deaths of despair in social and economic language: cases of suicide, drug and alcohol poisoning and alcoholic liver disease have been rising in certain Global North countries over the past 30 years, especially the United States and especially among working-class adults within deindustrialised areas. Between 1999 and 2017, the rate of drug overdose deaths among Americans between 25 and 64 increased fivefold, from 6.7 per 100,000 in this age group to 32.5 per 100,000 (Woolf and Shoomaker, 2019). Partly as a consequence, life expectancy in the United States began dropping in the 2010s, even before the pandemic began. Indeed, Berlant (2007: 775) offers a particularly bleak outlook, by arguing that we are all

> exhausted by neoliberal capitalism: we are all in slow death, but for some this is more apparent. This includes: the bodies of US waged workers will be more fatigued, in more pain, less capable of ordinary breathing and working, and die earlier than the average for higher-income workers, who are also getting fatter, but at a slower rate and with relatively more opportunity for exercise.

Needless to say, the social and geographic distribution of these deaths of despair are uneven; they particularly hit areas and people marked by high unemployment, ill-health and austerity, as well as social isolation and a lack of social infrastructure. As presented by Case and Deaton, deaths of despair is essentially a middle-life problem, but it is also a problem of public health – the frequently unsung safety nets (e.g. clinics, sanitation, treatment programmes, health insurance, availability of healthy food, etc.) that have contributed at least as much to growing life expectancies over the past century as heavily-funded and dramatic medical breakthroughs. Deaths of despair capture death on the instalment plan, which contrasts sharply with the rapid overdoses that can occur from a new wave of synthetic opioids that we examine in a few paragraphs.

A second way to frame the shift can be roughly illustrated by the notorious rise and slow decline of the 'War on Drugs' in the United States over the past 50 years. In the 1970s and especially 1980s and 1990s, the main response by the American federal state towards addiction was through punishment, imprisonment and interdiction. This implementation of the War on Drugs was very much filtered by race, in that the social problems raised by addiction were most acute in racialised inner-city areas of (large) American cities, especially within African-American communities (e.g. the American ghetto in Chapter 8). The punitive aspects of the War on Drugs were heavily influenced by dystopic representations of these places. Unlike the current responses to deaths of despair, which verge on the decriminalised and even supportive, the responses in the 1980s and 1990s towards racialised addicts were downright criminalising and carceral. This was based in a moral panic emerged around the emergence of crack cocaine in the 1980s. Highly addictive and socially disintegrating, the spread of crack across American cities was framed as a moral and criminal matter, rather than a public health crisis that required treatment. We can also add that this level of addiction can bring a *social death* before the actual physical death that comes from overdoses or the long-term effects of self-harm. That is to say, the addicted person dies in the eyes of society (and sometimes their family and friends) before they actually die as a physical process, especially if they are homeless and taking drugs in public. This is similar to the fates of individuals with full-blown AIDS in the 1980s and 1990s – their physical appearance, combined with the stigma of the disease itself as well as pervasive homophobia – meant that they died socially before their (inevitable) physical demise. The movie *Philadelphia* (1993) captures this moment, when Tom Hanks' character is asked to leave the library to use a private room because of his obvious symptoms, and because he is reading AIDS-related case law.

Since the 2000s, however, there has been a slow shift away from these tendencies, to see addicts in a less judgemental light. Just the same, one cannot stop the older system dead in its tracks, given it has coloured all state-based responses since the 1980s. More recently, there has been social experiments around loosening the connection between drugs and the criminal justice system. This has strained the system – already a patchwork at best. The system itself has also been strained by a new generation of synthetic drugs like Fentanyl that are many times stronger than heroin, and that can be lethal the first time they are taken. The situation in places like Philadelphia, a city suffering in 2025 from a full-blown opioid epidemic, is cautionary (FIgure 10.2). Addicts waver between the old punitive system and the promises of the new system of reducing harm and, perhaps, treatment. Overdoses have surged but criminalisation has faded, revealing gaps: implementing local solutions that decriminalise drug-taking remains

FIGURE 10.2 Philadelphia, 2020 © G DeVerteuil

difficult when the legal proscriptions of the War on Drugs remain firmly in charge at the state and national levels.

One way to focus on the waning War on Drugs is through the concept of *harm reduction*. Harm reduction holds a value-neutral view of the substance abuser and supports the flexible application of a wide range of effective treatment options that reduce the harm associated with substance abuse, including needle exchange, methadone programmes, safe injection sites and drug courts. Harm reduction has spread to various nations in the Global North, including Germany, the Netherlands, Switzerland, Canada and even has a toehold in the United States. The pros and cons of harm reduction are passionately debated. On the plus side, harm reduction is seen as moving away from a punitive model that bloats the criminal justice system with those addicted to intoxicants. This shift in resources also helps to reduce deaths from overdoses. Yet harm reduction might also lead some addicts to avoiding any kind of rehabilitation, allowing them to fully embrace a life on drugs. This leads to accusation that harm reduction without a robust treatment system is all carrot and no stick, offering self-harm instead. It also raises a social justice issue, in that some neighbourhoods are inequitably burdened by addicts doing drugs in public, while others need not worry. Moreover, these same neighbourhoods are also expected to host safe injection sites, given the co-location with users and community opposition almost everywhere else. More on this in Chapter 13, but suffice it to say that harm reduction is full of inequalities, and has become a divisive issue in places like Glasgow and Toronto (see Figures 10.3 and 10.4), both of which featured newly-minted safe injection sites in the early 2020s, but also in

FIGURE 10.3 Toronto, 2021 © G DeVerteuil

places like Vancouver, which has had a safe injection site for over 20 years (see Chapter 8 and Figure 8.10).

Finally, addiction treatment is a rich yet understudied area of social geography. In particular, social model programmes see drug/ alcohol dependence as a product of a lifetime socialisation process in a particular milieu that implicitly or explicitly encourages dependence. Recovering from dependence must involve facilities that are non-institutional, self-governing and based in mutual support. The best-known social model programme is Alcoholics Anonymous/12-step. These treatment programmes have a strongly masculine nature to them, which mirrors the strongly masculine nature of addiction, at least on the streets, as well as for other social problems (Chapter 1) like homelessness.

Overall then, and largely using American case study material, the spaces of addiction and treatment have gently rebalanced, from an obsession with inner-city America awash with crack cocaine to a renewed focus on small-town America with deaths of despair. Nonetheless, the urban (and largely male) nature of (street) addiction remains strong in the social imaginary of most Americans, as the Philadelphia case made clear. The spaces of treatment are usually in *close* proximity to the spaces of addiction, as in

FIGURE 10.4 Toronto, 2021 © G DeVerteuil

the case of harm reduction facilities that are almost always immersed in the same drug-plagued neighbourhoods. It is difficult for substance abuse treatment facilities to locate in areas without intoxicants, given the stigma attached to those addicted. The only exception to this is the very high-end market for treatment among well-off individuals, usually located in stunning natural settings (mountains, ocean-fronts) and build on the well-worn therapeutic value of the countryside compared to the city. In this sense, we have come full circle from the institutionalisation/deinstitutionalisation debates around mental ill-health and health in the 18th century.

10.6 Conclusions: Mundane Mental Health During and After the Pandemic

So far, we have focused mostly on stigmatising mental ill-health and the in-between situation of addiction, and how their encounters shifted from asylums to urban neighbourhoods. But this is but the tip of the iceberg of a much larger set of everyday issues with mental health in broader society. The pandemic in particular raised mundane issues around anxiety and societal malaise brought on by long periods of isolation. By loosening social ties and face-to-face meetings, the pandemic brought to the fore a variety of everyday concerns. Encounters shifted from face-to-face to online, from in-person to the computer or phone screen, with important implications. While for some, the lack of social contact was perhaps a welcome respite from the various pressures and anxieties around appearing and acting in certain ways, for others (particularly those in self-help and peer-support groups for mental health/addiction issues) the lack of interaction would prove catastrophic in terms of staying away from depression and/or intoxicants. For instance, more people died of drug overdoses in San Francisco in 2020 than from the pandemic, almost three times in fact (Fuller, 2021). Online gambling addiction also became a major issue over the pandemic, as people had more time on their hands and more time online.

Social capital in many Global North contexts was eroding even before the pandemic, and we could argue that social media and the virtual world may have something to do with this. For many individuals below the age of 40, social media has become the first and sometimes only way to present themselves to the world, rather than in-person interactions, the building block of social capital. This trend only accelerated during the pandemic. The isolating, competitive and individualising effects of social media, combined with its deleterious impacts on everyday self-esteem via a culture of insult and vulgarity, has itself become a mundane mental health issue. For some, mundane mental health is synonymous with digital mental health; not in the sense of using online and mobile technologies to provide mental health services, but the relationship between a person's digital life and their mental health. Does a person's entire mental well-being depend on the number of digital friends on Facebook, likes on TikTok, or favourable comments on Instagram? If so, then surely what makes good mental health is now as much digital as it is interpersonal and thus societal. Thankfully, the pandemic raised awareness about mental health across a broad spectrum of society. Society is now more invested in mental health well-being among the population, and we argue that this is progress.

Returning to the three strands of social justice, distribution may continue to lag in terms of moving resources to mental health treatment, but recognition grew substantially. Enormous strides have been made during and since the pandemic around issues of mental health and its de-stigmatisation, but so have larger societal feelings of brokenness, despair and general anxiety, particularly among young men. Of course, if young men in particular are more lonely and bitter than they used to be, surely it is up to them to heal themselves? Returning to the notion of social problems (Chapter 1) and masculinity, but also prefiguring material in Chapter 12 on gender, Wilton et al. (2014) noted that addiction, mental health and treatment are increasingly seen through the prism of failed (and even toxic) masculinity. Finally, procedures to obtain social justice remain murky. For instance, the halting move towards harm reduction clearly shows that the way addiction is managed remains fraught with opposing ideas, NIMBYism, ignorance and a distinct lack of patience.

Further Reading

American Psychiatric Association (2022) *Diagnostic and Statistical Manual of Mental Disorders (DSM-5-TR)* Psychiatry.org – DSM.

Boy, J.D. (2021) 'The metropolis and the life of spirit' by Georg Simmel: A new translation. *Journal of Classical Sociology*, 21(2), 188–202. https://doi.org/10.1177/1468795X20980638

Case, A. and Deaton, A. (2020) *Deaths of Despair and the Future of Capitalism*. Princeton: Princeton University Press.

Gilroy, P. (2004) *After Empire: Melancholia or Convivial Culture?* London: Routledge.

Imrie, R. (2003) Architects' conceptions of the human body. *Environment and Planning D*, 21, 47–65.

Imrie, R. (2014) Space, place and policy regimes: The changing contours of disability and citizenship. In K. Soldatic, A. Roulstone and H. Morgan (Eds.), *Disability, Space and Places of Policy Exclusion* (pp. 13–30). London: Routledge.

References

Anglin, D.M., Alberti, P.M., Link, B.G. and Phelan, J.C. (2008) Racial differences in beliefs about the effectiveness and necessity of mental health treatment. *American Journal of Community Psychology*, 42, 17–24.

Berlant, T. (2007) Slow death (sovereignty, obesity, lateral agency). *Critical Inquiry*, 33(4), 754–780.

Bourdieu, P. (1986) The forms of capital. In J. Richardson (Ed.), *Handbook of Theory and Research for the Sociology of Education* (pp. 241–258). Westport, CT: Greenwood

Boyer, L.B. (1978). Countertransference Experiences with severely regressed patients. *Contemporary Psychoanalysis*, 14(1), 48–72.

Dear, M. and Taylor, S. (1982) *Not on Our Street: Community Attitudes to Mental Health*. London: Pion.

Dear, M. and Wolch, J. (1987) *Landscapes of Despair.* Princeton: Princeton University Press.

DeVerteuil, G. (2003) Homeless mobility, institutional settings, and the new poverty management. *Environment and Planning A*, 35(2), 361–379.

DeVerteuil, G. (2022) Deaths of despair and the social geographies of health denial. *Geography Compass,* 16(2). https://doi.org/10.1111/gec3.12607

Fokht, E., Barabanov, I. and Ivshina, O. (2024) Ukraine war: No more easy deals for Russian convicts freed to fight. BBC News BBC 2024.

Foucault, M. (1977) *Discipline and Punish: The Birth of the Prison*. London: Pantheon Books.

Fuller, T. (2021) San Francisco contends with a different sort of epidemic: Drug deaths. The New York Times (nytimes.com).

Giggs, J. (1993) Schizophrenia in Nottingham: Lifelong residential mobility of a cohort. *The British Journal of Psychiatry: The Journal of Mental Science*, 163(2–3), 613–619.

Ignatieff, M. (1981) *A Just Measure of Pain: The Penitentiary in the Industrial Revolution, 1750–1850.* Chicago: University of Chicago Press.

Kelly, B. (2005) Structural violence and schizophrenia. *Social Science and Medicine*, 61, 721e730.

Knowles, C. (2000) Burger King, Dunkin Donuts and community mental health care. *Health & Place*, 6(3), 213–224.

Mental Health Foundation (2015) Mental Health Foundation I Everyone deserves good mental health.

Moon, G., Kearns, R. and Joseph, A. (2006) Selling the private asylum: Therapeutic landscapes and the (re)valorization of confinement in the era of community care. *Transactions of the Institute of British Geographers*, 31(2), 131–149.

National Health Service (2024) *Mental health*. NHS England » NHS mental health dashboard.

Philo, C. (1987) *'Hue and Cry from Town to Town': Towards a social geography of outsiders in Medieval England*. Working Paper. Department of Geography, University of Cambridge. (Unpublished.)

Philo, C. (1995) Journey to asylum: A medical-geographical idea in historical context. *Journal of Historical Geography*, 21, 148–168.

Power, A., Lord, J. and DeFranco, A. (2013) *Active Citizenship and Disability Implementing the Personalisation of Support.* Cambridge UK: Cambridge University Press.

Putnam, R. (2000) *Bowling Alone: The Collapse and Revival of American Community*. Boston: Simon and Schuster.

Rothman, D. (1971) *The Discovery of the Asylum: Social Order and Disorder in the New Republic.* New York: Walter de Gruyter.

Rothman, D. (1980) *Conscience and Convenience: The Asylum and its Alternatives in Progressive America.* Boston: Little & Brown.

Scull, A. (1984) Deinstitutionalization and public policy. *Social Science & Medicine*, 20(5), 545–552.

Takahashi, L. (1997) The socio-spatial stigmatization of homelessness and HIV/AIDS: Toward an explanation of the NIMBY syndrome. *Social Science and Medicine*, 45(6), 903–914.

Tracy, S. and Acker, C. (Eds.) (2004) *Altering American Consciousness. The History of Alcohol and Drug Use in the United States 1800–2000.* Amherst: University of Massachusetts Press.

Wilton, R., DeVerteuil, G. and Evans, J. (2014) 'No more of this macho bullshit': Drug treatment, place and the reworking of masculinity. *Transactions of the Institute of British Geographers*, 39(2), 291–303.

Woolf, S. and Shoomaker, H. (2019) Life expectancy and mortality rates in the United States, 1959–2017. *Journal of the American Medical Association*, 322(20), 1996–2016.

World Happiness Report (2023) Home I The World Happiness Report. happierlivesinstitute.org

World Health Organization (2008) *Principles of Drug Dependence Treatment.* Principles of Drug Dependence Treatment (who.int).

World Health Organization (WHO) (2022) Mental health. Mental health (who.int).

Yin, C. and Cohen, J. (2020) The Misuse of China's Mental Hospitals – The Diplomat.

11 THE HOUSING CRISIS

Homelessness, Gentrification and Generation Rent

11.1 Home, Housing and Security

Before entering into the social geographies of housing, we must underline that housing is a complex term, reflecting its attachments to elements of survival, the capitalist property system and the emotional connections to place. A parsimonious definition helps to distinguish housing from shelter or dwelling, which is seen as a fundamental human need along with food, hygiene and safety. Shelter or dwelling, however, are terms that lack the connection to current socio-economic contexts dominated by capitalist investment. Second, housing must be distinguished from 'home', where the latter is usually associated with 'a material and an affective space, real or imagined, that is generally formative of personal and national identity, shaped by everyday practices, lived experiences, social relations, memories and emotions' (Peil, 2009: 180). However, before we move on from home, it is worth pausing to consider how a sense of home can be made and unmade by certain housing configurations.

Home is central to our ontological security. Home is a place of refuge, providing respite from external concerns, allowing relaxation, quality time with family or housemates, a sense of control and an opportunity for self-expression (Peil, 2009), although home can also be a source of stress and even terror (Chapter 6, Chapter 12). Over the past couple of decades there has been a growing body of literature exploring the meaning of home (Blunt and Dowling, 2005; Brickell, 2012). This literature highlights how home is more than just a physical space made of bricks and mortar, home is also a feeling, an imaginative place, constructed through memories and attachments. Home-making, the practice of constructing a sense of home is a 'continuous process' through everyday routines (Easthope, 2004) and the non-passive role of objects (e.g. personal possessions) to create that sense of 'home' (Miller, 1998).

The sense of home for youth was already covered in depth in Chapter 6, but here we want to expand on the connection between home, housing, and security. A good example has been the rise, since the 1980s of *gated communities*. These are residential developments that are, for reasons of security and sometimes prestige, set apart from the rest of the city. Gated

DOI: 10.4324/9781003414889-13

communities tend to locate in areas of high insecurity. Many Latin American cities feature extensive gating, to the extent that there are even entire gated cities, such as Alphaville on the outskirts of Sao Paulo (Brazil), in reaction to high crime rates and the perceived lack of spatial separation between rich and poor. Alphaville contains homes for 32,000 residents, 1400 businesses, several shopping centres and parks, a private university and hotels. Alphaville is located 16 miles northwest of the Central Business District (CBD), on land once reserved for farming. Begun in 1974, Alphaville is a visible consequence of the continuing intensification of social disparities and spatial fragmentation, the increase in social conflict and violence, and of the growing differentiation of lifestyles in urban society. It is entirely built by private developers, right down to the infrastructure. For the middle- and upper-class people who live there, Alphaville represents a self-selected, exclusive community where they live in their own protected and segregated world and do not have to deal with the poor and violence. At the same time, Alphaville is an expression of the increasing concentration and self-segregation of wealth and the increasingly divergent lifestyles of urban society under the influence of globalisation. In effect, the people who have chosen to live there are following a globalised lifestyle, and in many ways mimicking a North American 'edge city', privatised and suburbanised approach to everyday life that is not so different from a place like Southern California. As such, the marketing strategies of the developers usually refer explicitly to North American models of suburbia. Further, the exclusive control of these spaces by private capital can be interpreted as an integral part of development patterns, characterised by deregulation and flexibility, which have emerged in Latin America in the face of reduced state spending.

However, this account of the home and security (and the detour around gated communities) is based on the normative assumption that we all *have* or *can gain access to* appropriate housing. Home is thus a place of contradictions that holds perpetual promise for self-care and can serve as a refuge, but it can also become the site where the relationship with precarity is most visceral and threatening. In some cases, Baxter and Brickell (2014) argue that a sense of home can be *un*made, where there is a perceived loss of independence, privacy and freedom of choice. This is often a fluid process, negotiated through daily encounters. We will now turn to housing, where over the next few sections it will be framed as a resource, a profit-making machine and a source of precarity and crisis for certain age segments of the population.

11.2 The Housing Crisis and Housing Precarity

Let us return now to housing, moving beyond notions of 'home' via emotional connections to place and survival. In particular, we want to connect to the capitalist property system, itself framed by a variety of legal containers and demographic conventions. First, there are the differences in tenure – from property owner (with or without a mortgage), to renter in the private sector, to renter in the social sector (e.g. social housing, from Chapter 4). Second, most cities in the Global North (and East) have some kind of zoning framework for land use, so that contradictory land uses are not in close proximity. For instance, polluting heavy industry located right next to residential uses. Third, the term 'household' is used, not in a legal sense but in a demographic sense. According to the UK census (2021), a household is a group of people who live at the same address and share a living room or dining area, and cooking facilities. A household can also be a single person living alone. Over the past 40 years, the number of households has increased fairly dramatically in the Global North, as more people choose to live alone rather than in family units. This has added to the pressure on the housing market in terms of needing more units (above and beyond just population growth), but as we shall see directly below, it is but one of many factors contributing to the housing crisis of the 2020s.

We frame the rest of this chapter through the idea of a pervasive housing crisis in many cities and in many nations, whereby affordable housing has become a precious resource in the face of gentrification, displacement, hyper-commodification and AirBnB via what is usually called 'platform urbanism'. Housing is a decidedly distributional matter of social justice, yet we need to recognise that the current housing crisis affects certain groups far more than others. The most negative outcomes of the pervasive housing crisis – which is one of precarity above all – fall on renters, the homeless, the young and those priced out of the housing market entirely. It is on these populations that we will focus our attention, marked by an emerging *housing precarity* ill-served by current configurations of housing supply. This population effectively fuses the distributional bottlenecks of the housing market with particular precarities.

The geography of the housing crisis is now increasingly global, and not just reserved to the largest cities. Take the UK for example. No longer is London the only expensive city – in fact, its very lack of affordability has made all other cities within commuting range more expensive, as well as those cities that receive priced-out (or cashing-out) Londoners. In effect, an entire 100-mile radius from London heading East, West and South has become increasingly expensive, the preserve of long-suffering commuters. The same can now be said, in greater or lesser degrees, for places as varied as Vancouver, Los Angeles, Toronto, Sydney, Dublin and Paris.

What can explain this current housing crisis? We emphasise above all the *over-consumption* and *over-valorisation* of certain housing stocks, the product of certain investment strategies that see housing entirely in terms of exchange value – the value of the building and the land – rather than on use value, that is the use of the building and land to owners and renters on an everyday basis. These trends first emerged in the 1980s, when investment started pouring into fixed assets such as housing, and away from productive assets like factories. This post-Fordist assetisation of housing increased markedly in the 2000s, leading to the sub-prime crisis of 2008 in the United States, which eventually criss-crossed the globe. Since then, there has been even more 'cheap' money (in the form of very low interest rates, which were lowered to almost zero in the wake of the 2008 financial crisis). This cheap money has been used by private equity firms to buy up even more fixed assets such as housing, as well as allowing well-off consumers to buy second and third homes as investment properties or as landlords. The consequence of this over-valorisation and over-consumption of housing is a bloated, unaffordable market by the 2020s. Many younger people (essentially those under 40) feel that the housing system is now rigged against them, in favour of the very old and the very rich (see Section 11.7). This has further led to frustration and the tendency to blame immigrants specifically (and over-population more generally, rather than over-consumption) for the current housing crisis, when in fact the causes are far more deep-seated and capitalistic in nature. One way to make sense of the current housing crisis is to tap into the concept of *rentier capitalism*, of how the owners of assets are able to extract rent from those who depend on those assets. This rent can take the form of, for example, payments for land, royalties for natural resources or interest payments on loans. This mode of capitalism is not productive in terms of growing the economy. Rather, it is based in monopolising ownership of key assets (like housing) that all people need to survive, an obligatory and lopsided relationship.

In the wake of the 2008 financial crisis and ensuing global recession, new avenues for investment emerged in the form of a more globalised real estate market, in which the very precarity generated by the financial crisis spurred the increasingly aggressive seeking of second- and third-home investments in so-called 'safe cities' (Ley, 2012; DeVerteuil and Manley, 2017). These safe cities were first very much at the upper echelons of the global city hierarchy – places such as London, New York, Paris and Hong Kong. But as the recession's effects began to lessen in the early 2010s, safe-haven overseas investment began to flow downwards to many other cities (e.g. Sydney, Vancouver, Auckland), as well as some of which were clearly well-established on the tourist map (e.g. Athens, Barcelona, Lisbon, Venice). Investment flows fed off uncertainty in financial markets in the Southern European Union (e.g. Spain, Italy, and Greece), the Middle East, Russia and China. Additionally, many nations whose financial health was shaken by the 2008 global recession began selling permanent residency to those who would invest into local real estate markets. These 'golden visas' served to prop up the real estate market and cash-strapped governments, but would eventually bloat the market and make it increasingly unaffordable for local residents on local salaries. Spain, Portugal and Greece used this system for over a decade, eventually ensuring housing inflation in which locals could not compete for housing with their low salaries.

We can illustrate some of these pressures by taking the extreme case again of London. The flood of overseas investment in the post-2008 period focused on areas of already-established wealth, such as Mayfair, Belgravia and Notting Hill. There are no bargains to be found there, but the point was to buy properties that would hold their value and see it increase over the long run. The goal was also not to spend the full year in London but use the property as an investment and occasional visits, and not even bother to rent it out. As such, the overseas investors connect to Vertovec's (2009: 3) concept of transnationalism, as a 'condition in which, despite great distances and notwithstanding the presence of international borders ... certain kinds of relationships have been globally intensified and now take place paradoxically in a planet-spanning yet common arena of activity', referring back to Chapter 3 and the limits to the grasp of the state. Looking to park their money in a safe-haven investment is paramount, rather than any real connection to the neighbourhood. A clear example is that of Russian expatriates massively investing in London real estate, as a bolt-hole in case of trouble at home (which came eventually in 2022 with the unprovoked war against Ukraine).

This kind of overseas investment very much parallels investment in tourist-heavy cities such as Amsterdam, Lisbon, Venice and Barcelona. In other words, overseas investment can produce a touristified city, emphasising transience and rootlessness above all, and putting more pressure on the local housing stock via over-consumption and over-valorisation. The issue of localised belonging versus catering to outsiders (e.g. tourists, investors) – of even being hospitable to them – was already discussed in Chapter 10. Here we can see the gap between the local and the global opening up around the distribution and allocation of housing in particular places – what housing should be allocated and even reserved for local inhabitants, versus how much should be open to outsiders? In other words, what loyalty is due to the local population who, random as it was, was born and grew up in these over-valued locales? These gaps matter when the local population no longer see the advantages of being on the investor/tourist 'map'. Belonging is stretched to breaking point when locals are blatantly ignored in the single-minded pursuit of outsider investment.

11.3 The Decline of Public (Social) Housing

A second reason for the current housing crisis is the decline in the availability of public or social housing. Social housing is the 'bricks and mortar' of the state brought up in Chapter 3, and can be used as way to show the crystallisation of the welfare state in particular places, in a material sense, but also of how the welfare state has shifted over time, from attempts at universalism to fragmentation in a more (but never completely) neoliberalised world. As the physical embodiment of the welfare state, social housing has loomed large in the academic imagination. Post-war social housing (from the mid-1940s) was bound up in modernism, social progress and emancipatory potential but also reflected a certain authoritarianism and top-down social engineering in both its capitalist and socialist forms. It is now a residual of a post-utopian world, belonging to a previous age, 'as remnants of hopeful but naïve thought, of depleted husks unable to respond to current demands' (Pinder, 2003: 12).

State interventions into the housing market have always been remarkably uneven within urban centres and across national contexts. To give but one example, the level of state support for affordable housing varies tremendously – in Singapore, close to 75% of the population lives in social housing (much of it owner-occupied), while in the UK the rate is closer to 15%, and in the US closer to 1% (Pacione, 2009). In the immediate post-war period, a large amount of social housing was constructed across Global North nations, as part of the expansion of the national welfare state. Behind the Iron Curtain, entire cities were remade along socialist lines, with vast swathes of government-build housing to replace the wartime destruction (e.g. Berlin, Warsaw). In Western Europe, extensive social housing estates were built, especially on the periphery of large cities such as Paris and Amsterdam (Dikec, 2007). In the UK, estates were both inner city and exurban, while in the US and settler nations (e.g. Australia, Canada, New Zealand) the location was decidedly inner city. Since the 1970s, however, some of these social housing stocks have come under increasing stigmatisation and threat as the state has simultaneously withdrawn its support and promoted more public-private regeneration and social mix. In the more neoliberal contexts of the UK and US, this has taken the form of allowing tenants to buy their apartments, or more radically, tearing down social housing entirely to make way for socio-economically mixed redevelopments that minimises affordable housing provision to former residents. In virtually all nations, the production of social housing has slowed to a standstill, leading to a crisis of affordable housing.

Depending on the national context, social housing has had a very uneven legacy in the so-called post-welfare period (explored in Chapter 4, Section 4.6). From the Red Road Flats in Glasgow to the Cité Pablo Picasso in Paris to the Baruch Houses of New York's Lower East Side, and Chertanovskaya in Moscow or Karl-Marx-Allee in Berlin, large-scale vertical social housing developments were crucial in providing affordable housing and social reproduction. Yet we also understand social housing, among other residuals of a previously more generous welfare state, as less a case of design and social failure and more a victim of general state neglect that is reflected in the wider stigmatisation of the foundational and everyday economy, increasingly seen as 'out of sync' with the accelerating neoliberalisation, financialisation and gentrification of cities. We want to show the experience of two different cities and their social housing, one American and one British.

At one extreme is Chicago, which effectively tore down most of its social housing, known as housing projects or just 'projects'. At the other extreme is London, which has kept most of its social housing estates but has redeveloped some of them, while building almost no new stock since the early 1980s. Let us begin with Chicago. Like many of Chicago's most

FIGURE 11.1 Robert Taylor Homes, 2002 © G DeVerteuil

FIGURE 11.2 Robert Taylor Homes, 2004 © G DeVerteuil

prominent housing projects, the massive Robert Taylor Homes were built specifically to segregate working-class African-American populations from White neighbourhoods, and over time catered increasingly to those on welfare. From the 1980s onwards, the Robert Taylor Homes suffered decline, abandonment, dispossession and eventual demolition in 2007, with images capturing just before their clearance across 2002 and 2004 that Geoff took. Looking back, he asked himself: What does it mean for something so large, so prominent to disappear, materially speaking?

The demolition of these massive social housing projects signified, in very material terms, the end of an era for social housing, at least in Chicago. If the welfare state in the form of large-scale social housing has faded in Chicago, in London the issue has been more neglect mixed with persistence and sometime blatant erasure, depending on which part of the city the housing was located. When Grenfell Tower burned overnight in June 2017, causing over 70 deaths, the fire materialised the chasm between private wealth and public squalor. Upon Geoff's first visit to Grenfell only two months after the fire, feelings in the area were still understandably raw. He was hardly the only photographer in the area, and several local residents were curious as to my business. Not one to exaggerate his importance, he said he was taking shots as an academic and a teacher of social justice. Sensing a purpose greater than gawking, the residents seemed pleased that someone could pass on the visuals to a greater number of (young) people, such as yourselves in these pages!

The juxtapositions became crystal clear. Grenfell Tower lies in close proximity to the wealthy enclave of Notting Hill. Grenfell's raw exterior has since been covered up, but the figures below show in stark detail the shell of the tower (Figure 11.3). By making the invisible fault lines visible in London, the interstitial space of Grenfell Tower underlines a conversation about the margins of state responsibility. However, in comparison to the housing projects of Chicago, which were close to 100% African-Americans, the Grenfell inhabitants were far more racially diverse (White, Muslim, Asian, Afro-Caribbean), which speaks to the less segregated nature of social housing in the UK.

There are now plans afoot in 2025 to tear down the tower, and make it a proper community memorial for the victims. Of course, some in the community see the tower's continued existence as a sort of memorial, while others are quite happy to see it torn down. Conversely, some social housing in London's inner city has long been targeted for redevelopment. The most prominent example is Elephant and Castle in South London. The Elephant and Castle area has been the object of redevelopment efforts since the 1960s, when it was radically altered through large-scale, modernist renewal to produce the Elephant and Castle Shopping Centre and Alexander Fleming House. Even before the 1960s redevelopments, there was always the sense that the area had great potential given its centrality, yet perennially undervalued (Campkin, 2013). Figure 11.4 shows the extensive redevelopment of the site between 2013 and 2015.

The Elephant and Castle site must be seen as a 'borrowed place... provisionally tolerated for the time being but without any future promise of state or market restraint' (DeVerteuil, 2015: 39). This case study juxtaposes the 'high and dry' nature of residual social housing and the increasingly dynamic and powerful city fabrics that threaten to overwhelm and overtake it, particularly in valuable areas like Elephant and Castle. Another vulnerable area in London is the East End, and especially Stratford, the site of the 2012 Olympics and massive redevelopment before, during,

FIGURE 11.3 Grenfell Tower, 2017 © G DeVerteuil

FIGURE 11.4 Transformation of the Heygate Estate in Elephant and Castle, May 2013, June 2014, August 2015 © G DeVerteuil

and after the event. Like Elephant and Castle, this was a largely ignored part of London until the early 2000s. The Carpenter Estate, pictured on page 234, consists of three large towers. Since 2012, however, they have been increasingly surrounded by new high-rise development that caters to a wealthier group of people. Now marooned, the Carpenter Estate symbolises the old working-class East End of London. The two material environments are quite different – on the left are the new-build towers that house professional workers or overseas investors, while on the right is the

FIGURE 11.5 Stratford City and the Carpenter Estate, 2024 © G DeVerteuil

old 1960s/1970s style of social housing, based on Brutalist architecture and a one-size-fits-all ethos.

In France, there has been even less drive to redevelop social housing estates, particularly those built in the 1960s and 1970s in impressively large concentrations in the suburbs of most French cities. These were the *grands ensembles* built to house the French working class, at least initially, but now largely house immigrants to France and their descendants. There has been some redevelopment, however, and some continued construction, but at a much smaller scale and more scattered. This physical integration into the existing built environment also appeals to a more middle-class clientele. The older *grands ensembles* are therefore at risk of being redeveloped, although to a lesser degree than in London, given the former's largely suburban locales. Figures 11.6 and 11.7 show the *Cité Gagarine* in the southeast corner of Paris (just outside the city limits but connected by the Métro system). Gagarin(e) was the first man in space in the early 1960s, which suggests the modernist age of the social housing and also the explicitly socialist pedigree, connecting to the then-Soviet Union. In the 2020s, the older high-rises were torn down and replaced by newer, shorter ones, part of a process of deconcentrating poverty:

The smaller-scale social housing is also vulnerable to what is called 'social mix', whereby some of the market-rate housing is inserted to subsidise the remaining stock. But there is no real discussion to dismantle the social housing system. This resembles the situation in some cities in the Global East, such as Hong Kong and Singapore, which built extensive stocks of social housing in the 1970s–1990s and have subsequently

FIGURE 11.6 Cité Gagarine, Paris, 2024 © G DeVerteuil

FIGURE 11.7 Cité Gagarine, Paris, 2024 © G DeVerteuil

safeguarded them. To this day, more than 80% of the total stock in Singapore is subsidised, and close to 50% for Hong Kong (Davidson, 2024; Dimsum Daily, 2024). In both places, and given extreme levels of unaffordability, housing is considered too important to be left entirely to the market. But, beyond issues of affordability is that of community – that social housing can act as a way to foster a multicultural society, albeit with complete governmental oversight into who lives where.

11.4 Housing Crisis in Action: Platform Economies and Mega-Events

Building on the previous section, the rise of AirBnB is instructive when considering the housing crisis in many cities, further cutting into the affordable housing stock. The math is quite simple – more money can be made by renting out housing by the night than by the month, up to ten times more. And for every houseowner who uses AirBnB, there are five corporations looking to buy rental housing to convert into AirBnB as a permanent moneymaker. This has led to a further over-consumption and over-valorisation of housing in tourist hotspots, to the detriment of previous renters in the area. In Europe, this version is ensured by *not* redeveloping the city, for the benefit primarily of tourists. From Berlin to Bologna, Dubrovnik, Lisbon, and Rome, over-tourism and touristification is slowly strangling the vibrancy of these

FIGURE 11.8 Over-touristed European cities. Barcelona (x3), 2016; Rome, 2005; Amsterdam, 2008; Berlin, 2011 © G DeVerteuil

cities for native populations, connecting the tourist economy to real estate (Cocola-Gant, 2018). We can therefore suggest that technologies such as AirBnB have enabled, along with lax conversion policies and cheap airfares, the over-valorisation of certain tourist cities, and have produced results not dissimilar to those produced by overseas investors.

The concatenation above shows a variety of European cities that have experienced an influx of tourist-based real estate investment via AirBnB and other platforms, including Barcelona, Rome, Amsterdam and Berlin. All these cities have always been central to tourism in Europe, but since 2008 this ongoing touristification has distorted the rental market and led to an all-out housing crisis for locals.

A second illustration of the housing crisis comes from the impetus of mega-events, with the clearest expression provided by the Summer Olympics. Mega-events are large-scale, systematically planned special festivals and special events as tourists attractions, but also as catalysts for rebuilding cities and rebranding them (Duffy, 2009). How do mega-events act as culprits for the housing crisis? Although very specific to the cities holding the mega-events, the blatant catering to outsiders only worsens housing affordability both during the event but also long after the event. The 2012 London Olympics were instructive – the Olympic Village that was built to house the athletes was privatised after the event. The promises of revitalising the Stratford area were ultimately hollow, as the Olympics only accelerated the upgrading of the area to the detriment of locals who lived in social housing prior to 2012. Other Olympics – Athens 2004, and Rio de

FIGURE 11.9 Anti-Olympics poster, Paris 2024 © G DeVerteuil

Janeiro 2016 – served less to upgrade the host cities and more to financially cripple them with unsustainable debts and 'White elephant' projects that are of little use to the local population, including private golf courses and cycling arenas. Finally, the run-up to the 2024 Paris Games led to the deportations of homeless individuals in the city to far-flung locations, either temporarily until the Games were over, or more permanently. During Geoff's visits to Paris in June 2024, he was made aware of how the Games were being used as an excuse for 'social cleansing', particularly of homeless migrants who congregate in certain high-profile parts of town, including in front of the Paris City Hall.

The poster says in French: *here, so that we can welcome all the nations we had to expel all those who were already here.* A clear expression, we would argue, of how mega-events (and AirBnB, over-touristification and overseas investment) increasingly cater to well-off outsiders, rather than already-existing inhabitants. The resulting housing crisis across many cities is intertwined with a crisis of belonging, of whose city is being created? The next section focuses on how the state has been managing this question, with largely unpromising results.

11.5 Signs and Outcomes of a Housing Crisis: The State and the 1%

The trends outlined in the previous sections combine with other, more latent reasons as to why the housing crisis is increasingly entrenched. One key reason is the growing impossibility of building new housing, which directly connects to land use zoning and the management of density. In effect, certain parts of cities are protected to the point where they do not need to change at all. These places – whether inner-city districts in European cities, gentrified zones or self-centred suburbs in North American cities – need never worry about encroachment or densification, and thus contribute to a growing lack of new housing. From the Bay Area to Nassau County, on Long Island to Southern California, there is a spirited and predictable defence of class privilege, particularly the untouchable status of single-family homes, where local governments act like 'private clubs' that keep out any class diversity or even population growth. So here we are interested in the 1% and what does not get built via NIMBY, as an important barrier to solving the housing crisis. The image below shows a mansion on a 'Great Gatsby' scale in the far west of Montréal Island, forever protected by the local municipality and its absurdly rigid land-use laws.

The opposition to any new housing obviously chokes supply. This artificial lack of supply on behalf of the 1% combines with in-migration to certain housing markets. Migration is frequently seen as a culprit, particularly if the incoming people are more monied. This has combined with a preference for urban living among younger generations that was lacking 30 or 40 years

FIGURE 11.10 Senneville, 2022 © G DeVerteuil

ago. In-migration of wealthier people, and the preference for urban living among a subset of the population, are frequently connected to *gentrification*, the subject of the next subsection.

The role of the state has been altogether complicit in these various facets of the housing crisis – in terms of promoting mega-events, overseas investment and 'golden visas', but also initially assisting the growth of AirBnB and keeping interest rates low for over 15 years. But the state also does little to intervene in terms of 'bread and butter' issues like ensuring housing affordability via the construction of new social housing. Even when governments acknowledge the housing crisis, action is usually quite slow. Few national governments want to re-invest in large-scale social housing, preferring that the private market develop housing with minimal oversight, and usually under the purview of local governments and planning regulations. When the national government does intervene, a key barrier is the need for long-term planning and infrastructure investment that usually outlasts the party in power, assuming the nation is democratic. Behold the 2024 proclamations by then Prime Minister Justin Trudeau, in the Canadian context (Canada's Housing Plan, 2024):

> One of the biggest pressures on people right now is housing. Young Canadians are being priced out of their communities. Families are finding it difficult to get a good place to settle down. Rising rents and the high cost of buying a home are making it more difficult for younger generations to find a place to call home. We need more homes in Canada and we need to keep them affordable.
>
> Right now, governments across Canada are sitting on surplus, underused, and vacant public lands, like empty office towers or low-rise buildings that could be built on. By unlocking more public lands for housing, we can lower the costs of construction and build more homes, faster, at prices Canadians can afford. *Canada's Housing Plan* lays out a bold strategy to unlock 3.87 million new homes across the country by 2031, including 250,000 homes through a new Public Lands for Homes Plan.

Canada is the world's second-largest country, and in theory there should be more than enough space to build new housing. Yet even there, housing costs have far outpaced salaries in cities like Vancouver and Toronto. All told, this has abetted a *squeezed middle class*, one that is increasingly unable to afford housing, but also for the working and lower-classes, who are prone to seeing the housing crisis in terms of far-right, anti-immigrant rhetoric that the country is 'full'.

This rhetoric was certainly mobilised in the case of the riots in Dublin in November 2023, fuelled by anti-immigrant sentiment tied to broader issues around violence and housing unaffordability, via social media e.g. #Irelandisfull. It also underlay some of the social unrest in Britain in August

2024 (see Section 7.4). These pressures can be further framed by two opposing (and sometimes, but not always, related) tendencies in many cities, some of which have more than 40 years of experience with both of them: gentrification on the one hand, and homelessness on the other.

11.6 Signs and Outcomes of the Housing Crisis: Gentrification

To many social and urban geographers, gentrification – a process that Hackworth (2002: 815) defined as 'the production of space for progressively more affluent users' – has now moved beyond a quaint, localised phenomenon to a systematic private–public urban strategy at the forefront of a globalised neoliberal urbanism (Smith, 2002). As gentrification has become increasingly globalised, there have been more comparative studies as it spreads to new locales in new guises. By the 2010s, the state had become front and centre in the drive to gentrify cities, particularly in the Global South and Global East, where the scale of redevelopment has been unprecedentedly large and complex.

Social geographers have especially sought to relate the process of gentrification to more fundamental social change in the Global North (inner) city, in some ways ignoring the much larger and dominant tendency towards suburbanisation that pervades most city-regions. In their book on gentrification, Lees, Slater and Wyly (2008) argued that the process of gentrification is now more comprehensive than its previous small-scale and ground-up neighbourhood guise, a generalised middle-class restructuring of place that features the following key elements: reinvestment of capital; social upgrading of locale by incoming high-income groups; landscape change; and direct or indirect displacement of low-income groups. Pre-eminent social geographers like David Ley have studied the social and geographic implications of the shift towards a 'post-industrial city', in which the service sector, young professionals and liberal local politics came to displace older modes of social organisation (1980). Using Vancouver as his case, Ley pinpointed the inner city as the key site for the emergence of this post-industrial city, with its attendant deindustrialisation and the conversion to sites of conspicuous consumption. Almost 30 years on, Hutton (2010) found a more mixed landscape of new economy sites of production (Internet and design companies especially), but one which bears the unmistakable stamp of post-industrial sites of consumption.

If one adheres to the post-industrial thesis, then gentrification is best explained not by demand-side accounts in which undervalued land attracts new capital that induces upgrading (the so-called 'rent gap'), but rather by a greater focus on the (increasing) supply and production of potential gentrifiers, the so-called 'new middle class'. To Ley and others, the emerging and expanding professional middle-class, along with more marginal gentrifiers who deemed the suburbs inadequate for their social needs (e.g. single mothers, LGBTQ+ groups, artists, and students), sought inner-city spaces with high levels of amenities, including architectural, and proximity to White-collar jobs in the CBD.

Along these lines, Sharon Zukin, in her book *Loft Living: Culture and Capital in Urban Change* (1982), chronicled this process in Soho in Lower Manhattan. The derelict manufacturing spaces in Soho were first converted to artists' lofts in the 1970s, and this cultural component formed the beachhead for attracting further capital to the area. The 'Soho' phenomenon would be replicated in other urban locales, although not without severe repercussions for the original population (nowadays, the model is Brooklyn – see Figures 11.6 to 11.8). For instance, Shaw (2007) examined how the systematic gentrification of Central Sydney has resulted in a monotonous 'whitewashing' effect, whereby all non-gentrifying groups are slowly forced out to make space for a largely White professional class. Those that have remained, such as the Aboriginal person (Indigenous) and Asian immigrants (e.g. Vietnamese and Cambodian), are severely contained to decrepit and stigmatised neighbourhoods.

Bloch and Meyer (2023) argue that by the end of the 1990s, as the violent crime rate in cities across the US declined and gentrification increased, new residents felt they could safely move into lower cost, 'up-and-coming' neighbourhoods. This is a theme we pick up again in Chapter 12 (Gender and Inequality), where young professional women are increasingly being targeted as viable rental consumers in these regeneration efforts. Graffiti also became more relatively mainstream as an outcome of the sterilising effects of gentrification. Once a hallmark of disenfranchised groups, former 'graffiti gangs' have become regarded as 'street artists'. Bloch and Meyer (2023) describes how in many cities, from Pittsburgh (Pennsylvania) to Pretoria (South Africa), street artists are invited to help brand neighbourhoods that are being revitalised and gentrified as legitimately hip destinations for business owners, home buyers and influencers. He provides some examples of some 'up-and-coming' neighbourhoods that offer street art tours and host graffiti festivals in cities like Bogotá, Colombia; Dakar, Senegal; Mexico City; Brisbane, Australia; and Seoul, South Korea. Although street art can be critical of the growing tide of gentrification (explored as a form of quiet activism in Chapter 13), the street-level creativity narrated as part of a gritty urbanism conversely has become eminently marketable on a global basis (Hubbard and Lyon, 2018).

Richard Florida, in his 2002 book *The Rise of the Creative Class*, pinpointed the importance of young gentrifiers (and especially LGBTQ+ people) working in creative industries such as design, as an essential tool for local economic development. This emphasis on attracting and retaining the creative class apparently resonated with many declining post-Fordist cities eager for revitalisation. We can underline the importance that Florida accorded to university students to show the contours of this creative class. As university students yourselves, you are considered educated, certainly more than the average person. Given your education, you are able to participate in the knowledge economy. Further, your experiences with living in dense, rental housing has perhaps pre-conditioned you to becoming a gentrifier later in life. Certainly, your predilection for living among your peers – other university students – has unintentionally attuned you to living in lopsidedly student-focused communities in which the interests of transient and mainly young students are prioritised over those of longer-term residents (Smith, 2008). This could be seen as similar to gentrification, displacing lower-income residents. This *studentification* is relatively common in UK cities. The example of Cardiff, where Geoff teaches, is instructive – the areas directly to the east of campus are up to 80% students, as the university itself has embarked on a long-term expansion in student numbers since the trebling of tuition fees in 2011.

While hesitant at first, local governments soon came to support these pioneer gentrifiers and their specific social and economic needs. By the 2000s, this recognition of the importance of gentrification had permeated across a wide swathe of policy maker toolkits. This was very much evident in Amsterdam during Andy's urban fieldcourses there. Gentrification had, according to Neil Smith (2002), moved from a locally-bound phenomenon to a more systematic, private–public urban strategy at the leading edge of neoliberal urbanism. Lees et al. (2016) talk about a state-focused *planetary* gentrification which now criss-crosses the globe, showing up not only in the Global North, but also the Global South and East in various guises, usually in the form of heavy-handed redevelopment to massively upgrade city spaces. In parallel, the long focus on central-city gentrification now includes smaller cities, suburbs and even second homes in rural areas.

In 2017, the debate came full circle when Florida acknowledged the detrimental effects of gentrification and the 'winner-take-all' urbanism that he helped to popularise (though did not originate), particularly the ill consequences to vulnerable populations (e.g. poor renters, public housing inhabitants) and the shifting social makeup of neighbourhoods. These deleterious effects constitute another area of inquiry for social geographers, albeit one attracting less attention when compared to explanations of gentrification and the experiences of the gentrifiers themselves. Tom Slater (2006) has been particularly vocal on this point, contending that far too much energy has been dedicated to explaining gentrification, rather than examining its outcomes for the most vulnerable, which include low-income renters, social housing tenants, single-room occupancy hotel dwellers and those double-up with relatives or friends.

Indeed, a wide array of research has shown gentrification to be harmful, with possible negative effects including loss of neighbourhood integrity, increased cost of living, loss of services and depletion of neighbourhood place-identity, homelessness, community conflict and of course displacement (Slater, 2006; Lees et al., 2008; DeVerteuil, 2015). Displacement of vulnerable groups – that is involuntary movement – has been particularly controversial in terms of how much is occurring, and the degree to which gentrification is causing it. One group that is particularly vulnerable are social housing tenants, especially when their estates are the target of redevelopment. This was shown in detail above, where places like the Heygate estate in Elephant and Castle, and the Carpenter Estate in Stratford, were torn down or surrounded by new developments that essentially represented state-sponsored gentrification. This tendency has been quite clear in London, but less so in cities like Chicago, where social housing was demolished but little development replaced it. In some cities like New York, social housing has not been redeveloped at all. The broader place-based stigma of social housing is certainly one reason why the state may find it easier to gentrify these estates, but their location also matters.

Another group that has been vulnerable to displacement has been immigrants. Immigration is one of the main generators of demographic change in cities – increasingly the main one for many large and global cities – while gentrification is one of the main paths of class-based restructuring and recomposition of urban space currently. Yet the two have rarely been counterposed or put alongside each other (DeVerteuil, 2018). In highly-successful Global North cities from Vancouver to Sydney, London to New York, immigrants increasingly avoid the gentrified central neighbourhoods, no longer flocking to the original immigrant-receiving areas, the Chinatowns and Little Italies. Rather, they are moving directly to suburban areas where living is cheaper, creating far-flung enclaves. At the very least, displacement may not be a widespread phenomenon but it still exacts a considerable toll on poor renting households, whether through eviction, demolition of public housing, or the excluding nature of new-build redevelopment (Newman and Wyly, 2006).

Let us now return to the four key elements of gentrification signposted previously: reinvestment of capital; social upgrading of locale by incoming high-income groups; landscape change; and direct or indirect displacement of low-income groups. One way to visualise these core elements is through two brief illustrative case studies of (1) Brooklyn, and (2) high-tech and the city. First off, Brooklyn has emerged as a global imaginary of gentrification, once peripheral to Manhattan but now the epitome of cool. Its name speaks to a brand. Once synonymous with decay and crime, certain parts of north-western Brooklyn – Park Slope, Downtown and Williamsburg – are now the epitome of upgraded areas (Figures 11.11 and 11.12).

These parts of Brooklyn have become a global shorthand for deep-seated and sophisticated upgrading, creating inspiration for cities as disparate as Leeds, Montréal and Vilnius (Figures 11.14, 11.15 and 11.16).

The narrative of a 'Brooklyn inspired' gentrification has travelled far and wide. So has the high-tech colonisation of certain key cities – especially but not exclusively San Francisco – that has engendered localised gentrification through high wages and high demand for living space. When Geoff first began photographing the early phenomenon back in the Summer of 2000 – the

FIGURE 11.11 Park Slope, Brooklyn, 2012 © G DeVerteuil

FIGURE 11.12 Williamsburg, 2020 © G DeVerteuil

peak of the first tech boom – he would have no inkling that this wave would come back stronger in the 2010s, version 2.0 if you will (Figure 11.16).

In 2000, this tech invasion of certain cities was considered more a curiosity. In 2025, it has been normalised. In effect, the distorting effects of over-reliance on one industry, and the catering of that industry by local officials,

FIGURE 11.13 Leeds, 2015 © G DeVerteuil

FIGURE 11.14 Montréal, 2016 © G DeVerteuil

has led to rapid change in neighbourhoods popular with tech workers, such as the various high-tech enclaves of Mile End (Montréal), the Mission District (San Francisco) and Old Street (London) (Figures 11.17 to 11.20).

And yet, highly unequal places like San Francisco are also marked by deeply-entrenched homelessness. This is not to say that gentrification (or the tech industry) is the only cause of homelessness, but it certainly puts additional pressure on the housing stock, contributing to an overall housing crisis whereby some (vulnerable) people will inevitably fall through the cracks.

11.7 Signs and Outcomes of a Housing Crisis: Homelessness

Like gentrification, there is also a longstanding tradition of social geographic research that has examined the multiple causes of homelessness, its experiences, its spatial aspects and possible exits. We understand

FIGURE 11.15 Vilnius, Lithuania, 2019 © G DeVerteuil

FIGURE 11.16 Mission District, San Francisco, July 2000 © G DeVerteuil

homelessness as a continuum from rooflessness (on the street – but sometimes with a roof, i.e. encampments) and houselessness/unhoused (in shelters) to insecure accommodation (informality) to inadequate accommodation (overcrowding) (ETHOS, 2005). Michael Dear and Jennifer Wolch (1987; 1993) proposed a multiscalar explanation of homelessness. In effect, homelessness is the product of housing unaffordability, combined with rising poverty and welfare state retrenchment, and exacerbated by personal vulnerabilities. Specific factors include: economic restructuring which has led to insecure and low-paid jobs; welfare state cutbacks for the unemployed and service-dependent (Chapter 4), including the deinstitutionalisation

FIGURE 11.17 'Silicon Alley', New York, May 2000 © G DeVerteuil

FIGURE 11.18 Montréal, 2015 © G DeVerteuil

FIGURE 11.19 Market Street, San Francisco, 2016 © G DeVerteuil

FIGURE 11.20 London, 2021 © G DeVerteuil

of psychiatric patients and individuals with mental illness (Chapter 10); and the reduced availability of affordable housing, through gentrification-induced displacement but also through declining provision of social housing (Chapter 11). This structural backdrop then interacts with personal vulnerabilities, such as mental illness, addiction and history of incarceration or foster care. Using Los Angeles as their template, Dear and Wolch illustrated their contentions through a series of locally-embedded case studies. Local characteristics – such as availability of food, emergency shelter, opportunities for panhandling and tolerance – were critical in both helping homeless people survive but also enabling them to possibly re-enter stable housing. Service-rich locales (e.g. service hubs, see Chapter 8 on Skid Row) could also serve as traps, however, making it easier to survive yet more difficult to exit homelessness altogether.

In the 1990s and 2000s, social geographers focused increasingly on how homelessness was managed in public spaces in the Global North. This connected to D.M. Smith's (1994: 151) insights that while 'human beings have no choice but to occupy a space: they just do, ... it does not follow that such a space allows for the full, adequate, and self-directed representation of human beings either to themselves or to others'. Much attention was paid to the escalating anti-homeless measures that framed the homeless as passive victims of a vengeful backlash against visibly poor people in prime urban spaces, as part of what Neil Smith (1996) called *revanchism*. The literature around revanchism, mostly from the United States, described a progressive breakdown of homeless spaces in the city under the pressure of anti-homeless laws that criminalise homeless activities such as scavenging, recycling, panhandling and public urination, as well as sweeps that are intended to clear parks, city centres and transportation nodes from their abrasive presence. An entire literature presented a dystopian, revanchist and rather lurid view of the city for homeless individuals. While not denying that punitive measures have increased in scale and scope since the 1990s, DeVerteuil et al. (2009) have argued against this trend that ignores the increasingly varied and complex geographies of homelessness that typify the contemporary city. First, they argued against that this collapse the supposed collapse of homeless space by drawing attention to the

FIGURE 11.21 Los Angeles, 2018 © G DeVerteuil

proliferation of homeless spaces such as shelters and hostels. Second, they drew attention to the more supportive measures in place in cities, such as soup runs and voluntary associations that belie the supposed vengefulness against homeless people. Third, they emphasised a more diverse homeless population (by gender and race) than suggested by revanchism, and one that is able to actively negotiate the dwindling access to public space.

Arguably, there is a need to move beyond these debates to think about homelessness in the current context, one that exceeds the Global North. The following images (Figures 11.21, 11.22 and 11.23) give a sense that homelessness has become more visible in the last few years, even before the pandemic. These are a direct result of the housing crisis for those at the margins – precariously housed, precariously employed, the disabled, people with stigmatising mental ill-health and so forth, returning us to the deep consequences of marginality we saw in Chapter 8.

As the images show, homelessness is a global issue, as much as gentrification. Many large cities in the Global North suffered from record-high levels of homelessness in 2025, in the wake of the pandemic and loosening of social ties and routines alongside the accelerating housing crisis. As such, homelessness acts as the tip of a much larger iceberg. This contrasts with, for instance, squatting as a safety valve in the housing market – abandoned

FIGURE 11.22 Tokyo, 2005 © G DeVerteuil

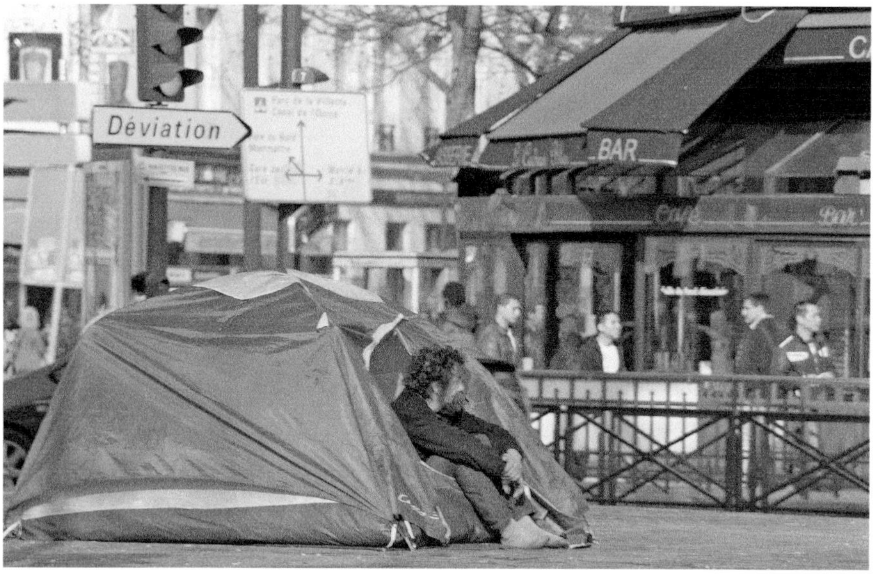

FIGURE 11.23 Paris, 2012 © G DeVerteuil

homes could be occupied (see Chapter 13). In some ways, this is akin to the auto-construction that has become a permanent housing solution in many cities of the Global South.

While the global scale of homelessness is not up for debate, the scale of homelessness (in terms of not having a permanent shelter) is that much

FIGURE 11.24 Kamagasaki, Osaka 2014 © G DeVerteuil

more prevalent in the Global South. It manifests itself across various forms of largely tolerated forms of informality, auto-construction, and squatting (see images in Section 2.4 on inequality). In cities like Lima (Peru) or Jakarta (Indonesia), upwards of 30% of the population live in auto-constructed neighbourhoods. Homeless issues also exist in the Global East. Geoff first visited Kamagaski in Osaka in 2014, and then revisited three more times (2016, 2018, 2022) as part of a larger research agenda on comparing service hubs in the Global East and Global North (DeVerteuil et al., 2022, DeVerteuil & Kiener, 2022). Figures 11.24 through 11.26 show how that even in a highly coherent society like Japan, some people (mostly men) can fall through the cracks of the welfare state and the housing system.

The images capture some of the key visual cues of the area: the predominance of older men (former day labourers), the concentration of services for them and the apartness of this area from the rest of the city. Like many homeless individuals around the world, the tendency to be male and middle-aged is rather pronounced, akin to the 'deaths of despair' we saw in Chapter 10.

With four in-depth visits as well as guided tours from service providers, residents and academics, Geoff came to know this area quite intimately. Even in translation, the visits allowed him to appreciate the idiosyncrasies of the place – the tiny rooms, pawn shops and the cheapest food-dispensing machines in the city. The 2022 visit found Kamagasaki in a state of change. The heart of Kamagasaki – the landmark Airin Welfare Center – had been closed for demolition, apparently no longer seismically safe (Figure 11.26).

FIGURE 11.25 Kamagasaki, Osaka 2014 © G DeVerteuil

FIGURE 11.26 Kamagasaki, Osaka 2022 © G DeVerteuil

Of course, the Japanese context contrasts sharply with other Global North nations, in that there is no pervasive housing crisis, as well as a robust welfare state. In fact, as Japan's population slowly drops, there is already too much housing in most cities save Tokyo. There is, however, evidence of gentrification in many inner-city areas, including Osaka that threaten service hubs like Kamagasaki.

Whether in the Global North, South or East, gentrification and homelessness are two sides of the same housing crisis. Yet for some readers, these sides might seem rather remote to their own experiences. In the next section, we set about to further ground the housing crisis in terms of what it means in generational terms, especially for those generations born after 1980.

11.8 'Generation Rent' and Wealth Inequality

Talking about the extremes of gentrification and homelessness as two sides of the same coin does not, however, capture a lot of the middle ground when it comes to the current housing crisis – the so-called squeezed middle-class from earlier in the chapter. In this penultimate section, we propose that the housing crisis is especially difficult for those young people in their 20s and 30s (as of 2025) when compared to the Baby Boomer generation, who bought at a time of much greater housing affordability. Stepping back, we need to speak about the idea of a 'generation' within society. In the anglosphere at least, seven existed at the writing of this textbook (Pew Research Center, 2020):

- The Greatest Generation – born 1901–1924.
- The Silent Generation – born 1925–1945.
- The Baby Boomer Generation – born 1946–1964.
- Generation X – born 1965–1979.
- Millennials – born 1980–1994.
- Generation Z – born 1995–2012.
- Gen Alpha – born 2013–2025.

Here we can also talk about 'Generation Rent', where it becomes impossible to buy a house in many cities and where lifelong renting (or living with one's parents) looms large. Indeed, those encountering difficulties in the housing market are mostly in the Z and Millennial generations, which together are demographically the largest as well. They suffer from housing insecurity, rather than poverty or marginality, but neither do they constitute gentrifiers for the most part. In places like Canada, those born before 1980 have had ample opportunities to buy into the housing market at relatively low prices (although not low interest rates) in cities like Montreal, Toronto and Vancouver. Since 2008, however, and abetted by historically low interest rates, prices have risen far faster than wages, to the point where the only option for home ownership is far-off suburbs or a move to cheaper cities entirely, or a deposit paid by parents (from the Boomer or X generations).

We cannot say that this is strictly about ageism, but there is also no doubt that the Baby Boomer generation in general and the 1% in particular are in no rush to relinquish their over-valued assets while defending them with great vigour (see Section 11.4 on opposition to new housing that would depress their own house's value). One future mechanism of transfer might be the inevitable sale of Baby Boomer-owned housing to younger generations, either through downsising or death. But even this process will be fraught with inequality – for those Millennials and Zs who do not have well-off, home-owning parents, the payoff will be zero. We spoke about inequality in Chapter 2, mostly framed around income. But of course wealth – and housing wealth in particular – is even more unequally distributed than income, and even less liable to be taxed by the state, even as inheritance.

But in this regard, let us look at the UK tax on inheritance to give some insight into wealth inequality, which is basically a tax on housing wealth (and of course many nations do not even have inheritance tax). As of the early 2020s, the first £325000 is untaxed when a child or children inherit from their parents. Anything over that amount is taxed at 40%; if one is married the threshold is set at £650000 before the 40% kicks in. If the parents deem the house to the children, the 40% only begins at £425000. For the least wealthy households, there is no inheritance tax at all because there is no inheritance – no housing ownership in particular. For those in the middle-class, the 40% threshold depends on where in the UK the house was located. In the 100-mile radius around London, £425000 may not actually be all that expensive, while in other parts many houses would fall well below the threshold. As a side note, inheritance tax did not apply when Queen Elizabeth passed away in 2022. King Charles paid no inheritance tax on the over 1 billion Pounds worth of property that passed to him.

Part of the discussion around the inequalities of being Generation Rent connects to larger inequalities around age and demographic realities in society (Chapter 6 on youth), which can be difficult to perceive when it comes to slow-moving change, as in the housing markets whose encounters are elusive and slippery. These demographic realities become more apparent, however, in times of crisis. Thinking back to the pandemic, it became obvious that the young and the old were most impacted. For those under 25, and this could include future readers of this textbook, in-person education was severely curtailed for almost two years. For those over 70, the pandemic was a lethal threat. The ability to shift these demographic realities is nearly impossible, and so their slow-moving yet relentless advance must be acknowledged and 'lived with' until new opportunities arise as the older generations fade, a sort of passive social justice that we do not promote! (but see Chapter 13 on quieter modes of activism).

11.9 Conclusions

This chapter highlights the ambiguous nature of the encounters when it comes to housing. There is an uneasy mix of social and private housing renters, homeowners, state planners, developers, AirBnBs, foreign interests vs. local interests, tourists and visa buyers versus locals, and so forth. It can be hard to delineate who contributes to the housing crisis when there are so many different players. Conversely, private sector investors, real estate agents (realtors) and landlords also have played their part. Given the complexities, it is certainly irresponsible to blame the housing crisis on just one group – for instance, scapegoating immigrants or tourists, or just AirBnB, or older generations hoarding their own housing. It is the combination of many factors that we have outlined here that have led to the hyper-commodification of the housing market in many nations over the last two decades. In turn, this has compromised the ability of many younger and less well-off people to afford to buy or even rent, and so the relationalities of housing are generational as much as they are global/local.

This chapter has mostly raised social justice issues of age-based distribution, or should we say mal-distribution. The housing crisis, gentrification, generational divides and homelessness are all manifestations of not enough housing in the hands of the people and especially those with less power in society. There are strands of procedural social justice that can be unpacked around the housing crisis. These can relate to how community opposition to new housing is managed by planning systems – are there provisions to impose a 'fair share' model where all communities must do their part? This is not an academic question, since it directly implicates who benefits and who is burdened by the housing crisis across society, in real material terms. Finally, and as in Chapters 6 and 9, this chapter raises an important social identifier, that of age. Age is more supple than other social identifiers – especially race and gender – as people age continuously, moving from one category to the next (children, youth, adult, elderly). Yet the fixed notion of generations adds friction to this continuous aging process. The combination of a moving age and fixed generation certainly complicates matters in the future. As time goes on, your generation's priorities in life will change, and this includes the key resource of housing. Will there be enough? What sort of legacy will the Baby Boomer generation leave on the overall housing market? Will your generation behave in the same way as previous ones? Will the state reconsider its role in helping to buttress housing markets?

Further Reading

Florida, R. (2002) *The Rise of the Creative Class: And How it's Transforming Work, Leisure, Community and Everyday Life*. New York: Basic Books.

Florida, R. (2017) *The New Urban Crisis: How Our Cities Are Increasing Inequality, Deepening Segregation, and Failing the Middle Class, and What We Can Do About It*. New York: Basic Civitas Books.

Holloway, S.L., Hubbard, P., Jöns, H. and Pimlott-Wilson, H. (2010) Geographies of education and the significance of children, youth and families. *Progress in Human Geography*, 34(5), 583–600. https://doi.org/10.1177/0309132510362601

Ley, D. (1980) Liberal ideology and the postindustrial city. *Annals of the Association of American Geographers*, 70(2), 238–258.

Prime Minister of Canada (2024) Building homes on public lands I Prime Minister of Canada (pm.gc.ca).

Rosenbaum, J.E. et al. (2002) How do places matter? The geography of opportunity, self-efficacy and a look inside the black box of residential mobility. *Housing Studies*, 17, 71–82.

References

Baxter, R. and Brickell, K. (2014). For Home UnMaking. *Home Cultures*, 11(2), 133–143.

Bloch, S. and Meyer, D. (2023). Displacement beyond dislocation: Aversive racism in gentrification studies. *Dialogues in Urban Research*, *1*(3), 206–225.

Blunt, A. (2005) Cultural geography: Cultural geographies of home. *Progress in Human Geography*, 29(4), 505–515.

Brickell, K. (2012) Geopolitics of home. *Geography Compass*, 6(10), 575–588.

Campkin, B. (2013) *Remaking London: Decline and Regeneration in Urban Culture.* London: Tauris Press.

Cocola-Gant, A. (2018) Tourism gentrification. In *Handbook of Gentrification Studies* (pp. 281–293). London: Edward Elgar Press.

Davidson, M. (2024) *Justice and Cities: Metro Morals*. London: Routledge.

Dear, M. and Wolch, J. (1987) *Landscapes of Despair.* Princeton: Princeton University Press.

DeVerteuil, G. (2015) *Resilience in the Post-Welfare Inner City: Voluntary Sector Geographies in London, Los Angeles and Sydney*. Bristol: Policy Press.

DeVerteuil, G. (2018) Immigration and gentrification. In *Handbook of Gentrification Studies* (pp. 428–443). London: Edward Elgar Press.

DeVerteuil, G. and Kiener, J. (2022) The commons, the counterpublics and dissident urban space. *Geography Compass (Urban)*, 16(8). DOI: 10.1111/gec3.12654

DeVerteuil, G. and Manley, D. (2017) Overseas investors into London: Imprint, impact and pied-a-terre urbanism. *Environment and Planning A*, 49(6), 1308–1323.

DeVerteuil, G., Marr, M. and Kiener, J. (2022) More than bare-bones survival? From the urban margins to the urban commons. *Annals of the American Association of Geographers*, 112(7), 2080–2095.

DeVerteuil, G., May, J. and von Mahs, J. (2009) Complexity not collapse: Recasting the geographies of homelessness in a 'punitive' age. *Progress in Human Geography*, 33(5), 646–666.

Dikec, M. (2007) *Badlands of the Republic: Space, Politics and Urban Policy.* London: Wiley.

Dimsum Daily (2024) Public housing in Hong Kong and Singapore: Lessons across the straits 17 February.

Duffy, M. (2009) Festival and spectacle. In N. Thrift and R. Kitchen (Eds.), *International Encyclopedia of Human Geography*. New York: Elsevier Press.

ETHOS (2005) *ETHOS Typology on Homelessness and Housing Exclusion*. Brussels: FEANTSA.

Government of Canada (2024) *Solving the Housing Crisis: Canada's Housing Plan.* Ottawa: Housing, Infrastructure and Communities Canada – Solving the Housing Crisis: Canada's Housing Plan.

Hackworth, J. (2002) Postrecession gentrification in New York City. *Urban Affairs Review*, 37(6), 815–843.

Hubbard, P. and Lyon, D. (2018) Introduction: Streetlife – the shifting sociologies of the street. *The Sociological Review*, 66(5), 937–951. https://doi.org/10.1177/0038026118771281

Hutton, T. (2010) *The New Economy of the Inner City.* London: Routledge.

Lees, L., Shin, H. and Lopez-Morales, E. (2016) *Planetary Gentrification*. Cambridge UK: Polity Press.

Lees, L., Slater, T. and Wyly, E. (2008) *Gentrification*. London: Routledge.

Ley, D. (2012) *Millionaire Migrants: Trans-Pacific Life Lines*. London: Wiley.

Newman, K. and Wyly, E. (2006) The right to stay put, revisited: Gentrification and resistance to displacement in New York City. *Urban Studies*, 43, 23–57.

Pacione, M. (2009) *Urban Geography: A Global Perspective*. London: Routledge.

Peil, T. (2009) Home. In R. Kitchen and N. Thrift (Eds.), *International Encyclopedia of Human Geography* (pp. 180–184). London: Elsevier Press.

Pew Research (2020) Where Millennials end and Generation Z begins I Pew Research Center.

Pinder, D. (2003) *Visions of the City: Utopianism, Power and Politics in Twentieth Century Urbanism*. London: Routledge.

Shaw, W. (2007) *Cities of Whiteness.* London: Blackwell Press-Antipode Book Series.

Slater, T. (2006) The eviction of critical perspectives from gentrification research. *International Journal of Urban and Regional Research*, 30, 737–757.

Smith, D.M. (1994) *Geography and Social Justice*. Oxford: Blackwell.

Smith, N. (1996) *The New Urban Frontier: Gentrification and the Revanchist City*. London: Routledge.

Smith, N. (2002) New globalism, new urbanism: Gentrification as global urban strategy. *Antipode*, 34, 427–450.

Smith, D.P. (2008) The politics of studentification and '(un)balanced' urban populations: Lessons for gentrification and sustainable communities? *Urban Studies*, 45, 2541–2564.

Vertovec, S. (2009) *Transnationalism*. London: Routledge.

Wolch, J. and Dear, M. (1993) *Malign Neglect: Homelessness in an American City*. San Francisco, CA: Jossey-Bass.

Zukin, S. (1982) *Loft Living: Culture and Capital in Urban Change*. New Brunswick NJ: Rutgers University Press.

12 GENDER AND SEXUALITY

12.1 Introduction

As a core social identifier, gender has become a key social justice topic in contemporary times. A recent rise in populist right-wing political leaders and influencers have sought to establish a 'culture war' over how gender is defined and how it relates to definitions of sex. Debates often revolve around how to define women and the rights of transgender individuals, including access to gender-affirming medical care, the use of gender-neutral bathrooms, and the inclusion of transgender people in various social settings. Across this spectrum, we appreciate the varying degrees of gender normativity and diversity that different people may embody and experience. The chapter is situated within a longer historical narrative, exploring how people of different gender identities have faced and overcome many barriers in gaining full citizenship. It draws on a range of literature and theoretical fields in this context, including trans theories, queer engagement, feminist and masculinist geographies and gender geographies. It includes a focus on persistent inequalities, for example, in domestic responsibilities, informal care, access to public space and political engagement and how these have been exacerbated in the wake of austerity and post-welfare contexts.

Likewise, sexuality has come under intense scrutiny, with disagreements over the extent to which lesbian, gay, bisexual, trans and queer+ (LGBTQ+) individuals and issues should be included in schools, workplaces, and public spaces. While the chapter cannot do justice to the full breadth of scholarship across both the geographies of gender and sexuality, or purport to cover both equally, we are nonetheless aware that the two axes intersect in complex and multifaceted ways which are important to disentangle. While the chapter primarily focuses on gendered inequalities, we highlight how gender intersects with sexuality to create unique experiences of privilege and oppression. As we have done elsewhere in the book, we seek to foreground this intersectionality, where relevant, as it can have a significant impact on the lives of individuals of any gender or sexual orientation. This is also a testament to the fact that feminist geographers have contributed a great deal to both the geographies of gender and sexuality.

DOI: 10.4324/9781003414889-14

We also must acknowledge our own positionality here, as two White male scholars, working in a discipline that has had a troubled past with gender. Up until the 1970s, the discipline has been long associated with masculine ideas of expeditions, with pervasive sexism, and with women largely excluded from geographic scholarship and the wider discipline. Tivers (1978) provides an important early contribution with the aptly titled, *How the other half lives*. This was the first published critique in British geography of the male-oriented view of society at the core of a discipline dominated by men and its complete absence of the study of women.

Much of this emerging scholarship was historically focused on the longstanding nature of patriarchy within the earlier part of the 20th century and since (Bowlby et al., 1982; Rose, 1993; McDowell and Sharp, 1997). This followed the emergence of feminist geography and the creation of the Women's Geography Study Group (WGSG, 2004). Within this 'first wave' of feminist literature, there was a core focus on the gendered geographies of labour and social reproduction. Building on the geographies of production outlined in Chapter 2 and our focus on ordinary citizenship in Chapter 3, we examine the patriarchal order largely institutionalised by men that had effectively tied women to the home and excluded them from the labour market. This related to cultural expectations around (feminised) domestic responsibilities within the home, the labour market and wider publics. While these represent the various relationalities for women, this chapter also considers how women commonly occupy a position on the 'server' side of various contact zones for others, in both unpaid roles such as mothers and informal carers, and paid roles such as care staff, teachers, welfare officers, and civil servants (Chapters 5 and 9).

Before we proceed, it is important to emphasise how gender is socially constructed, like race. This is not to demote the idea of gender; as we contend in Chapter 2, a social construction is not something purely subjective or ideal, or that sits outside of the material world. As we explore, the social construction of gender can manifest in powerful ways through shaping patterns of habitual behaviour, sanctioning, and institutionalised practices and rules. Here, gender is to be understood as formed through understandings of how biological and social processes influence identity formation. Put another way, it provides a relational blueprint for what it means to identify as a man, a woman or a non-binary person, consistent with our relational framing elsewhere. It thus places cultural expectations onto people based on gendered norms and assumptions. These norms, as we shall see, can create boundaries based on masculine and feminine expectations regarding certain jobs or ways to behave for example. In some cases, these cultural norms can become narrowly defined and interwoven with 'heteronormativity', a term used to highlight the often-invisible norms of heterosexuality between men and women, which can inadvertently exclude LGBTQ+ people.

Geographers have sought to destabilise the presumed naturalness of the binary man-masculinity and woman-femininity, since Judith Butler's (1990) book *Gender Trouble* prompted scholars to disrupt such norms and critique stable or essentialist understandings of gendered subjectivity. Shortly after this publication, geographers also began to trace out a fledgling sub-field on the geographies of sexualities. *Mapping Desire*, edited by David Bell and Gill Valentine (1995), was the first book on geographies of sexualities, and included many feminist geographers who sought to challenge normative ideas of gender and sexuality. Here, Butler's earlier call prompted echoing voices keen to challenge how rigid ideas of 'masculinity' and 'femininity' had become neatly embodied onto sexed bodies. We shall look at these debates in the latter half of the chapter.

However, as noted, there has been a recent push back against critical gender scholarship, with some women strongly defending their position of being solely understood as a biological sex class. This culminated with the 2025 Supreme Court ruling in the UK that 'sex' should be interpreted as biological sex in the Equality Act 2010. This means that, under the Act, a woman is a biological woman or girl (a person born female), and a man is a biological man or boy (a person born male). This has led to concerns that it could significantly affect legal protections for transgender women and non-binary people and more broadly the future of the LGBTQ+ community. This has revealed once again how potent these debates are, and so we start our chapter from a position of taking an inclusive and respectful way forward, and remain committed to the values of inclusion. The chapter begins with a focus on gender inequality, focusing on three core areas, politics and work, care and domestic labour and sexual violence.

12.2 Gender Inequality: Politics and Work

Women have long experienced unequal access to political rights as briefly noted in Chapter 3 (which for the most part pre-dated geographical interest in this area). Women's call for political rights emerged from the late 19th century, with New Zealand being the first to extend the franchise to women citizens in 1893 and most Western nations extending the right to vote (suffrage) from the early 20th century. In the US, the 1920 Nineteenth Amendment granted women the right to vote. English women won limited suffrage in 1918, and then the right to vote on the same terms as men in 1928. Women's political rights were also granted in many newly emerging states e.g. Pakistan and Bangladesh (1947), Indonesia (1941) and most North Africa countries (by mid-1950s), where women participated in the first national elections or soon following. By 1960, this right had extended to women in most other world regions, with more than half of the countries and territories granting women the right to vote by this time. More recent additions have included Kuwait (2005), United Arab Emirates (2006) and Saudi Arabia (2011).

This extension of suffrage was typically preceded by a hard-fought achievement, where suffragette movements around the world often faced significant resistance. In meeting this resistance, women activists gradually began to use more militant tactics and civil disobedience to mobilise for support and gain political recognition to extend the right to vote. As noted, much of this earlier period of change pre-dates the interests of geographical scholarship on the gendered spatial experiences of women, but it provides an important backdrop to the later interests which largely developed from the late 1970s. At the time, there was a lack of visibility of women in wider public debate, including the labour market.

Given this historical context and lack of a political voice, women's rising labour market participation in the late 20th century was a major societal and economic revolution – in the Global North at least. Here, much like achieving political rights, this labour market transformation was preceded by a women's movement in the 1960s and 1970s, which was bolstered by feminist writers critiquing the patriarchal order. This movement was in fact a series of social movements and political campaigns across many industrial countries (that would soon begin de-industrialising!) for liberal and radical reforms on women's issues created by inequality between men and women. These issues extended to reproductive rights, maternity leave, equal pay, as well as sexual violence. This movement signified a moment of wider social change, with greater college enrolment and graduation among women from professional and graduate schools, and a later age at first marriage (Goldin, 2006). The 1960s is also remembered as a time when big shifts in social freedoms occurred, with abortion and homosexuality (for those over the age of 21, at least) being partially decriminalised in 1967 in the UK, the same year the contraceptive pill was first made available to unmarried women. The Stonewall Riots also occurred shortly afterwards in 1969 after a gay bar was raided, prompting a riot, and galvanising activism and raising national awareness, and fostering a more visible and assertive LGBTQ+ community. This shift was by no means universal though and not without opposition from some segments of society. Indeed, in many countries where Catholic and Muslim faith dominates to this day, women and gay people often continue to experience firm boundaries based on gender and sexual orientation.

As well as marginalisation from the labour market, women also sought to challenge exploitation and powerlessness *within* the labour market (to use Young's faces of oppression), which stemmed from the high levels of gender-based segregation which persisted in employment sectors. Women have been disproportionately represented in service-sector employment and in low-paid, low-status work. Certain sectors had become female-dominated, including caring jobs, clerical employment and work in the retail sector (EHRC, 2011). Within these sectors, women have fared worse, experiencing relatively low pay, inferior career prospects and precarious patterns of employment. By contrast, men have enjoyed greater freedom and recognition within the labour and higher salaries. Thus, women have had to struggle for equal pay for work of equal value. McCann (1994) believes that much of this disparity was due to deeply-rooted discriminatory practices, whether through deliberate exploitation of women workers or through an unwitting acceptance of traditional norms.

By contrast, the Nordic countries have made remarkable progress in eliminating gender pay gap differences. This has been somewhat due to the compressed wage structure in these countries. However, despite the smaller gender pay gap, a study on wage differentials in Norway, Pal Longva and Steiner Strom (2002) concluded that 25% of the remaining gender pay gap could be attributed to sex segregation by industry and 71% to discrimination. Nowhere has reached complete equality and parity, but some places are closer than others.

This discrimination has been persistently overlooked by men since the 1980s, as one common justification for the gender pay gap was that women should naturally anticipate shorter and discontinuous periods of paid employment due to their reproductive, caring, and housework roles and so invest less 'human capital' in education and training than men (Breen and Cooke, 2005). This reinforces how prevalent heteronormativity was, and moreover how this was underpinned by a *hegemonic masculinity*. Hegemonic masculinity refers to a perceived ideal masculine identity, typically associated with traits like strength, courage, aggression, and the ability to provide, and protect. Hegemonic masculinity refers to the culturally dominant form of masculinity that is socially prized and upheld, often at the expense of other forms of masculinity – and women. Indeed, much of this research has sought to examine the privileged status of masculinities and how this has evolved over time. For example, McDowell (1995) traces the shifting cultures of work and employment and how this is mapped out through masculine bodily performances in the workplace (McDowell, 1995, 2005). This includes observations about the change from a 'loud, aggressive, macho version of masculinity' that was often required when trading and dealing – for instance, in the merchant, banking and finance sector – to a computer-savvy and fast-thinking masculinity (McDowell, 1997, 2005).

The gender division was also reinforced by assumptions of 'traditional' roles for women, reflected in the quote below by Berger and Piore (1980: 50):

> They are willing to take jobs because they see their commitment to these jobs as temporary and they are able to bear the flux and uncertainty of the industrial economy because they have their traditional activities on which to fall back.

From this reasoning, women's lower pay has been assumed to be a result of their lower productivity – a point that feminist scholars in geography (e.g. McDowell, 1991) have sought to challenge. Based on these assumptions, we ask you to consider the following questions.

Box 12.1 Engagement Questions

1. Why did you go to University?
2. What jobs do women commonly work in that do not involve a University degree?
3. What jobs do men commonly work in that do not involve a University degree?
4. Now, compare the typical incomes in these jobs, and discuss the implications on University participation.

Is this a reason why many degree programmes now have higher rates of women students?

By contrast, in the Global South, women have continued to be mainly self-employed in the informal sector, providing unskilled labour on farms or as micro- and small- entrepreneurs in home-based work (producing goods for sale), and wholesale or retail vending. Work in the informal sector is typically low-wage, insecure, and often seasonal. Employment in the formal sector, on the other hand, is better paid and more secure; but a smaller proportion of women have such jobs (Mehra, 2021). Traditional gender roles and societal expectations can also create barriers for gay people in seeking employment and advancing in their careers. Due to fear of stigma and discrimination, some gay individuals may work in the informal sector or in jobs that do not require them to disclose their sexual orientation (Cortés, 2023).

In the Global North, lesbians and gay men still receive fewer invitations for interviews than their heterosexual counterparts. Furthermore, they also receive fewer invitations to interview for positions in which masculine or feminine personality traits are highlighted in job applications and at firms that do not provide written equal opportunity standards. This suggests that the level of discrimination depends partly on the personality traits that employers seek and on organisation-level hiring policies. Heteronormative discourse thus continues to prevail in the labour market, but this operates in more subtle ways (Drydakis, 2015).

Returning to the case of women, they now attain more high-paying professional and managerial positions. A shift in theorisation has emerged to account for this, away from a focus on the marginalisation of women *from* the labour market towards precarity *because of* the labour market, which is the essence of Young's (1990) exploitation and especially powerlessness. This emerged because despite becoming increasingly involved in waged labour, women remained responsible for most of the tasks of *domestic* labour, which we turn to next.

Gender, Care and Domestic Responsibilities

In McDowell's (2014) review of her own earlier writing from the late 1980s/early 1990s, she reflects how she had originally considered that something would have to give if the burden of paid work and domestic responsibilities continued to land on women. However, as it turned out, she found that women have merely shouldered more responsibility and now juggle more in their private life; '*optimistically, I imagined the downfall of the capitalist system rather than fewer immaculately clean homes*' (p. 5). In this quote, she recognises the juggle of everyday life women face, and how these tensions are managed in everyday, mundane ways within the household. Fagan and O'Reilly (1998) provide some further insight here by reminding us that the decisions between men and women over the type of work participation assume consensual decision-making within households, ignoring unequal power relations based on gender. In short, they contend that men have enjoyed greater bargaining power within (heteronormative) households. Assumptions are often made over care work, which has remained an entangled and pervasive social practice associated with women.

In LGBTQ+ households, Barrett (2015) explored how couples divide and understand domestic labour (and childcare, where relevant) in their homes. The perceived normativity of coupled domesticity and childrearing meant that on the one hand the LGBTQ+ participants could be seen to fit in with normative ideals of domestic family life. On the other hand, the couples challenged heteronormative ideas about gendered household practices. They disrupted traditional divisions of parenting and domestic labour, questioning the assumed 'natural' connection between women, childcare, and specific household tasks. They enacted diverse roles and meanings and used their agency and creativity to assign particular preferences and sometimes entangle meanings to their roles and division of labour. This also included the very process of starting a family, which was something of a communal endeavour, involving third parties such as donors, surrogates and social workers.

This entangling of childcare and work can also be entangled in the workplace, as illustrated in the US context, where parental leave is minimal and women have been forced to combine wage work and care work in novel ways. Specifically, Kate Boyer (2014) explores how workplaces have had to adapt to mothers breastfeeding or expressing breast milk *at* work. Boyer's research centres on US federal rights legislation which mandates that workplaces of over 50 employees must provide lactation rooms and breaks during the workday to express milk. This example serves as a fitting illustration of *neoliberal motherhood,* which Boyer (2014) argues describes how women must contend with the growing expectation of working motherhood. While juggling domestic responsibilities, women also now are more likely to work part-time, which offers greater flexibility on the one hand but has been regarded as a trap for women on the other. Breen and Cooke (2005) describe the basis of this trap, as part-time work can marginalise women's position in the labour market, can reduce their chances for economic independence, and lets men 'off the hook' in sharing domestic labour responsibilities more equitably.

To better understand the roots of the concept of neoliberal motherhood, studies have shown that women's growing motivation to challenge dominant, oppressive social attitudes which underpinned their earlier 'stay at home' status, coincided with state policies and institutional arrangements (Scharff, 2016) that *demanded* their labour market involvement. Neoliberal policies since the 1980s, on the back of de-industrialisation, sought to 'encourage' participation of women in the emerging service economy and thereby oriented labour market participation as a key to the benefits of citizenship. While this undoubtedly opened up opportunities for women, this political imperative initially coincided with a period of economic recession, fiscal deficits and unemployment (although the latter primarily affected many male-dominated industrial and mining sectors). Men were also no longer entitled to a 'family wage' (a wage generous enough to support a family) which had characterised the Fordist mode of production. This also paved the way for a shift in hegemonic masculinity away from a focus on traditional male roles like provider and protector to encompass a more nuanced and diverse range of expressions, which we explore in Section 12.3.

One of the main achievements of this body of literature and feminist theorising about labour and employment has been to insist that (and expose how) unwaged work in the arena of social reproduction (Chapter 3), 'from cleaning to childcare, sex, love, and companionship' – is essential to the maintenance and continuation of a capitalist mode of production' (McDowell, 2014: 1). This achievement has been about recognising and valuing women's important role which had been overlooked. Similar approaches have also been taken by Boyer et al. (2017) and Mckie et al. (2001) in their work on the changing gender dynamics of household care. The other interrelated social justice achievement was the contribution to and amplification of debates over fairer pay and conditions for women, linking back to the previous section. In more contemporary times, it is reassuring to see how the gender pay gap has narrowed, and that women's involvement in employment has widened to other professional sectors.

Women's continued and disproportionate engagement with care work though (Chapters 5 and 9), and the building of social capital (Chapter 10), has persisted to this day. This is not necessarily in the same way as before their marked rise in labour market participation. Women remain central to the provision of care, albeit now organised on a different social and spatial basis. To meet the demands of labour market participation, women in many countries now must make their own private arrangements for childcare. Women thus commonly occupy a position on both sides of the contact zone, in both unpaid roles such as mothers and informal carers, and paid roles such as care staff, personal assistants, nurses, social workers and community support workers. This explains why they appear so frequently in other chapters in this book – often appearing as the 'other' in zones of contact (Chapters 4 and 5) with those seeking access to state or voluntary sector support, or in terms of social capital that was raised in Chapter 10 on mental health. All these forms of care involve vulnerabilities, needs and dependencies at the individual level. However, these forms and practices of care are themselves caught up in wider structures and institutions that may be caring or uncaring. The gendered norms associated with care work should therefore inform how we think about these contact zones. The relative (low) valuing of this work is based on the attitudes of people requiring care work (see Chapter 8), the agencies care staff work for, and the state which often determines how such work is resourced.

This has been the core focus of some recent geographic scholarship on the impacts of austerity on women both seeking and delivering care. Ellie Jupp (2023) has traced how mothers' access to care has been particularly affected in the wake of austerity cuts to welfare services. In Jupp's work, she considers what kinds of care are being offered or withdrawn by the welfare state and what does this mean for caring at home. Her focus is on the heavy budget cuts in the UK which led to a significant withdrawal of early-years services – that formerly offered places of 'rest' or 'time out' for parents (overwhelmingly women) and to support those who needed 'help'. Middleton, with Semanani (2020), echo these impacts, with a focus on the experiences of first-time parents (almost exclusively mothers), and those working in early-years drop-in services. In both cases, this work shows how women are unequally impacted by benefit changes, loss of employment in the public sector, and the withdrawing of resources from public services. Consequently, Hall (2023) explores how this has manifested in more personal ways with women specifically pausing and delaying decisions to have any or more children as a direct result of changes in the name of austerity.

Gender and Sexual Minority Based Violence

Another form of inequality based on gender and sexuality is the unequal levels of violence people face based on their gender identity and/or factual or perceived sexual orientation. Violence against LGBTQ+ people typically falls within definitions of gender-based violence. The forms of violence can include any harmful act that is perpetrated against a person's will including threats of such acts, coercion or arbitrary deprivation of liberty, whether occurring in public or in private life. The mention of private lives is to signal that gender-based violence also includes domestic and intimate partner violence rather than solely through external victimisation in public. Furthermore, men can also be targeted with gender-based violence, although statistically, the number of such cases is much smaller, in comparison with women, but it should not be neglected. Few studies of gay populations have been sufficiently large to yield useful information regarding the incidence or prevalence of anti-gay victimisation, but in one study (Huebner et al., 2004), a third of gay men reported some form of violence or other criminal activity directed at them due to their sexual orientation. Gay men are also far less likely to report incidents to the police than lesbians, so it may well be higher.

For women and girls, Pain (1997) describes how the fear of violent crime can have profound effects on their lifestyle, mobility and behaviour. It can affect countless decisions during the day, including whether a woman chooses whether to go out or to go out alone and whether to avoid certain streets and certain types of transport. It can also extend to specific sites including spaces of leisure and social activities. Each of these micro-decisions map onto certain times of the day and are never fixed but can be fluid and based on fleeting moments and recollections. Moreover, fear affects women differently across social identifiers such as class, race, sexuality, ability and age. Taken together, this fear has been understood as a central dimension of women's subordination. Gender-based violence against women is argued to be a legacy of patriarchy (Day, 2001), where traditional notions about women's roles and the places which are considered appropriate for them are reproduced (i.e. outside of the public sphere).

Pain urges us to think critically about the social construction of space into 'safe' and 'dangerous' places. There have been various macro mapping approaches and public engagement efforts to 'design out' crime and fear. In one empirical study based in Toronto, Kern (2010) explores how property developers recognise that 'the social imaginary of urban fear' potentially inhibits their ability to sell urban property to certain groups, including single women, the elderly, and families. While Toronto's image has never been primarily defined by a 'scary city' rhetoric, and the city itself is experienced by many as a safe, clean and friendly place to live and visit, Kern notes the increasingly panicked discourse that had arisen around public violence, which the media dubbed 'the summer of the gun' in 2005. Within this context, new spaces must be 'purified' and maintained as clean, safe and attractive to make the city more inviting, echoing themes we explored in Chapter 11. The security features of condominiums formed a key part of a marketing strategy to help the middle classes manage the perceived risks of city living. For most of the women in the study, the security features were important in their daily lives. Security helped these women to feel comfortable in their homes, attenuating the perceived risks, at least in the private realm.

Another tactic to tackle gender-based violence against women has been through campaigns that advocate that women travel together at night to avoid potential violence. These campaigns have been criticised for focusing attention away from the social and political causes of crime (Valentine, 1992) and placing responsibility on women. Efforts to 'reclaim the Night' have been a vocal response to gender-based violence since the mid- to late 1970s around the world, including Philadelphia (1975), Leeds (1977), Soho, London (1978) and Perth and Sydney (1978). Typically, these have emerged as a response to violent crime (e.g. 'Yorkshire Ripper' murders around

Leeds), either directly or to the police responses which instruct women to stay out of public spaces after dark. These marches have persisted to this day and have extended to the Global South, with marches in India in 2017.

While these marches have made public the subordination of women in public life, it is important not to forget that the home remains the location of most incidents of violence against women and gay men. This means that the primary focus on 'hate crime' and whether 'misogyny' or 'anti-gay' rhetoric is included in government policy is somewhat mis-weighted. While this seems motivated by well-meaning, Vera-Gray and Fileboard (2023), argue that this would merely serve to acknowledge 'the tip of the iceberg'. They contend that promoting hate crime legislation merely serves to undermine the extent of the problem and fails to address the prevalence of the more mundane, everyday forms of harassment many women and gay men experience within the home.

Meanwhile, spaces shared and reappropriated by LGBTQ+ communities across the globe known as 'Queer spaces' or 'gay villages' are perceived to constitute a safe haven at the margins of society for queer people (Hubbard, 2012). Examples include Key West, Tel Aviv, Toronto, Brighton & Hove and the Castro District, San Francisco. The latter became known as America's first and best-known gay neighbourhood. Its origins can be traced back to the Second World War when gay people and lesbians serving in the armed forces were often discharged from military service in San Francisco. Yet, Hubbard argues that this inclusion in these 'queer spaces' is not automatic, and must be negotiated. Indeed, these spaces may be unwelcoming for gender variant and transgender people, as shown in Browne and Lim's (2010) research in Brighton, 'the gay capital of the UK'. They suggest that trans lives are both excluded from and inextricably linked to geographical imaginings of these 'gay capitals'. This can give rise to feelings of having no place to be – a theme we pick up again in Section 12.4. For now, we turn to explore the contested geographies of masculinity.

12.3 Hegemonic and Contested Geographies of Masculinity

In contrast to the long standing geographical interest in feminist geographies since the 1970s – which includes critiques of hegemonic masculinities which underpinned patriarchal gender norms – research on 'the experience of men *as men* has scarcely yet been addressed' (Jackson, 1991: 209). Since Peter Jackson's initial observation, this topic has evolved within geography but remained as a relatively small field. A few reviews have sought to collectively map out the emergence of this sub-field (for a fuller historical account of this scholarship, see Hopkins and Gorman-Murray, 2019). This geographic interest contrasts with its more extensive coverage as a subject of enquiry within the social sciences and humanities. Its coverage in the geographies of social justice has been even less evident, most likely due to experiences of masculinity rarely being considered as a form of cultural imperialism and violence. Much of this research has sought to examine the privileged status of hegemonic masculinities or indeed White (male) supremacy and how this has evolved over time, as we saw earlier in the workplace. Hegemony, according to Connell (1995) is likely to be embodied through having a culturally ideal body type and institutional power. However, we recognise that 'hegemonic' is not a fixed character type, always and everywhere the same. Rather, it varies by place, based on local patterns of gender relations and is always contestable.

One area of common ground among scholars of social geography has been the acknowledgement of men's intersectionality with marginalised male identities. Indeed, we have already made the argument that certain social problems – namely mental ill-health, addiction and homelessness – are strongly associated with a certain failed masculinity, of becoming (through disability or addiction in particular) marginal in society, outside of the labour market entirely, and therefore, missing a key element of what it means to be a 'breadwinner' (see Chapters 1 and 10 on the strongly masculine nature of social problems). The study of masculinities, as with feminism, has been shaped by a greater engagement with the ways in which gender intersects with other social categories – in particular, those associated with racial and cultural diversity – as well as broader questions around intersectionality and class (Hopkins and Gorman-Murray, 2019).

One common intersectional strand has been a focus on masculine youth and working-class identity, which was initially explored in Chapter 6 on youth. With the decline of traditional employment opportunities for young working-class men, McDowell (2005: 25) shows that issues relating to 'weight, style, clothes, and accent' have increased in significance, as has the requirement of these young men to 'learn to serve' customers politely and empathically. In a sense, the distributional strand of social justice has shifted to the recognitional. These changes have all meant that young men have been forced to think carefully about how they present their bodies and how they construct their masculine identities. Since this initial work, McDowell returns to these communities in the wake of austerity to find that precarity has also loomed large in young working-class men's lives.

This research recognises that there are competing masculinities which can also be structured in varying degrees of dominance. Connell (1995) proposes three (competing) forms of masculinities: subordinate, complicit, and marginal, with each defined in terms of their relative positions of power. *Subordinate* refers to the political and cultural exclusion some men face, particularly gay men. This can extend to cultural abuse (in the United States gay men have now become the main symbolic target of the religious right), legal violence (such as imprisonment under sodomy statutes, still prevalent in African and Middle Eastern nations), street violence (ranging from intimidation to murder), economic discrimination and personal boycotts. *Complicit* refers to the problem that most men do not embody or rigorously practise the hegemonic pattern in its entirety, yet gain from its hegemony, since they benefit from the patriarchal dividend. This is an advantage that men in general gain from the overall subordination of women. Marginal refers to masculine identities that may be exemplars for hegemonic masculinity but are at the intersection of other structures such as class and race, which creates marginalised positions. While the idea of competing masculinities is welcomed, Hopkins and Gorman-Murray (2019) point out that it is important not to reduce the complexity and nuances of what men actually do. Rather, masculinities are constituted by an amalgam of practices, values and meanings which are realised in particular places and different contexts. So, while some men may feel subordinate in certain wider publics, they may be complicit with gendered hierarchies within their own families or within their own communities.

The nature and meaning of male embodiment can also play into specific men's fears. Longhurst (2005) for example explores about the marginalised experiences of men who have so-called 'man-breasts'. She notes how these men can feel uncomfortable in many public places (swimming pools, beaches, changing rooms) and feel subject to derision; simultaneously, men's lifestyle magazines promote a male body that is muscular and tough, marginalising the types of men who deem that their bodies fail to match up to these norms. Such changing demands have led to a 'rise in gym culture, fitness and the pursuit of the buff, toned body' (Tanner, Maher and Fraser 2013: 67), and a situation where 'men now have to work at being men' (68). As Hopkins and Gorman-Murray (2019) contend, this work is interesting because it represents a shift away from solely masculinities considered minorities, like gay men or Black men, and towards so-called mainstream White men. As we have highlighted elsewhere (e.g. in Chapter 11 on the 'squeezed middle' and Chapter 8 on White communities 'left behind'), we can also study, in interesting ways, what's going on in how masculinities are constructed in the mainstream, not just at the margins.

Returning to the wider social and economic imperatives that inform labour market participation within post-industrial economies (explored in Chapter 2), these have prompted younger generations of men to be more reflexive about their bodily presentation. This has led to debates about the crisis of masculinity and changing models of acceptable masculinities. Two conflicting manifestations have emerged: a softer and more emotionally astute masculinity and a more toxic, libertarian, far-right, populist rendering. In the first manifestation, men are engaging to a greater extent in the everyday tasks of social reproduction, including the increased involvement in care giving in domestic/familial environments (Boyer et al., 2017). Boyer reminds us that the proportion of men relative to women who carry most social reproduction remains low – but is rising. Given this rise, Boyer calls for greater attention from geographers to understand the factors, motivations and institutions that facilitate and constrain men's involvement – in the context of stubborn gender inequalities.

Building on this, Bourantani (2018) addresses three spaces of encounter which she describes as obstacles for men who take on primary caring work. First, men carers need to confront and mediate common assumptions about work-based masculinities, based on the strong association between masculinities and paid work, and the neoliberal work ethic, and to carve out their own spaces that allow them to prioritise and value childcare. Second, men carers need to challenge binary gendered distinctions in parenting to allow their male identities to emerge as part of their embodied and intimate care work, as something distinct from mothering. Third, men need to navigate childcare places that are deeply feminised; these encounters prove to be both a blessing and a curse, as men are received with overwhelming praise and, at the same time, treated as 'aliens'. The lack of connection to both mothers and other fathers is indicative of their in-betweeness, of their liminal position that challenges binary spaces and urges them to create new ones.

Men have also sought to navigate more emotional entanglements alongside more 'blokey' masculinities. The latter term 'blokey' refers to a masculine archetypal identity particularly in Australia, to describe an 'ordinary', strong, White and healthy male that's good for a laugh. Warren (2016), for instance, examines the masculinities of male surfboard makers (a particularly White industry and sport). Within this particularly niche market in the geographies of production, men weave between masculine work cultures and 'blokey' pretensions, such as exhibiting physical strength, being self-reliant and stoic, along with a more personally attuned, emotionally rich, passionate and 'soulful' craft-worker identity. This demonstrates how waged work can comprise embodying a blend of masculinities (that may sometimes be complicit in terms of playing it both ways), shaped by working cultures, relations and labour processes.

Building on the theme of emotional masculinities, this can also extend to men's embodied and emotional experiences of religion, faith and spirituality. The relational gendering of Muslim identities has already been covered in Chapter 6. Beyond this, Hopkins and Gorman-Murray (2019) describe how Christian men are more likely than women to claim that none of their male friends are religious, thereby suggesting that faith is more privatised for men than it is for women. From a social justice perspective, this point is important for considering the gendered framing around how men might make sense of or act upon the injustices facing them or in the world around them.

In the second manifestation, involving more far-right masculinities, Luger (2022) explores the geographies of *toxic masculinity*, specifically within far-right White male communities in the US and UK. He acknowledges that most men in gyms are not extremists, nor are they necessarily right-wing. However, he explores how one of the key facets of the new far-right in these countries has been the positioning of the strong, muscular, usually White, male body, as a crucial aspect of identity, culture and ideology. Luger draws on the example of the male bodybuilding community from various gym settings in the US and

UK, and identifies how it was largely united against COVID-19 related public health measures (like masks, gym closures and vaccines). It also leaned toward far-right libertarian ideals and sought to construct notions of 'freedom' and the 'lone wolf'. This is contrasted alongside efforts to demasculinise leftists, portrayed as 'soft' men (with descriptors like 'woke' and 'modernist'). This is reinforced by 'heroic' masculine media influencers, like Joe Rogan, Andrew Tate, Russell Brand and Jordan Peterson, who cultivate a brand of hyper-masculine 'gang masculinity'. This serves to escalate and police boundaries between 'masculine' men and other perceived masculinities. This trend can be partly understood as a symptom of global neoliberal structures, where men feel threatened by a perception that western culture is increasingly 'anti-men' and feel the 'elites' want to push women into the job market at any cost.

The populist radical right in the Netherlands, explored by Linders et al. (2023), serves as another example of how stronger, more determined and more audacious masculinities are juxtaposed against the threat of 'elites', scientists, the media, mainstream politicians and immigrants. This also plays out in men seeking to police borders and nationhood. We should also underline that toxic masculinity is not always perpetrated by well-known figures, Hollywood gurus, pop stars and such – it is also found at an everyday street level (Wilton et al., 2014).

Toxic masculinity can also be understood relationally, specifically the extent to which men are connected to wider social communities, how porous these communities are and where these communities meet and spend time. During the pandemic and since, a trend has emerged for young men to spend inordinate amounts of time online – playing video games, posting on social media, gambling, but also being exposed to (or seeking out) more radical messaging around Incel (involuntary celibate), misogynistic or racist subcultures. This dropping out of mainstream, in-person society is potentially troubling. As we have seen with social problems, young men who feel excluded tend to lash out or self-destruct, which harms society as a whole. As Luger (2023: 56) argues, these virtual online constructions are spatialised through produced territories and local community infrastructures (homes, churches, schools): 'The patriot-bro, in other words, moves from online personality to spatial fixture, with implications (and perils) for democracy and community cohesion'.

Relatedly, the theme of religious engagement explored above can also play out in more toxic ways. Returning to Luger's work, he explores the link between White, far-right masculinism and muscular Christianity, especially in the US, where a proliferation of non-traditional churches (e.g. in US malls) have reinforced 'strong' male body cultures in their religious rhetoric and imagery. He points to a growing synergism between right-libertarianism and fitness and wellness cultures, which he characterises as Christianism. This is reinforced by messages that the LGBTQ+ community is not welcome within these religious communities, with one church's website stating plainly how: 'Marriage is a sacred covenant made between one man and one woman, before God'. Given the degree of variation in the gendered experiences of men outlined in this section, largely due to the intersectionality of different identities, embodied norms, and positions, the gendered geographies of social justice for men are far from straight forward. The various contact zones are embedded in people's relational networks which are policed through enforced norms for male communities.

12.4 Gendered and Sexed Geographies Beyond Binaries

While the sections above consider the various social geographies of women and men, and intersected with different sexual orientations, as noted in our introduction, Judith Butler's (1990) book *Gender Trouble* provides a call to scholars to disrupt the accepted 'naturalness' of the binary man-masculinity and woman-femininity. Since this publication, geographers have been 'troubling' gender (see a themed issue in *Gender, Place and Culture* entitled 'Towards trans geographies'; Browne et al., 2010) and have recognised and sought to challenge the high levels of violence, social stigma and social injustice attached to people whose gender is understood as non-normative. Hostile reactions towards 'gender transgressions' – where people are perceived by others not to conform to their expected genders – bring into sharp relief the performative and material consequences of binary gender norms. Such transgressions can become acute or heightened in certain places, illustrated by the 'bathroom problem' (Browne, 2006) where women are mistaken as men and face hostile reactions. Browne et al. (2010) echo this call and seek to push for new and innovative understandings not only of the spatialising of gender and the creation of gender through socio-spatial relations, but also to more fully engage with the challenges and resistances trans people experience in the spaces and places they use, create and reject. This scholarship is therefore concerned with the 'real', often marginalised materialities of bodies that are lived and felt by non-binary people. This mirrors the emancipatory lens found elsewhere in this book.

Rather than being fixed, gender terms are reflective of their personal, geographical, historical and disciplinary context. Multiple genders, and terms that trouble gender binaries, exist. These include numerous gender minorities, for example, gender diverse, transsexual, transvestite, cross-dresser, genderqueer, non-binary, gender fluid, neutrois, demiboy, agender, non-gendered, third gender, trans woman, trans man, drag king and drag queen (Johnston, 2016). Moving beyond Anglo countries, the breadth of gender terms widens again, including Two-Spirit in many Indigenous cultures to encompass those who embody both masculine and feminine spirits, and third gender concepts in many countries (Muxe in Mexico; Hijra in India), as local associations have evolved over time.

The term 'transgender' or 'trans' is usually understood – at least in the Global North – as an umbrella term to refer to people who move away from the gender they were assigned at birth (Stryker, 2008). Some trans people though may believe in a binary conception of gender based on their perception that their identities are framed by a change from one sex to the other. Here, gender is seen as binary, where individuals see themselves as merely moving from one binary category to another. By contrast, many align with a non-binary or queer subjectivity, based on the belief in multiple genders that do not adhere to a biological body. This is typically joined to a commitment to breaking down any binary thinking over gender, especially that would render any group as inferior to another (Miller, 2021). A person may identify as 'genderqueer' when they do not align with normative or conventional binary gender identities (Nestle et al., 2002).

Irrespective of these differences, for trans people, access to their preferred gendered identities is often fraught. Trans people are often subject to discriminatory attitudes and sometimes violence. They are commonly bullied at school, denied access to employment, sometimes rejected by family members, and pushed out onto the streets. As noted elsewhere, intersectionality is a factor, as the fear of violence has for a long-time closely followed lines of disadvantage in society (see, for example, Crawford et al. 1990), based on social, political and economic inequalities. Pain (1997) argues that discrimination, isolation or a lack of social acceptance follows those who feel a lack of integration into their neighbourhoods and those who have little control over resources.

People who transgress gender norms can experience discriminatory attitudes that shape nearly every aspect of their public and private lives. Drawing on ethnographic research, Doan (2010) traces the various geographies in a continuum of public to private spaces, including: parking lots, public restrooms, shopping malls, the workplace and the home. Across these various places, people experience various relational encounters with their family, local community, work colleagues and wider publics. At times, hostile encounters may be fleeting, and yet can mirror the harmful experiences felt by other groups explored in this book.

Like these other marginalised groups, trans people often adapt to wider hostility through occupying 'safe havens' which are spaces they may develop a sense of belonging within broader contexts of hostility (see Chapter 9 for the parallel example of disabled people). Moving beyond physical encounters, Rooke (2010) highlights the importance of trans virtual cultural spaces as sites for the development of trans young peoples' ontological security and their navigations and negotiations of a gendered social world. More on these safe spaces will be taken up in Chapter 13, as quiet forms of activism but also survival and regrouping. However, as noted above, 'gay villages' may be unwelcoming for gender variant and transgender people (Browne and Lim, 2010). Not 'fitting in' is illustrated in the high levels of homelessness for transgender youth (Nash, 2010), and this has been noted even in gay neighbourhoods such as The Castro in San Francisco and Toronto's gay village.

Here, the spaces of encounter are multi-layered. Complex sexual and gender solidarities are simultaneously created and contested within and

FIGURE 12.1 Pride, Rennes (France), 2024 © G DeVerteuil

across trans and LGB communities, where people of different genders and sexual orientations navigate an ambivalent space. In this way, Browne and Lim (2010) signpost the paradoxes of the hopes and solidarities that co-exist – and are held in tension – with experiences of marginalisation. To this end, Doan (2010) calls out how, for both trans and LGB communities, gender norms produce gender oppression. Doan argues that the strongly gendered dimensions of these spaces suggests that a discursive revisioning of gender is needed to create more transgender friendly urban spaces. Pride events, part of 'Pride month' (usually June or July) have played an important role propagating a more varied and extensive gay geography, and more varied spaces of encounter (Figure 12.1 above). Consider the Engagement Questions below about its role.

Box 12.2 Engagement Questions

How does 'Pride Month' work to create a more varied and extensive gay geography, and more varied spaces of encounter?

The Pride movement has sought to advance the recognition of LGBTQ+ people by creating more inclusive spaces of celebration. This movement has also been motivated by efforts to *maintain* progressive gains, and to ensure these are not eroded by oppression. Moreover, Pride is about *not* being invisible in society. International examples serve as reminders of how political regimes can undermine previous advances. In Hungary, for example, Prime Minister Viktor Orbán led an effort in April and May 2020 to end legal recognition of trans people. These experiences point to the need to explore a more particular historical and political conceptualisation concerning the formations of gendered subjectivities, identities and forms of embodiment in these spaces.

12.5 Conclusions

This chapter provides a starting point to understanding the wide breadth of social geographical scholarship exploring gendered geographies, as well as a way to connect to and gather up previous material in the first 11 chapters on gender and social geography. It also explores the close association with people's experience of their sexual orientation. As such, it offers a reflection of the evolution of these often intertwining sub-fields, from the first wave of feminist scholarship challenging the marginalisation of women, and with (gay or straight) men and non-binary people being relatively late as subjects of enquiry. We primarily focused on the spaces of encounter with the wider publics that shape norms and expectations and determine interactions, including the geographies of production and social reproduction, which frame them. These encounters often encompass attitudinal barriers from wider mainstream patriarchal and heteronormative gender norms.

The various encounters based on gender and sexual orientation that people experience were acutely shaped by specific historical and geographical contexts, and over time become codified in statutory regulatory framings. According to Brown, Nash and Hines (2010), the political, social and economic constitutions of place offer different historical contexts in the practising of gender. In non-binary and trans scholarship, these experiences have been particularly focused on urban metropolitan and Anglo-American contexts, with researchers drawn to 'LGBTQ+ friendly' cities.

However, as we saw, these spaces of encounter are complex and at times contested between different non-hegemonic gender groups. These spaces within and across different gendered communities have therefore been ambivalent. We also acknowledge the way that various struggles for social justice based on gender identities are commonly fragmented, held occasionally in solidarity and frequently in tension, as different calls for rights sometimes impinge on others. Consequently, complex solidarities based on gender and sexuality are simultaneously created and contested, with overlapping and sometimes complicit hegemonies. In this way, Browne and Lim (2010) signpost the paradoxes of the hopes and solidarities that co-exist – and are held in tension – with experiences of marginalisation. This chapter brings Part II to a close, and we pick up on this theme of hope in Part III.

References

Barrett, C. (2015) Queering the home: The domestic labour of LGBTQ couples in contemporary England. *University of Southampton, School of Geography, Doctoral Thesis*.

Bell, D. & Valentine, G. (1995) *Mapping Desire*. London: Routledge.

Berg, L. and Longhurst, R. (2003) Placing masculinities and geography. *Gender, Place and Culture*, 10(4), 351–360. doi: 10.1080/0966369032000153322.

Berger, S. and Piore, M. (1980) *Dualism and Discontinuity Industrial Societies*. Cambridge: Cambridge University Press.

Bourantani, E.A. (2018) Re-gendering Care in the UK: The Experiences of Male Primary Carers, PhD Thesis, University of Southampton. https://eprints.soton.ac.uk/419533/1/Eleni_Bourantani_PhD_thesis_final_copy.pdf

Bowlby, S.R., Foord, J. and Mackenzie, S. (1982) Feminism and geography. *Area*, 14, 19–25.

Boyer, K. (2014) 'Neoliberal motherhood': Workplace lactation and changing conceptions of working motherhood in the contemporary US. *Feminist Theory*, 15(3), 269–288. https://doi.org/10.1177/1464700114545321

Boyer, K., Dermott, E., James, A. and MacLeavy, J. (2017) Regendering care in the aftermath of recession? *Dialogues in Human Geography*, 7, 56–73.

Breen, R. and Cooke, L.P. (2005) The persistence of the gendered division of domestic labour. *European Sociological Review,* 21(1), 43–57.

Browne, K. (2006) 'A right geezer-bird' (man-woman): The sites and sights of 'female' embodiment. *ACME: An International E-Journal for Critical Geographies*, 5(2), 121–143.

Browne, K. and Lim, J. (2010) Trans lives in the 'gay capital of the UK'. *Gender, Place and Culture*, 17(5), 615–633.

Browne, K., Lim, J. and Brown, G. (2007) *Geographies of Sexualities: Theory, Practices and Politics.* London: Routledge.

Browne, K., Nash, C.J. and Hines, S. (2010) Introduction: Towards trans geographies. *Gender, Place & Culture*, 17(5), 573–577.

Butler, J. (1990) *Gender Trouble*. New York: Routledge.

Claudia G., Lawrence F.K., and Ilyana K. (2006) The Homecoming of American College Women: The Reversal of the College Gender Gap. *Journal of Economic Perspectives*, 20(4), 133–156. Fall 2006.

Connell, R. (1995) *Masculinities*. Oxford: Blackwell.

Cortés, P. P. (2023) Sexuality and migration in the Global South: An overview. GDI Working Paper 2023-062. Manchester: The University of Manchester.

Crawford A., Jones T., Woodhouse T. and Young J. (1990) The second Islington Crime Survey Centre for Criminology, Middlesex Polytechnic, Enfield, Middlesex.

Day, K. (2001) Constructing masculinity and women's fear in public space in Irvine, California. *Gender, Place and Culture*, 8, 109–127.

Doan, P.L. (2010) The tyranny of gendered spaces: Living beyond the gender dichotomy. *Gender, Place and Culture*, 17, 635–654.

Drydakis, N. (2015) Sexual Orientation Discrimination in the United Kingdom's Labour Market: A Field Experiment. IZA Discussion Paper No. 8741, Available at SSRN: https://ssrn.com/abstract=2544805

EHRC (Equality and Human Rights Commission) (2011) *Sex and Power 2011*. London: EHRC.

Fagan, C. and O'Reilly, J. (1998) *Part-Time Prospects: An International Comparison*. London: Routledge.

Hall, S.M. (2023) A Pregnant Pause? Reproduction, waiting and silences in the relational endurance of austerity, *Geoforum*, 142, 103755.

Hines, S. (2010) Queerly situated? Exploring negotiations of trans queer subjectivities at work and within community spaces in the UK. *Gender, Place and Culture*, 17(5), 597–613.

Hird, M. (2003) Considerations for a psychoanalytic theory of gender identity and sexual desire: The case of intersex. *Signs*, 28(4), 1067–1092.

Hopkins, P. and Gorman-Murray, A. (2019) Masculinities and geography, moving forward: Men's bodies, emotions and spiritualities. *Gender, Place & Culture*, 26(3), 301–314. http://doi.org/10.1080/0966369X.2018.1552560

Hopkins, P. and Noble, G. (2009) Masculinities in place: Situated identities, relations and intersectionality. *Social & Cultural Geography*, 10(8), 811–819.

Hubbard, P. (2012) *Sexualities and Cities*. London: Routledge.

Huebner D.M., Rebchook G.M. Kegeles S.M. (2004) Experiences of harassment, discrimination, and physical violence among young gay and bisexual men. *American Journal of Public Health*, 94(7), 1200–1203. doi: 10.2105/ajph.94.7.1200. PMID: 15226143; PMCID: PMC1448421.

Jackson, P. (1991) The cultural politics of masculinity: Towards a social geography. *Transactions of the Institute of British Geographers*, 16, 199–213.

Johnston, L. (2016) Gender and sexuality I: Genderqueer geographies? *Progress in Human Geography*, 40(5), 668–678. https://doi.org/10.1177/0309132515592109

Jupp, E. (2023) *Care, Crisis and Activism: The Politics of Everyday Life*. Bristol: Bristol University Press.

Kern, L. (2010) Selling the 'scary city': Gendering freedom, fear and condominium development in the neoliberal city. *Social & Cultural Geography*, 11(3), 209–230. https://doi.org/10.1080/14649361003637174

Linders, N., Dudink, S. and Spierings, N. (2023) Masculinity and sexuality in populist radical right leadership. *Politics & Gender*, 19(3), 653–674. doi: 10.1017/S1743923X22000265

Longhurst, R. (2005) 'Man-breasts': Spaces of sexual difference, fluidity and abjection. In B. van Hoven and K. Hörschelmann (Eds.), *Spaces of Masculinities* (pp. 165–178). London: Routledge.

Longhurst, R. (2008) *Maternities: Gender, Bodies, Space*. London: Routledge.

Longva, P. and Strøm, S. (2002) Wage differentials and gender in Norway. In C. Jonung and I. Persson (Eds.) *Women's Work and Wages* (pp. 1–17). London: Taylor & Francis.

Luger, J. (2022) Celebrations, exaltations and alpha lands: Everyday geographies of the far-right. *Political Geography*, 96, 102604.

Luger, J. (2023) 'Patriot bros', the fascist creep, and the spatial fantasies of white nationalist masculinity. In J. Mercer and M. McGlashan (Eds.), *Toxic Masculinity: Men, Meaning, and Digital Media* (1st ed., pp. 56–71). (Masculinity, Sex and Popular Culture). London: Taylor & Francis. https://doi.org/10.4324/9781003263883-5

McCann, M.W. (1994) *Rights At Work: Pay Equity Reform and the Politics of Legal Mobilization*. Chicago: University of Chicago Press.

McDowell, L. (1991) 'Life without Father and Ford: The New Gender Order of Post-Fordism'. *Transactions of the Institute of British Geographers*, 16(4), 400–419. *JSTOR*, https://doi.org/10.2307/623027

McDowell, L. (1995) Body work: Heterosexual gender performances in city workplaces. In D. Bell and G. Valentine (Eds.), *Mapping Desire* (pp. 75–95). London: Routledge.

McDowell, L. (1997) *Capital Culture: Gender at Work in the City*. Malden: Wiley-Blackwell.

McDowell, L. (2000) Learning to serve? Employment aspirations and attitudes of young working-class men in an era of labour market restructuring. *Gender, Place and Culture*, 7(4), 389–416.

McDowell, L. (2003) *Redundant Masculinities? Employment Change and White Working Class Youth*. Oxford: Wiley-Blackwell.

McDowell, L. (2005) The men and the boys: Bankers, burger makers and barmen. In B. van Hoven and K. Horschelmann (Eds.), *Spaces of Masculinities* (pp. 19–30). London: Routledge.

McDowell, L. (2014). Gender, work, employment and society: Feminist reflections on continuity and change. *Work, Employment and Society*, *28*(5), 825–837. https://doi.org/10.1177/0950017014543301 (Original work published 2014).

McDowell, L. and Sharp, J. (1997) *Space, Gender, Knowledge: Feminist Readings*. London: Arnold.

McKie, L., Bowlby, S. and Gregory, S. (2001) Gender, caring and employment in Britain. *Journal of Social Policy*, 30, 233–258.

Mehra, R. (2021) What works for low income working women: Review of evidence, Co-Impact Research Working Paper. https://co-impact.org/wp-content/uploads/2021/06/Evidence-Full-Women-ECON.pdf

Miller, L. (2021) Queer versus Trans: A false binary. *QED: A Journal in GLBTQ Worldmaking*, 8(2), 87–95. https://muse.jhu.edu/article/851614

Namaste, K. (1996) Genderbashing: Sexuality, gender and the regulation of public space. *Environment and Planning D: Society and Space,* 14, 221–240.

Nash, C.J. (2010) Queer conversations: Old-time lesbians, transmen, and the politics of queer research. In K. Browne and C.J. Nash (Eds.), *Queer Methods and Methodologies: Intersecting Queer Theories and Social Science Research* (pp. 129–142). Burlington, VT: Ashgate.

Nayak, A. (2006) Displaced masculinities: Chavs, Youth and Class in the Post-Industrial City. *Sociology*, 40(5), 813–831.

Nestle, J., Howell, C. and Wilchins, R. editors (2002) *GenderQueer: Voices From Beyond the Sexual Binary*. Los Angeles: Alyson Books.

Pain, R. (1991) Space, sexual violence and social control: Integrating geographical and feminist analyses of women's fear of crime. *Progress in Human Geography*, 15, 415–431.

Pain, R. H. (1997). Social geographies of women's fear of crime. *Transactions of the Institute of British Geographers*, 22(2), 231–244. www.jstor.org/stable/622311

Rachel, H. (1997) Pain social geographies of women's fear of crime. *Transactions of the Institute of British Geographers*, 22(2), 231–244.

Richard, T. (2006) Ely lecture the quiet revolution that transformed women's employment, education, and family by Claudia Goldin. AEA Papers and Proceedings, MAY 2006.

Rooke, A. (2010) Trans youth, science and art: Creating (trans) gendered space. *Gender, Place and Culture*, 17(5), 655–657.

Rose, G. (1993) *Feminism and Geography. The Limits of Geographical Knowledge*. Cambridge, UK: Polity.

Rose, G. (1995) Tradition and paternity: Same difference? *Transactions of the Institute of British Geographers*, 20, 414–416.

Scharff, C. (2016) *Repudiating Feminism: Young Women in a Neoliberal World*. Routledge.

Stryker, S. (2008) *Transgender History*. Berkeley, CA: Seal Press, 1.

Tanner, C., Maher, J.M. and Fraser, S. (2013) *Vanity: 21st Century Selves*. New York, Palgrave Macmillan.

Tivers', J. (1978) How the other half lives. *Area*, 10(4), 302–306.

Valentine, G. (1992) Images of danger: Women's sources of information about the spatial distribution of male violence. *Area*, 22-29.

Vera-Gray, F. and Fileboard, B. (2023) Hiding the harm? An argument against Misogyny hate crime. In Hall, E., Clayton, J. and Donovan, C. (Eds.), *Landscapes of Hate, Tracing Spaces, Relations.* Bristol: Bristol University Press.

Warren, A. (2016) Crafting masculinities: Gender, culture and emotion at work in the surfboard industry. *Gender, Place & Culture*, *23*(1), 36-54.

Williams, E.L. (2013) Sex work and exclusion in the tourist districts of Salvador, Brazil. *Gender, Place and Culture*, 21(4), 453–470.

Wilton, R., DeVerteuil, G. and Evans, J. (2014) 'No more of this macho bullshit': Drug treatment, place and the reworking of masculinity. *Transactions of the Institute of British Geographers,* 39(2), 291–303.

Women and Geography Study Group (2004) *Gender and Geography Reconsidered*. Glasgow: Women and Geography Study Group.

Young, I.M. (1990) *Justice and the Politics of Difference*. New York: Princeton University Press.

PART III

GEOGRAPHIES OF HOPE FOR SOCIAL JUSTICE

If you have come this far, there is a bit more to go – and after the somewhat gloomy material around inequality, social injustice and the various social problems that assail the downtrodden, we can now safely proceed to higher ground in the final part of the textbook. We use the next two chapters to advocate for alternative, more progressive ways of understanding the cross-cutting threads (relationality, encounter), the core social identifiers already considered in the previous chapters (dis/ability, race, age, gender, class), as well as the three key themes of social geography: the social, inequality and especially social justice and its three strands (recognition, distribution and procedure).

For social justice in particular, we seek to reframe it for the emerging alternative and even utopian possibilities in the 21st century. This must be cast in terms of the more recent surge in interest in social and spatial justice in the early 2010s. For example, two well-established urbanists – Ed Soja from UCLA, and Susan Fainstein from Harvard University – both came out with books on socio-spatial justice in 2010: *Seeking Spatial Justice* and *The Just City* respectively. Taken together, these books were at the forefront of a movement to place socio-spatial justice at the core of understanding (urban) space within a period of profound and growing inequality. This resonated

DOI: 10.4324/9781003414889-15

with the rise of 'critical geography', which calls for both interpreting injustice but also challenging it. The 2010s also featured an extravagant amount of social justice agitation, whether by social media (#) or in-person via protest movements. Across the globe, protesters flooded city streets, camped out, occupied, but also lobbied and hash-tagged. In some places – Egypt, Sri Lanka, Ukraine – rulers were toppled. In other places, and abetted by social media, the social justice agenda became more entrenched, bleeding into politics at the national levels – Canada, Germany, New Zealand. The 2008 global economic crisis and the 2020 pandemic added fuel to the fire, heightening the sense that capitalism was only working for a small (and increasingly smaller) group. Even more pressing are the calls for environmental justice, within a larger context of climate change that adversely and disproportionately impacts more vulnerable groups. This was entirely evident in the wake of Katrina in New Orleans (2005).

The protest groups that emerged during this tumultuous decade included Occupy, Black Lives Matter, #MeToo and Extinction Rebellion. To some, social justice can only be achieved by contesting and appropriating space, which empowers activists to push for social and political change via visibility. The ability to control and marshal resources within a territory, and to regulate the flow of resources across borders, is also of critical importance to social movements. This is especially true if they want to move beyond the local scale. In the case of the Occupy Movement, the taking of space was its defining act, and it has become a central feature of diverse uprisings around the world in the 2010s and beyond, including the Arab Spring (2010–2012) but also protests against the war in Gaza in 2023+. Yet at the same time, there has been a sense of fatigue around the utility of constant and highly visible protests. Sometimes, people get involved in quieter forms of activism, and seek to operate more behind-the-scenes; strategies we will consider in Chapter 13.

13 GEOGRAPHIES OF ALTERNATIVE SPACES, RESILIENCE AND SLOW/QUIET ACTIVISM

13.1 Social Encounters with Quiet Activism

We understand that the world has developed in ways that are inequitable and unjust, marked by ableism, ethnocentrism, racism, sexism, ageism and classism, and these practices are intimately tied to our spatial world in ways that mean we are not always allowed to celebrate our differences positively. Rejecting the status quo involves exposing the marginalisations, exclusions and injustices of the world, deciphering how they are inscribed and naturalised within our existing societies but also with a view of strategising a more progressive, socially just, emancipatory and sustainable formations of social life.

This creation of alternatives is very much linked to the idea of hope, utopias and *otherwise worlds* (King et al., 2020; DeVerteuil and Kiener, 2022). For them, it is important to first imagine otherwise possibilities, so as to then create an otherwise world. In other words, creating alternatives requires utopian thinking to eventually bypass the dominant understandings of the (social) world entirely, rather than working within their structures. Utopian thinking is essential in foregrounding a better future, but one that cannot always be predicted!

How to get there though? To begin, we need to consider activism as a practice and an outcome, and always connected to social justice in its various strands (distribution, recognition, procedure) and inflections: contested; spatial; practical; and utopian; linked to social injustice, and especially, though not exclusively, urban. We acknowledge that different forms of activism do not fit neatly into discrete categories. Moreover, there is often a lot of overlap between different forms of activism, even within the same social movement. contemporary activism takes place across a spectrum ranging from 'inside the system' to 'outside the system'. Groups often utilise more than one tactic in an effort to address a particular personal or social concern. Sometimes, a form of resistance is only labelled as such after a particular action, for example when interpreted through a media or political lens to try and discredit and undermine the action. As we have considered the relational nature of power (Chapter 3) elsewhere in this book, we are also aware that for

DOI: 10.4324/9781003414889-16

every action, there are consequences in terms of a counter force, seeking to stymie or divert change, maintaining the status quo at all costs.

Another caveat is that some people choose not (or inadvertently fail) to recognise their actions are a form of activism, with some preferring other terms to describe their political positions and actions. Resistance seems to imply *intention* (Cresswell, 1992: 333): 'a purposeful action directed against some disliked entity or situation with the intention of changing it or lessening its effect. The resistance of an action therefore appears to be in the intention of the perpetrator, in the eagerness to overcome or change some obstacle'. But what constitutes meaningful and effective *intentional* action? A notable example of the difficulty in discerning this question is the first time the 'F-word' was used on television by The Sex Pistols in an interview on British Television in 1976. This caused a media-fuelled moral panic at the time, and was interpreted as a form of symbolic resistance against 'decent' moral standards, embodied by the style and demeanour of the band in their interview. However, this was a spontaneous moment which unfolded in the live context of a challenging interview and was only defined as resistance after the fact (Mitchell, 2003).

The above provocation forces us to consider the question of what we actually might consider as forms of activism. Disability-justice scholars highlight that taking care of oneself or others through activism, mutual aid or family or other social and economic arrangements is always an act of resistance in neoliberal capitalism (Schalk, 2022). This broadening of the concept of activism has been driven by feminist scholars who have sought to demonstrate how small, modest everyday acts of kindness, kinship and creativity can transform social relations in ways that can foster social change (Abrahams, 1993), re-imagining everyday urban landscapes for more care-centred futures.

Building on these more modest forms of activism, some may feel that they are just 'getting on' or 'coping' with circumstances that they feel are unjust, rather than seeking to resist or transform these same circumstances. Others may feel that they are merely making small decisions to live alternatively to the mainstream or to utilise some resourcefulness to live with significant challenges. Pottinger (2017) suggests that these more modest acts can be regarded as a form of *quiet activism*, which she defines as small, every day, embodied acts, often of making and creating, that can be either implicitly or explicitly political in nature. These contrast with more resistant, unruly, unbowed, vocal, antagonistic and demonstrative forms of protest. We will cover these in Chapter 14, and which have sustained more attention, and arguably have been more romanticised, by geographers. So despite the overlapping nature of the myriad forms of activism, we explore the quiet and the vocal separately across two discrete chapters. We recognise that this is a largely superficial demarcation, and so we continue to emphasise how each form sits within a broader set of positions and practices, and how they can build on each other across the two chapters.

We draw obvious connections between alternatives to the current social world and the social geographies of resilience and reworking, as well as slow and quiet activism. These activities relate directly to social justice, especially the long-term, incremental and patient kind that aims to create alternatives to the current status quo. But we are interested in more than just social change, capturing the role of geography in the process. Indeed, social change must take place somewhere. In what follows, we outline four kinds of quiet, incremental and alternative place-making practices that aim for social justice: creating commons; counterpublics and counter-places; abolition geographies; and practicing resilience. The first (commons) aligns most closely to Cindi Katz's idea of reworking as it is about improving conditions within existing structures. Creating counterpublics and abolition geographies can both be understood as having a blend of reworking and resistance within Katz's framework, depending on the context and intent. Because they are quieter forms of resistance, we include them in this chapter, however as noted elsewhere, they can be linked to forms of protest. For the latter (resilience; Katz's third response), we explore how coping and waiting on change is a temporal practice. Yet, even with resilience, it can overlap with protest, in the form of the resilience of protesters, that is explored in Chapter 14. Thus, all practices can have different angles with very overlapping practices.

Each of these place-making and temporal strategies necessarily involve envisioning and realising otherwise worlds, utopian ideals and both anti-capitalist and alongside/beyond-capitalist practices. They also create somewhat different relationalities and encounters, to which we will return at the end of the chapter, ranging across all three strands of social justice – distribution, recognition and procedure. Finally, the four practices are very much part of the buffet approach that this textbook takes – we allow you as the student to choose what you think works best in your own worldview. It is up to you to choose how to engage with the world based on your personal beliefs. You may feel a kindred connection to one or more of these responses (or none!). However, we should caution about the mixing of the various practices. They do not necessarily always work well together. The insistent nature of abolition geographies, for instance, might not merge well with the more patient and perhaps timid nature of resilience and the commons. All the same, the very existence of these four place-making practices further points to the incomplete nature of capitalism, neoliberalism and the state-maintained status quo.

13.2 The Commons

The concept of the commons represents the promise and practice of 'life beyond marketization, privatization and commercialization' (Jeffrey, McFarlane and Vasudevan, 2012: 1249), creating and holding on to space that is 'both collective and non-commodified – off-limits to the logic of market exchange and market valuations' (Harvey, 2012: 73). The market logics are usually characterised as 'enclosures', the opposite of commons, seeking to close off and commodify spaces. The origin of the term 'enclosure' derives from the fencing off and marketisation of agricultural lands in the 18th and 19th centuries in many parts of the Global North. It has since become a term to characterise all types of encroachment via property, accumulation, appropriation and dispossession, all hallmarks of capitalism, not just in the countryside, but also in the city. As we saw in Chapter 11, equity firms can easily outbid individuals when buying homes, homes which are then rented back to individuals outpriced by the market. This is a clear example of enclosure by and for the private market. An even stronger example includes the proliferation of AirBnBs in many cities at the expense of other uses, such as affordable housing that is not designed to make a profit. By always prioritising exchange value over use value, neoliberalism operates as a mode of enclosure. Just the same, Harvey (2012: 80) claimed that although 'capitalist urbanization tends to destroy the city as social, political and liveable commons', commons are always being rebuilt within the in-between spaces of the city.

Conversely, an example of the commons is that of most public space (defined in Chapter 1, Section 2), such as green areas, sidewalks and streets and community gardens. Technically, these belong to everyone, and they are not easily commodifiable. The same can be said of social housing, but also of more radical examples of squatting in abandoned buildings. Alex Vasudevan (2017) talked about squatting (e.g. illegally trespassing and living in abandoned properties) as part of the larger commons of the 'autonomous city', where alternative forms of living exist. These forms of living are structured by the inhabitants' own rules, and usually revolve around sharing, solidarity and place-making. Squatting is also in direct response to the housing crisis outlined in Chapter 11, moving away from the state and the private market and towards a radical sense of community and civil society (Chapter 5). Of course, in the Global South this kind of place-making remains quite well-established with large informal settlements (e.g. Favelas in Rio De Janeiro; Dharavi in Mumbai), given the inability of the private and state sectors to provide affordable housing. But in the Global North, squatting as a form of commons remains rather tenuous.

From the squatting example, commons tend to provide public goods (via the state or the voluntary sector, or even self-provided), from health care to food to simply having a place to exist and interact without the spectre of profit-making. Community-based food aid is an example of a type of commons, focused on providing food security and access (and minimising the harsh effects of austerity on poor people), often in contrast to traditional state or market-driven systems. It emphasises collective action and mutual support in ensuring that people experiencing food insecurity have access to sufficient food. Food commons initiatives can include community gardens, food sharing networks, and local food production systems.

Another example, that relates to Andy's work is the peer-support networks which adults with intellectual disabilities co-create with local grassroots voluntary organisations. These can be considered a type of commons as they create spaces (e.g. in community centres, parks) where shared resources and knowledge are accessed and utilised by a community. More broadly, these share some characteristics with online forums (e.g. Mumsnet; subreddits) where individuals with shared experiences come together to support each other and sometimes offer their time to help others (e.g. pro bono legal work). A final example is housing co-operatives, where individuals or groups collectively manage and reclaim resources, including land, housing, knowledge and care, to meet their needs. This approach emphasises shared ownership or control, collective decision-making and a focus on community needs over individual profit.

Together, these types of commoning can halt or avoid entirely the forces of commodification – in the form of not just buying and selling land, but also not 'renting' green spaces and streets for profitable ends (e.g. festivals, Pride marches); more on public space as contested space in Chapter 14. Further, the commons tend to exist alongside, beyond or against capitalist relations, but not overturning them necessarily (DeVerteuil et al., 2022). Rather, they offer some relief from the neoliberal hyper-commodification of everything in the city, and by extension everything in society. They are less visible to the point of being overlooked even. And yet, they are omnipresent all across the globe.

At the very least, commons support subsistence outside of the property and commercial system. By not charging for certain life-sustaining services, the commons involve a commitment to everyday survival of marginal populations, such as service hubs in Chapter 8 and homeless individuals as discussed in Chapter 11. Commons can also sustain spaces of care (see Chapter 5). As Gibson-Graham, Cameron and Healy (2016: 195) noted, 'We characterize commoning as a relational process – or more often a struggle – of negotiating access, use, benefit, care, and responsibility. Commoning thus involves establishing rules or protocols of access and use, taking caring of and accepting responsibility for a resource, and distributing the benefits in ways that consider the wellbeing of others'. The commons, therefore, entail a sustained engagement to people and places, of caregiving and place-making. Finally, the commons enable spaces of solidarity, which hints at the potential for something more transformative as opposed to just acting as a placeholder. Commons can involve self-governance and collective ownership that undermine accumulation, and can act as springboards to collective action. In this way, traditional critiques of the commons – that it inevitably leads to a free-for-all with regards to consumption, because everything is technically free – can be avoided as long as no one in particular can use up all of the (non-commodified) resources.

13.3 Counterpublics, Safe Spaces and Counter-Places

Counterpublics are part of a larger socio-spatial framework to seek social justice for stigmatised and ignored populations. Nancy Fraser (1990) is particularly associated with this concept, where she recognised that rather than one public in society, there are many and that some counter the dominant public domain. By this she meant that subordinate groups express themselves and resist notions of one public, strategically using specific *safe spaces*: 'subordinated social groups ... have repeatedly found it advantageous to constitute alternative publics ... parallel discursive arenas where members of subordinated social groups invent and circulate counterdiscourses, which in turn permit them to formulate oppositional interpretations of their identities, interests and needs' (1990: 67). Sziarto and Leitner (2010: 383) put it this way: 'Counterpublics are spaces with a dual character: they are spaces of withdrawal and regrouping, and spaces of publicity'. We have already considered publics as multiple in Chapter 2. With the practice of counterpublics, encounters become even more fragmented, potentially open to many different groups and divergent aspirations. These counterpublics can include immigrant hubs as per Chapter 4, or for disabled people in Chapter 9 to battle pervasive exclusion. The point for Fraser was that these counterpublics, embedded in particular spaces, were both a setting for re-groupment and withdrawal, but with the aim of eventually rejoining larger society to challenge it. To this end, Fraser saw counterpublics as civil society 'in training', based on awareness of being subordinate vis-à-vis the dominant public, a productive tension that articulates a particular identity and, as we will propose, a particularly grounded spatiality.

Examples of counterpublics are numerous, and they usually refer to parallel social and spatial groupings that are ideologically and culturally opposed to the dominant public, a source of conflict and rivalry (DeVerteuil and Kiener, 2022). Subordinated groups – ranging from homeless individuals to racially-marked minorities to LGBTQ+ people (deemed 'sexual dissidents' in homophobic societies) – create spaces that can range from encampments, barber shops, and hair salons, and clubs – that sustain contrarian identities but also seek to develop oppositional politics. As such, counterpublics are both insular and outward-looking, avoiding fringe or permanently separatist spaces. Here we can consider the overlap with safe spaces, where vulnerable groups can protect themselves with the eventual aim to reconstitute a more socially just society that values them. An example that Geoff uses in class is that of Grindr, a gay male dating app that is the progenitor of the heterosexual Tinder dating app (although a site that Nancy Fraser would be unlikely to deem a safe space!). For a gay man, it can be safer to use Grindr than to approach men in public, not knowing if they are gay or even homophobic. Of course, not all counterpublics are productive, especially if their activism is particularly quiet, based in its very existence rather than any kind of oppositional intent.

Kurt Iveson (2017) gives an example of counterpublics with the 'Aboriginal Tent Embassy' in Canberra, the capital of Australia. Over 50 years old now, the Embassy occupies a patch of land in a highly symbolic part of Canberra, with the activists mobilising and sustaining a counterpublic. This counterpublic has worked against the violent subordination of the continent's Indigenous peoples for over 250 years now, but also acts as a space of withdrawal that approaches an indigenously sovereign space. Counterpublics can be remarkably durable, a point we will take up under resilient practices and relationalities in a few pages when we consider the aspect of duration.

Yet we must admit that some counterpublics are also formed by groups who are reactionary, seeking to use vigilante violence to bolster some kind of established order or an idealised, established order that previously existed or does not yet exist, usually in concert with the state and not in 'isolation from the ongoing violence produced by the...state' (Goldstein, 2005: 391). From this perspective, vigilante violence seeks to maintain rather than overthrow, becoming a conservative force, essentially negative but superficially disorderly. Even in the case of frontier vigilantism in the United States, Belew (2014: 92) found that it served existing power: it is 'group violence that serves systemic power ... subtle power exerted by many informal structures that uphold it. When the state is weak, systemic power – often patriarchal, racial, or religious – commonly prevails'. So by these notions, vigilantes may prop up and fill the gaps of a weaker, less 'vigilant' state or even extend a more powerful one, even when it appears that there is no law at all – this very vacuum is what motivates vigilante violence.

In sum, this is the preferred sequence for counterpublics to follow: exclusion leads to the creation of safe spaces; safe spaces are used to promote tolerance and fight against exclusion; eventually the safe space rejoins the larger society, perhaps no longer even needed. But this might not happen when society is altogether unyielding, such as in the case of disabled people who must navigate the city within a broader context of exclusion. Or it could create spaces for reactionary groups whose politics are completely at odds with changing the current status quo maintained by the state.

Like the commons, counterpublics are place-making practices. The combined result could be termed *counter-places.* Counter-places can be considered the spatial overlap of commons and counterpublics – places of solidarity and de-commodification, but also sustaining alternative groups and their eventual challenge to larger social structures. While some counter-places are quite subtle, some are more high-profile and vilified as transgressive zones, and their very visibility makes them targets for re-commodification through redevelopment. These are extreme cases, places that are contained, self-exiled and alternative places like Christiania in Copenhagen, Uzupio in Vilnius, Stokescroft in Bristol, Exarcheia in Athens and People's Park in Berkeley. They are self-styled liberated zones, with obvious links between social justice and everyday city living. Citizens of such places like to think of themselves as beyond the rest of the city, living

FIGURE 13.1 Christiania, Copenhagen September 2008 © G DeVerteuil

FIGURE 13.2 Uzupio, Vilnius, July 2019 © G DeVerteuil

in zones unto themselves. Vradis (2020) talks about Exarcheia as part of a rough spatial contract that limits protest to very particular places, thereby avoiding overspill into more powerful and protected parts of the city. This spatial containment strategy ensures that these exceptional places survive, although it also makes them easy targets for redevelopment.

The two images above show counter-places as they existed when photographed: Christiania and Uzupio. Each counter-place shows off a particular facet: Christiania in terms of being insular and separate from the rest of Copenhagen, and Uzupio in terms of being welcoming to alternative lifestyles and ways of making a living.

When Geoff visited Athens in November 2012, during a particularly difficult time for the Greek nation, the economy had contracted by one-third, and basic services were being withdrawn. He was hardly the only visitor – many radical scholars wanted to see the nefarious effects of full-strength austerity, of how a Global North nation could slide into economic chaos, albeit capturing only the 'helicopter' version of the crisis. With a Greek colleague (Dr Dionysia Lambiri), he was shown the area between the liberated zone of Exarcheia and the powerful city fabrics of Syntagma Square, which was covered by extensive political graffiti. The messages speak directly to everyday concerns, pointed critiques of the system, and suggestions for transformational change. Figure 13.3's message translates as 'to oppose the new measures' (*metrah*) and 'get a stone' (*petrah*) instead. The message

FIGURE 13.3 Exarcheia, Athens, 2012 © G DeVerteuil

FIGURE 13.4 Exarcheia, Athens, 2012 © G DeVerteuil

in Figure 13.4 calls for a people's uprising, in the people's hands, and that we won't pay the heavy taxes (the word is Turkish (*haraç*) and originates from the Ottoman Empire era when non-Muslims had to pay land taxes. The term is still used in Greek when the government introduces new forms of taxation and is usually discriminatory – i.e. targeting those who earn less), signed 'Athens Squatters'.

Sometimes these counter-places are temporary rather than permanent, more pleas about improving everyday life around Lefebvrian *auto-gestion* (e.g. self-management, as seen in Chapter 3 in Paris and Montreal) or making sarcastic, Banksy-like observations about powerful forces acting upon everyday (Figure 13.5).

The LA mural in particular Figure 13.6 represents future potential, that like-minded people can join together to bring change at a societal scale but starting from a particular locality.

Murals can be understood as a way of confirming and communicating a counterpublic's identity, giving a sense of belonging and territorial identity. Despite their increasing mainstreaming, murals still hold some potency in disenfranchised urban areas, as a tool that disempowered and stigmatised groups can raise social justice issues (e.g. recognition) and memorialise individuals who were unjustly killed. Murals afford a visual introduction to the values of smaller subgroups with their own distinctive cultures.

FIGURE 13.5 London, 2008 © G DeVerteuil

FIGURE 13.6 Los Angeles, 2013 © G DeVerteuil

13.4 Resilience

Up to now, we have focused on very specific spatial manifestations of quiet activism. For this section, we are as much interested in the temporal manifestations of this quiet opposition – how long do they last, and can they actually last? What concept can best encapsulate the slow and quiet, rather than the fast and loud, activism and opposition? In this regard, resilience seems an appropriate but somewhat controversial choice. At its most basic, resilience relates to the ability to bounce back after a shock of some kind, enabling people and places to survive without changing the circumstances that challenge their survival. This 'getting by' is essential when it is not possible, or preferable, to enact system overhaul, better known as 'resistance', or changing the system to better suit one's needs, better known as 'reworking' (Katz, 2004). Resilience usually involves: (1) a particular shock to a relatively stable system; (2) the capacity to bounce back to some degree of stability and previous identity after the shock; and (3), in bouncing back, perhaps showing some adaptability and gaining capacity for innovation (DeVerteuil, 2015; DeVerteuil and Golubchikov, 2016). Resilience shifts attention from risk and vulnerability to something more positive and prospective, on analysing the capacity of people either as individuals or groups (social resilience) and places (spatial resilience) to anticipate, persist with, adapt to and minimise the damage caused by change, risk, and adversity.

We understand social and spatial resilience to be different from engineering or ecological resilience, whereby the preferred outcome of a shock is the direct bouncing back to the original position prior to the shock, which of course can lead to the very same level of vulnerability pre-shock. A series of globally-felt shocks have occurred in the past 15 years – the 2008 economic crisis, which has been signposted across various chapters, including Chapter 11 on the current housing crisis, as well as the 2020–2022 pandemic, which have left its legacy in a more unequal world. For both of these shocks, our resilience outlook would dictate that society does *not* return to the status quo pre-crisis, because this would put it in a similarly vulnerable situation. Instead, our resilience perspective would valorise adaptation over a return to the status quo, the sustaining of new directions that are not necessarily transformative but certainly running against the pre-shock settlement.

And so while resilience has been critiqued for maintaining the (neoliberal) status quo, which is usually understood by critical geographers as unjust and certainly unsustainable, they are not inevitably linked. If anything, neoliberalism erodes and destroys social resilience, leaving people and places vulnerable. And while resilience is certainly not immediately transformative, by surviving the present it can provide a platform to call for eventual social change later on. Resilience is more of an incremental, long-term strategy to bypass powerful social structures. In particular, social resilience involves a response to ensure (collective) well-being when faced with the socially corrosive effects of market-led neoliberalism. This can be in the form of institutional and community resilience, but also the role of the state as an important counterweight to market-driven forces (Hall and Lamont, 2013). Resilient place-making practices can help ensure that some of the counterplaces illustrated above can survive in the meantime.

FIGURE 13.7 Sydney, March 2011 © G DeVerteuil

FIGURE 13.8 Sydney, June 2014 © G DeVerteuil

FIGURE 13.9 Sydney, January 2018 © G DeVerteuil

To illustrate resilience, we show a variety of place-making outcomes. The first is how resilience is never guaranteed, based on the case study of Redfern in inner-city Sydney. Below are three images taken by Geoff in 2011, 2014, and 2018 of this Indigenous counterpublic and counter-place (Figures 13.7, 13.8 and 13.9). The Block in Redfern was a highly-conspicuous concentration of housing, care and services, created by Aboriginal people, that emerged as part of intense political contestations during the 1960s and 1970s over the creation of specifically Indigenous space in inner Sydney that provided refuge for the most dispossessed, hungry and threatened (DeVerteuil, 2015). Back in the 2010s, the Block represented a resilient product of self-help and survival, and the images show, at least in 2014, a tent city to protest imminent redevelopment of the site. In this case, resilience lasted for 40 years with the Block remaining the same, demonstrating its longstanding presence, in spite of the Sydney skyline continuing to bloat. However, with this persistent pressure on this counterpublic and counter-place, by 2022, the site had indeed been redeveloped, with little of the original community hub remaining.

A second, potentially more resilient example is Skid Row Los Angeles, visited back in Chapter 8. Although eroded in a few places, especially on its

FIGURE 13.10 Skid Row Los Angeles, May 2009 and March 2023 © G DeVerteuil

western edges, Skid Row has proven remarkably resilient as a commons, counterpublics and even counter-place. The images below, taken by Geoff in 2009 and 2023, show the same block in Skid Row (on 5th Street). In 2009, the sidewalks were relatively empty, given the strongly interventionist policing of the place back then. In 2023, the sidewalks were full of tents, and a construction crane in the background points to more affordable housing being built. Resilient but also representing a consolidation and densification of the homeless over the 14 years. As a counter-place and a service hub, Skid Row is going nowhere, despite its saturated nature.

The resilience of Skid Row parallels the resilience of much social housing left remaining in cities like London, Paris and Vienna. These resilient 'past futures' now co-exist with newer, more commodified landscapes, their destruction prevented by state inaction, state support or even grassroots movements that Manuel Castells talked about in 1983. Conversely, socially destructive practices that sustain the current status quo can be remarkably resilient, such as neoliberalism. This parallels the creation of safe spaces for reactionary and vigilante groups.

Resilient places can also be combined with particularly resilient messages from well-known proponents of social justice. The words of Martin Luther King Jr – largely around social justice – is enshrined in a variety of places, officially in terms of monuments, parks and street names, including the national memorial in Washington DC, the site of his assassination in Memphis, and his burial place in Atlanta. But it is also an unofficial affair, woven into the everyday fabrics of African-American neighbourhoods in visible ways. These images on the next two pages show Martin Luther King Jr as deeply part of the African-American experience in places like Los Angeles, but especially in Memphis.

To sum up, resilience is not particularly radical or even transformative, but it does provide a conceptual platform for survival 'in the meantime' when transformation is not possible or even desired, a kind of 'breathing space'. Beyond the examples provided here, resilience is a concept that you as a student will likely encounter and may need to practice over your lifecourse, irrespective of your level of political engagement. This resilience may relate

FIGURE 13.11 Memphis, 2015 © G DeVerteuil

FIGURE 13.12 Memphis, 2015 © G DeVerteuil

to housing, mental health and caregiving. Resilience is often made possible with the voluntary sector and through the support of others. Resilient practices and place-making are by definition slow and steady, but able to adapt to a variety of shocks, including neoliberalism, capitalism and the state-maintained status quo. The next section, however, is far more radical

FIGURE 13.13 Los Angeles, 2005 © G DeVerteuil

and transformative, and perhaps more utopian, projecting a better world so that change can become possible (more quickly).

13.5 Abolition Geographies

The final means of quietly creating an alternative world is the practice, relationalities and encounters of *abolition*, engaging particularly with the work on race by the African-American geographer Ruth Wilson Gilmore (2022: 3). To her, abolition geography is nothing less than 'how and to what end people make freedom provisionally, imperatively, as they imagine *home* against the disintegration grind of partition and repartition through which racial capitalism perpetuates the means of its own valorization'. Unpacking this statement, her immediate focus is on abolishing the dominant use of police and prisons as modes of social control, within the American context of pervasive racialisation and criminalisation. Over-reliance and overspending on policing and prisons are but the tip of the iceberg in terms of a larger, more insidious mode of American-style capitalism that is based on racial exclusion. For many American cities and states, spending on the police and prison far outweighs that of welfare and community support, part of a state-building exercise through prison-building, rather than providing a functioning welfare state for the people. America also has by far the highest rate of imprisonment in the Global North (as well as by far the most guns per person).

Abolition geographies basically work against these forces that are arrayed to impinge on the freedom of everyday survival, places and place-making among African-Americans in particular. This radical reworking appears *utopian*, by design – perhaps too utopian for some, and certainly a stretch beyond resilience and 'in the meantime'. Utopias are about future worlds that improve on the one we have now. This is based in the idea that in order to improve the world, one must first think up a perfect world and proceed backwards from there. Abolition as utopian invokes the everyday transformation away from incarceration as the catch-all solution to social problems. Social problems can be avoided in the long-term by investing sufficiently before they become social problems, taking money from the bloated police and prison system. The mural below, in Brooklyn, sums this up using a quote from the abolitionist Frederick Douglass, who fought against slavery in the pre-Civil War United States.

Better to invest in children who will grow up to be strong adults, than trying to retrospectively fix grown-up men who have spent long periods incarcerated!

Abolition geographies require not just the end of harm, but also repair and reparations. In 2014, Ta-Nehisi Coates wrote a powerful piece in *The Atlantic* on the case for reparations in the United States. The case focused on demands for restitution for the 'multi-century plunder of Black people in America', who have faced economic dispossession not only because of the ongoing legacy of slavery but other legal and extra-legal racist practices such as incarceration (as the abolitionists argue), but also historical real estate practices designed to confine and disinvest African-American areas in cities, in terms of redlining (i.e. legally discouraging investment by making mortgages in these areas un-insurable) that made it impossible to build up

FIGURE 13.14 Brooklyn, 2019 © G DeVerteuil

land-based wealth like Whites did. Coates frames this call for reparations less as a payout, and more as a reckoning for America's past sins. Of course most Americans never owned slaves, so why should all Americans reckon with the past? Coates counters that America's current prosperity – the very prosperity that has brought millions of immigrants from around the world to its shores for hundreds of years – was partly built on the unpaid labour of slaves and the wealth they created for the nation before 1865. This gave America a huge head start on many different nations that did not benefit from slavery (nor the massive amount of emptied land stolen from the Indigenous peoples, an advantage only found in other colonial settler societies – see Chapter 7). Moreover, the century-long legal submission of African-Americans after the abolition of slavery allowed the White population and subsequent immigrant groups to face less competition and thereby move up the social ladder, albeit slowly and indirectly.

Since 2014, Black Lives Matter has also taken up 'reparations for past and continuing harms' as one of its core demands, to remedy the poverty gap that they identify as arising from colonialism, slavery, redlining, mass incarceration and surveillance. However, the demands for reparations by groups who have been marginalised, oppressed and subject to social, political and economic violence, stands in stark contrast to the actual payouts. In the early summer of 2020, in the wake of the George Floyd killing and ensuing unrest in many American cities, there were inklings of an opening in the discussion around not just reparations but also abolition.

But summer 2020 also marked a brief period when abolition became a serious conversation, especially through the talk of defunding the police in

FIGURE 13.15 Cardiff, 2020 © G DeVerteuil

many localities as a way to begin reducing the bloated carceral state. The interlude was indeed brief, in a nation with so much desperation and within a context of the pandemic, it was not long before crime began creeping up just enough to call for more policing, not less. Some feared a return to the 1980s, when many of the most dangerous neighbourhoods were under-policed rather than over-policed, leading to situations where 5% of the population terrorised the remaining 95%. 'Defund the Police' also largely lost its appeal through its choice of messaging. At the time, former President Barack Obama warned social justice activists that using 'defund the police' messaging likely derailed their attempts for change and alienated potential allies – thus demonstrating that optics matter in efforts to gain wider buy-in. As such, the interlude in 2020 was fleeting, but still managed to attract attention, with some more receptive to it than others (Figure 13.15).

13.6 Conclusions and Shifting Encounters with Activism and Social Justice

Let us conclude this chapter with a journey around the relationalities and encounters created and sustained by the four different place-making practices under review: commons; counterpublics and counter-places; resilience; and abolition. For the commons, encounters range widely, basically including any de-commodified space, as well as the boundary between it and the rest of the commodified world. But the boundary can be blurred, as areas that are technically commons, especially green spaces, can be 'taken over' by city-sponsored festivals which charge for admission. An example here is LGBTQ+ Pride marches – while originally free of charge, over the last decades some now charge for access. Counterpublics are very much ground-up and DIY, so that the encounters are within the safe spaces themselves, at least until such time as the subordinated group in question decide to engage with wider society. Counter-places range as widely as commons, and usually build on de-commodified places like commons, or state-sponsored sites like social housing or service hubs. The encounters for resilience are as much about time as they are about place-making – as long as certain practices can be sustained, there is always the chance for transformation (e.g. resistance) at a later stage. As for abolition geographies, the state is the frontline – of how prisons and the police insinuate themselves into everyday life in certain communities. Abolition is by far the faintest encounter, but also the most radical potentially, acting as a bridge to the more transformative (even militant) forces to be unleashed in Chapter 14. Further to this, we also consider in Chapter 14 the politics of visibility and invisibility in terms of social justice and public space.

Now that the relationalities and encounters have been settled, the next and final issue in the chapter is the nature of creating them, i.e. how can you as students activate these various practices? This moves you as the student from passive learner to active campaigner or activist citizen by engaging with this chapter (and the next one). Quiet activism might strike some of you as rather quaint, or more ready-made for older adults who have more patience and measure (or weariness). Yet a variety of strategies can be deployed, from hashtag protest to harnessing the system to help enable transformation, and the rejection of the status quo. Quiet activism is necessarily slow and measured, which might not appeal when you are 20 years old and want things to change now. That said, many forms of quiet activism can have low barriers to entry and may be practiced by many of you already, e.g. through peer-support and local volunteering.

Quiet activism can raise serious and contradictory issues, so an open mind is better than a closed one. For instance, the example of harm reduction was raised in Chapter 10 as an alternative to incarceration and abstinence with regards to intoxicants, and could even be seen as a fledgling form of abolition (via de-criminalisation). Yet harm reduction in itself is not a magic pill, as already mentioned, without a robust treatment system already in place, the temptation for some addicts will be to continue to use intoxicants without fear of being arrested or sobering up. In other words, harm reduction needs to have appropriate guardrails given the extreme level of addiction with certain intoxicants, especially opioids. Some will question harm reduction as a way to enable some people to do drugs full-time, with no aim of ever getting off them, and as such perhaps leading to even more overdoses on the street (or more long-term deaths from intoxicant abuse). Simultaneously, a looser approach to intoxicant use that comes with harm reduction also allows vulnerable people to do hard drugs in public, so as not to further stigmatise them or send them underground. But this has a negative impact on public space. In many nations, such as Canada, it is now forbidden to drink or even smoke in public space, but apparently doing hard drugs is (sometimes) tolerated. What message does this send to people trying to get off drugs, or impressionable members of the public? Suffice it to say that activists for harm reduction do not always have the big picture that is needed to advocate responsibly for major social changes.

A further example, and drawing on the discussion of race in Chapter 7, revolves around what constitutes anti-racist policy, that is opposing and eradicating racism rather than just acknowledging that it exists. An abolitionist stance would surely argue that the hyper-incarceration of African-Americans is racist, in that they are far more likely to be criminalised for lesser crimes than other groups, especially Whites. One obvious solution is to de-criminalise, for instance, low-level drug use and shoplifting under a certain amount of value, of treating it more as an issue of survival and social work rather than a matter for the police. This was instituted in some American states, including California. One unintended consequence, however, is that many small grocers struggle with increased levels of theft from all groups. Many of these small grocers are themselves immigrants from the Middle East or Asia. This certainly brings up the notion that racialised people are not always in solidarity with each other, but rather in competition or operating on entirely different trajectories.

The same goes for universities. Just as the authors cannot validate all of the contradictions and unintended outcomes that plague social problems and the socially-based activism around them, they also cannot validate everything students feel about the world and about issues of social justice in particular. This became apparent during the student occupations of various campuses in 2024 as a way to show solidarity with the Palestinian people; universities could not at the same time support these occupations but also appear to be in support of Israel, or even of the Jewish students on campus. This topic will be raised again in the next chapter on loud activism. More positively, quiet change is *always* possible, despite the obvious cautions above. The demolition of Apartheid in South Africa, for instance, was as much a product of slow-burn activism and playing the long game, as it was loud and spectacular unrest. It is to these more direct actions that we turn to in Chapter 14.

References

Abrahams, N. (1993) *Bricks that hold the building up: Women volunteers, political action, and the negotiation of family and community responsibilities*, University of California, Santa Barbara ProQuest Dissertations & Theses, 9405802.

Belew, K. (2014) Lynching and power in the United States: Southern, western, and national vigilante violence. *History Compass*, 12, 84–99. https://doi.org/10.1111/hic3.12121C

Cresswell, T. (1992) The crucial 'where' of graffiti: A geographical analysis of reactions to graffiti in New York. *Environment and Planning D: Society and Space*, 10(3), 329–344.

DeVerteuil, G. (2015) *Resilience in the Post-welfare Inner City: Voluntary Sector Geographies in London, Los Angeles and Sydney*. Bristol: Policy Press.

DeVerteuil, G. and Golubchikov, O. (2016) Can resilience be redeemed? Resilience as a metaphor for change, not against change. *City*, 20(1), 143–151.

DeVerteuil, G. and Kiener, J. (2022) Commons, counterpublics and dissident urban space. *Geography Compass*, 16(8), e12654.

DeVerteuil, G. Marr, M.D. and Kiener, J. (2022) More than bare-bones survival? From the urban margins to the urban commons. *Annals of the American Association of Geographers*, 112(7), 2080–2095.

Fraser, N. (1990) Rethinking the public sphere: A contribution to the critique of actually existing. *Social Text*, 25/26, 56–80.

Gibson-Graham, J.K., Cameron, J. and Healy, S. (2016) Commoning as post-capitalist politics. In A. Amin and P. Howell (Eds.), *Releasing the Commons: Rethinking the Futures of the Commons* (pp. 192–211). London: Routledge.

Gilmore, R.W. (2022) *Abolition Geography: Essays Toward Liberation*. New York: Verso.

Goldstein, D.M. (2005) Flexible justice: Neoliberal VIOLENCE and 'self-help' security in Bolivia. *Critique of Anthropology*, 25(4), 389–411. https://doi.org/10.1177/0308275X05058656

Hall, P. and Lamont, M. editors (2013) *Social Resilience in the Neoliberal Era.* Cambridge, UK: Cambridge University Press.

Harvey, D. (2012) *Rebel Cities*. New York: Verso.

Iveson, K. (2017) 'Making space public' through occupation: The Aboriginal Tent Embassy, Canberra. *Environment and Planning A: Economy and Space*, 49(3), 537–554.

Jeffrey, A., McFarlane, C. and Vasudevan, A. (2012) Rethinking enclosure: Space, subjectivity and the commons. *Antipode*, 44(4), 1247–1267.

Katz, C. (2004) *Growing Up Global: Economic Restructuring and Children's Everyday Lives*. Minneapolis: University of Minnesota Press.

King, T., Navarro, J. and Smith, A. editors (2020) *Otherwise Worlds: Against Settler Colonialism and Anti-Blackness*. Durham NC: Duke University Press.

Mitchell, D. (2003) *The Right to the City: Social Justice and the Fight for Public Space*. London: Blackwell.

Pottinger, L. (2017) Planting the seeds of a quiet activism. *Area*, 49, 215–222. https://doi.org/10.1111/area.12318

Schalk, S. (2022) *Black Disability Politics.* Durham: Duke University Press.

Sziarto, K.M. and Leitner, H. (2010) Immigrants riding for justice: Space-time and emotions in the construction of a counterpublic. *Political Geography*, 29(7), 381–391.

Vasudevan, A. (2017) *The Autonomous City: A History of Urban Squatting*. London: Verso.

Vradis, A. (2020) Spatial politics and the spatial contract in Exarcheia, Athens, Greece (1974–2018). *Transactions of the Institute of British Geographers*, 45, 542–558.

Young, I. (1990) *Justice and the Politics of Difference*. Princeton: University of Princeton Press.

Zizek, S. (2006) *In Defense of Lost Causes*. London: Verso.

14 GEOGRAPHIES OF RESISTANCE AND DIRECT ACTION

14.1 Introduction

The second chapter within Part III builds on the earlier themes from the previous chapter on quiet activism via creating commons, counter publics and counter-places, and abolition geographies *and* practicing resilience. It considers how groups have more directly challenged the status quo through acts of resistance that seek to overturn the system, rather than living with it or striking a better deal (e.g. resilience, reworking). These moments frequently occur through a sense of desperation tinged with hope, where a utopian ideal of human rights takes a collective turn, as when the rights of labour, women, LGBTQ+ and other minoritised groups come to the fore and, on occasion, have yielded results.

While some scholars have chalked up other, more quiet, subtle, everyday micro-politics as 'resistance', we return to Cindi Katz's (2004) three-part definition of change via activism first introduced in Chapter 13 – resilience, reworking and resistance – and direct our primary focus to the last of these (resistance) to help frame the chapter. Echoing Katz's understanding of resistance, in this chapter, we take it to mean a kind of upheaval and transformation within and across social structures, thereby removing the status quo. Katz argues that resistance draws upon 'oppositional consciousness' that aims to bring about 'emancipatory change'. For people taking part in resistance, we acknowledge that different forms of activism often do not fit neatly into discrete categories and are less easy to disentangle in practice, as the relationships between people and place shift and unfold over time, nonetheless the above definitions offer a useful starting point. We also acknowledge that activists frequently employ the other two forms of change/activism outlined by Katz (acts of resilience and remaking) and repeat our reminder that Chapter 13 and Chapter 14 offer two different angles on very overlapping practices. For instance, activists often shift between public, private and voluntary sectors over the course of their activist careers, and depending on the circumstance can advocate resilience (waiting on change) or reworking (not changing the system but seeking a better deal from it). Protest can also take its toll, given its embodied form, so we seek to challenge the heroic narratives frequently associated with it.

DOI: 10.4324/9781003414889-17

Both forms of overt resistance (public protest, social unrest) involve the use of public space to make claims for recognition and in the process making social injustice visible. Here we connect to the politics of visibility and consider how in some cases, making something visible can be 'radical' and transformative. We explore *how* public space is used in the practice of resistance – as a predetermined space of public address, akin to an open staging ground or something more complex. We trace some of the key geographical thinking on this from both a Global North and South perspective. So, in the section below, we first consider the wider question of what is public space?

14.2 Public Space, Protest and Occupation

Public space is often the most celebrated space in social, cultural and political life. It is a social space that is generally regarded as open and accessible to people (see Section 1.2 for an introductory definition and discussion). How we think of public space, at least in Europe and North America (we return to consider less 'parochial' accounts later in the chapter), is informed by the ancient Greek notion of Agora (18th century–8th century BC). The Agora was a central place in ancient Greek city-states, particularly in Athens, which was the best-known example. The literal meaning of the word is 'gathering place' or 'assembly'. People classified as free-born citizens would gather in the Agora for military duty or to hear statements of the ruling king or council. Later, the Agora became the centre of athletic, artistic, spiritual and political life of the city. It also served as a marketplace where merchants kept stalls or shops to sell their goods. While it has been critiqued for positioning women more at the margins, nonetheless, it has been influential in informing contemporary spatial imaginings of public space as an open, civic space for political address and protest. So, before we can begin exploring protest, we must consider the wider question of what is urban public space in today's context?

Box 14.1 Engagement Questions

What is public space? What keywords would you use to define it?
What is included in public space? Is it only parks and squares, or are there other venues that we think of as public space?
What can and do people use it for?
How is it controlled?

In the Engagement Questions above, we encouraged you to think of the idea of public space, which has been the subject of extensive critical inquiry in geography (Barnett and Low, 2004; Iveson, 2007; Mitchell, 2003; Staeheli and Mitchell, 2007). Some commonly used words to describe public space are free, open, accessible, visible, dynamic, heterogenous, mixed. It can be used for visiting, watching, travelling through, dog walking, meeting friends and unexpected others, exercising, eating and drinking, gardening, resting, taking part in recreation and performance (e.g. busking), shopping or browsing, taking photographs, working – and 'living'. The latter serves as a reminder that for homeless people, public space is a necessity to survive. Vidler (2001 cited in Mitchell, 2003: 3) argues that an urban life worth living consists in 'public-ness' itself:

> the street as a site of interaction, encounter and the support of strangers for each other; the square as a place of gathering and vigil; the corner store as a communicator of information and interchange. These spaces, without romanticism or nostalgia, still define an urban culture, one that resists all effort to 'secure' it out of existence.

Returning to our earlier reference to the Greek Agora, public space has also been associated with forms of protest and collective action – in the Global North at least. Here, public space can also be used to challenge, however briefly, the monopolisation of powerful city actors in the city, through speaking up and protesting (Iveson, 2007). In this account above, Kurt Iveson classifies our initial rendering of public space as being understood in a *topographical* sense, to refer to particular places in the city that are (or should be) open to members of 'the public', such as streets, footpaths, parks, squares, for circulating ideas and making claims for social justice. Historically, there is certainly plenty of evidence to support this view. The Protestant Reformation in 1517 is a poignant reminder that sometimes all it takes is one spark to start a fire. Martin Luther is widely credited for being the spark that started the Protestant Reformation, a movement that completely upended the Catholic Church in many parts of the world and changed the way that people practiced religion on a global scale. One man's voice in a public space, in the form of a list of '95 Theses', nailed to church doors was enough to create an entirely new religious sect. Indeed, the origin of the word Protestant comes from the Latin words *pro* and *testari*, to testify forth, or to hold forth a position on something. Its primary historical meaning *(pro testari)* has been to assert, to maintain, to proclaim solemnly or state formally. Another example was the Arab Spring of 2011, which overturned regimes in Egypt, Libya and Tunisia, and shook others (Morocco, Iraq, Syria) to the core. It all began with one man in Tunis, who immolated himself in protest after being harassed by the police amid a general atmosphere of corruption and economic stagnation.

During Victorian times in England, Hyde Park and Trafalgar Square were frequently disturbed by mobs and rioters, until it became the recognised place in which to air popular discontent or grievance. Indeed, Hyde Park became a key staging ground for protest in 1866 by working-class men seeking to secure the right to vote. Up until this time, only male property owners had the franchise. Other historical examples worth noting are the Temperance Movement, which sought to impose a more puritanical society in the US. This movement serves as a useful reminder that the right to the city can be invoked by groups that share views that do not necessarily resonate with our own. Mitchell (2013) argues that only through researching these grounded and historical examples can we understand how the seemingly serendipitous and haphazard emergence of public space is tied into historical contexts and broader social, political and legal agendas.

Protest has been the subject of much geographic enquiry ever since the 1970s, particularly within Marxist scholarship, where it is typically presented as a public, 'spectacular' and antagonistic practice (see Harvey, 2012; Mitchell, 2013). Protests have traditionally been regarded as high intensity and fast-paced (Routledge, 2019), relying upon numerous bodies occupying and moving through space in tandem. A more recent large and spectacular example of protest includes Extinction Rebellion. Participants in these protests have been bent on the 'real world' act of claiming and organising iconic public spaces. These dramatic events in squares, parks, streets and university campuses, according to Iveson (2007), redefine the importance of (largely urban) public space as a centre for social encounter and interaction, a forum for discussion and dissent, an interface of virtual and material connections and a stage for the reinstatement of democratic practice and resistance in the face of social injustice.

In contrast to the more antagonistic forms of protest above, and the heroic narratives that are romanticised within them, some forms of protest in public space are less vocal. Quiet marches and vigils are positioned somewhere in-between quiet activism and public protest. One recent example has been the Grenfell Tower silent vigils (see Back's 2021 paper *Hope's Work* for an account). The global Occupy Movement also sits somewhere in-between, as an example of non-violent encampments, accompanied by public talks and local marches. Following the 2008 global financial crisis and subsequent bailing out of banks, questions emerged over the inequitable redistribution of resources. Their calling card, 'We are the 99%', aimed primarily to advance social and economic justice and different forms of democracy. This led to the occupations in over 950 cities around the world including cities in North America, Latin America, Egypt, Nigeria, India, Asia, Oceana (New Zealand and Australia), UK and Europe. In social movement theory, taking space can be understood through what McAdam and Sewell (2001) call 'transformative events', or Tarrow (1998) terms a phase of 'heightened conflict', a particular temporality of protest that provides a turning point in the development of struggles. Such accounts are distinctively aspatial, however. Arguably one of the most cited (by activists and academics) conceptualisations of taking space from an anti-capitalist perspective is Hakim Bey's notion of the 'temporary autonomous zone' (TAZ), a 'moment of intensity' (1991: 98) that 'like an uprising … liberates an area (of land, of time, of imagination) and then dissolves itself to re-form elsewhere' (1991: 99) to

FIGURE 14.1 Edinburgh, 2011 © G DeVerteuil

avoid state capture. Political protest, wherever it takes place, thus creates context-specific spaces for representation.

This movement was allied to a sense of hope based on the utopian vision of a future world that would be more equitable. The geographical focus of the protests were on highly visible locations in strategic cities, including St George's Square in Edinburgh (Figure 14.1) and St Paul's Cathedral in London (Figure 14.2).

While temporary, both of these occupations were highly visible and strategic in 2011, but perhaps did not last long enough to produce durable change. In 2024 and 2025, the student occupations calling for a Free Palestine used similar tactics to the Occupy Movement back in 2011, but the focus was more on university campuses than on strategically visible parts of the city, in this case Cardiff University (Figures 14.3 and 14.4).

Protest in all its guises can be conceptualised as staging a deliberate 'act of citizenship' (Isin and Nielsen, 2008). Indeed, the Supreme Court in the US bases its public forum doctrine on the notion that since 'time immemorial' people have used the public spaces of the city – the streets, parks and squares as gathering places for 'communicating between citizens' and 'discussing public questions (Hague v. CIO 1939). The importance of public space in protest cannot therefore be overstated. According to Monbiot (2013: n.p.),

> Scarcely a human freedom has been obtained without the help of public protest. Scarcely an inch of social progress has been achieved without the same. Scarcely any effective movement in pursuit of this progress has remained within the bounds of the law.

FIGURE 14.2 London, 2011 © G DeVerteuil

FIGURE 14.3 Cardiff, 2025 © G DeVerteuil

FIGURE 14.4 Cardiff, 2025 © G DeVerteuil

Monbiot's key point is that protest operates as a form of suspending and disengaging from state-endorsed modes of responsibility and moral order. Suffragettes for instance used every militant tactic they could think of to get the women's vote. They shouted down parliamentary debates and forcibly entered government. They also undertook various forms of social unrest, which we return to later in the chapter, including throwing eggs and flour in courtrooms, smashing windows of government offices, setting light to post boxes and a West End theatre, burning slogans into putting greens, and destroying paintings in the National Gallery and the Royal Academy in London.

These tactics concern the 'practice' of public space itself and serve to illustrate the vital relationship between struggles over public space and movements for social justice. Within this topographical definition of public space, Kurt Iveson (2007: 8) outlines how there are three distinct aspects of 'publicness':

- A context for action (public space)
- A kind of action (public address)
- A collective actor (a/the public)

This neat classification is useful as a starting point, as it disentangles the three key components of public protest and allows us to consider where encounters are likely to appear. However, as Iveson himself concedes, this definition is based on a solely topographical understanding of public space – i.e. parks, streets and squares. As he and others have argued, the notion of public space carries many ambiguities which in turn make the job of conceptualising protest more complex, as we will unpack in the section below.

14.3 Unpacking Public Space and Protest

One ambiguity with public space is rooted in the very assumption that areas of the city are necessarily always 'public' (as understood as an open, accessible, civic space). Within urban and social geographic scholarship at least, there is widespread concern that public spaces in contemporary cities are becoming more exclusionary, and hence less accessible to those seeking to put them to work in circulating ideas and making claims for social justice. One only has to look at the exclusionary practices which seek to remove the public presence of homeless people as an example:

> While it is true that human beings have no choice but to occupy a space: they just do, ...it does not follow that such a space allows for the full, adequate, and self-directed representation of human beings either to themselves or to others.
>
> (Smith, 1994: 151)

Public space as we saw through Part II is increasingly characterised by stringent, austere and punitive encounters, where people are being controlled through surveillance cameras, police and security personnel, national and local laws, curfews, and more indirectly through urban regeneration and gentrification processes. These factors effectively create more concrete and stark spaces that often stall or reverse publicisation processes. Unsympathetic media, including social media, has also frequently been used to discredit the use of public space for protest. Extinction Rebellion protests for instance have been accused of blocking motorised traffic, and consequently have provoked forceful responses. Politicians also enact measures to restore 'order' and 'quality of life' in public space on behalf of a public that they claim is intimidated by political gatherings (Iveson, 2007).

As we saw in several chapters in Part II, different groups including young people and homeless people also face various *corrective* spaces with wider publics, mediated through the 'eyes on the street' from other members of the public in everyday encounters. These can also be hardwired into the design of the built environment, thus inadvertently causing barriers to public space for disabled people. Through processes of marginalisation and the enforcing of narrow social norms, these groups have frequently been confined to the margins of public life and kept in subordinate positions. Studies on publicity thus focus on the question of how negated access to public space and the public sphere has undermined both the principle and the practice of citizenship.

The tendency to privatisation and securitisation spurred by (incomplete) neoliberal capitalism has also chipped away at the integrity of public places that most scholars had come to associate with an active public sphere. Harvey (2023) argues that city authorities willingly encourage private sector investors to invest in public regeneration projects, in an effort to soak up their surplus capital. Consequently, those taking part in protest movements frequently confront encounters from urban authorities (both state and corporate) whose modes of policing public space mobilise new technologies and increasingly militarised forms of surveillance and control (Graham, 2010). With face-recognition cameras on lamp poles, private security officers, militarised vehicles in urban areas and the increase in privately owned public spaces (POPS) (Garrett, 2015), these throw into sharp relief questions about the nature of public space, how it is to be used and under what conditions.

These efforts to undermine the 'public-ness' of public space have led Iveson (2007) and others to interrogate taken-for-granted relationships between public space and public address. Moreover, urban scholars have been asked to ponder the veritable 'end' of public space as we know it (Mitchell, 1995). The above account of the perceived decline in the vitality, freedom and openness of public space offers a particularly bleak assessment. However, these accounts are often based on a nostalgic perception of a loss of a golden era for times and places, like the Agora, that were deemed more welcoming of political struggles. According to Iveson (2007: 6),

> When concerns about exclusion are articulated through narratives of loss, they imply that public spaces used to be more inclusionary – more 'public' – before their contemporary degradation.

In reality, public space has always been contested, as noted in Chapter 13. The publicness that we are supposed to have lost is in fact a 'phantom', based on a nostalgia for times and places that were purportedly more welcoming of political activism. In fact, public spaces have never been open to all. As we saw earlier in this book, historically, the encounters people have faced have been long standing and persistent (e.g. The Ugly Laws in Chapter 9), thwarting people's efforts to take their place as part of the public, to do their 'work of citizenship'.

While the earlier three-part framing of public protest by Iveson (i.e. as a context for action (public space), a kind of action (public address) and a collective actor (a/the public)), offers a helpful initial framing of the various components involved in public protest, it does not account for how inherently unstable 'public space' can be. Geographers have thus sought to move beyond a topographical understanding which solely thinks of public sphere as bounded in public urban spaces (i.e. in parks and streets). Rather, public space can be *any* space in which a certain kind of action – public address – takes place at a given time. For instance, one can undertake forms of public address in a private space – e.g. participating on social media from a private house. Don Mitchell (2023) argues that:

> what makes a space public is not its preordained 'publicness'. Rather, it is when, to fulfill a pressing need, some group or another takes space and through its actions makes it public. (35)

Mitchell's provocation offers a novel way to consider how public space is created through engagement rather than as something predetermined. At the point of taking space, Mitchell argues, we can see 'the beginning of public space' unfold.

Thus, on the other side of the controlling encounters, forms of mass protest are evidently growing to historically unprecedented levels in frequency, scope, and size. The number of protests around the world has more than tripled in less than 15 years up to 2021 (Younge, 2019; Ortiz et al., 2022), with increases across all regions of the world, with the largest concentration of activity in the Middle East and North Africa, and the fastest rate of growth in sub-Saharan Africa. As well as the increased frequency and scope, we are also seeing some of the largest protest movements ever recorded — including the farmers' protests that began in 2020 in India, the 2019 protests against President Jair Bolsonaro in Brazil and the Black Lives Matter protests between 2013–2014. We ask in our Engagement Questions below, what is driving this?

Box 14.2 Engagement Questions

What do you think are the key demands of protesters in our contemporary world?

Beyond the key demands, what enabling factors are relevant in fuelling mass protest movements?

The major demands driving this growing trend around the world include calls for democracy, increased access to jobs, public services, social protection, civil rights, global justice and against austerity and corruption (Ortiz et al., 2022). Other factors that have fuelled these protests have been slowing global economic growth and the worsening effects of climate change. The use of technology by protesters, and the foreign meddling in internal politics via disinformation and other tactics, has led to quicker development of protests. Of course, there are limits to protests. Although still early days, there is a sense that the second Trump administration (2025–2029) might lead to fewer protests than the first (2017–2021), and more space for legal activism, and political action rather than the mass protest model.

When this rise is positioned within the context of the increased securitisation of public space, we can see how arguments over the perceived 'end' and 'beginning' of public space is always in a dialectical relationship. Iveson (2007) refers to this alternative understanding of public space (as *any* space that is made public) as a procedural approach. We build on this understanding by arguing that public space is relational, that it is embodied, practiced and felt by people interacting through it – and must continue to do so for its publicness to remain. Therefore, when considering public space as an elusive political ideal, it is only realised by enduring struggles at various encounters for power, distribution of resources, procedural justice and recognition.

While we recognise that public space is not pre-ordained, that is not to say that all space holds the same symbolic value. It is important not to assume that any space can become equally and unproblematically as public as any other simply by being used for public address. Spaces develop symbolism and cultural codes over time and become emblematic of various struggles. When we look at some of the defining social movements in the US from the 1960s to the present day for example (many of which are noted in earlier chapters) – including the women's movement, free speech movement, civil rights movement, Anti-Vietnam war protests and more recent Palestine protests – one site comes to prominence: Berkeley Campus, University of California.

As we have learnt, the significance of this staging ground cannot be explained by it being open and inclusive to protest. In fact, the anti-Vietnam war protests in Berkeley campus were accused by faculty of using the university as a 'staging ground' for subversive forays into the larger community. The otherwise liberal philosopher and sociologist Lewis Feuer in particular, was deeply outraged by the Free Speech Movement and its effects on campus. He blamed the faculty for refusing to 'properly' limit the rights of the students. Also, in some protests, there was a teargassing of students on Berkeley Campus. So, as Mitchell points out, 'the American past was anything but inclusive – and public space was always a source of conflict' (1995: 121). But, as we have also seen, just how and where people are to meet, under what conditions they are to do so, and what they are able to discuss are all themselves points of struggle.

So, what enabling factors may be at play? Principally, new mobile media and computing technologies have enabled large and diverse groups of people to communicate with each other and to plan social activities and political uprisings (Vigneswaran et al., 2017). For example, during Occupy Wall Street in 2011, there was 200 unique Facebook and Twitter accounts, 10,000 videos on YouTube, and 10,000–15,000 posts on Twitter/X every hour (Preston, 2011). Indeed, in some cases, social media activism has taken place entirely online. However, geographers have argued that 'real world' activism remains vital for political struggles particularly when the motivation of this resistance concerns the *recognition* of political groups and to reverse the invisibility that some groups feel. Here, Gill Scott Heron's famous song title, *The Revolution will not be Televised* serves as a useful critique of armchair activists. The specific experience of disabled people provides a useful lesson here for two reasons. First, their core motive for taking space was to challenge the various relationalities that had excluded them from (topographical) public space, including institutionalisation, the design apartheid of the built environment and the attitudinal barriers they faced. They therefore sought to *take* space in a very literal sense. Second, disabled people have challenged the ostensible 'heroic', romanticised and ableist accounts of protest, by showing how vulnerability can also be used as a tool *in* direct action rather than an opposing feature to be excluded from it. We therefore feel it justified to look at this specific case separately as a case-study for exploring the key conceptual engagement above.

14.4 Case Study: Resistance and the Emergence of the Disabled People's Movement

When looking at disability activism, there are three routes to reform which we distil. The first was the initial emergence of a disability rights movement. Echoing the point earlier about how specific spaces develop symbolism and cultural codes over time and become emblematic of various struggles, the disability movement was no different. In Berkeley, the first *Centre for Independence Living* opened in 1972 after a group of disabled students at the University of California came together to lobby the university for support services for independent living on the campus. The disabled students had previously been confined to the Cowell Hospital which was situated on the campus. Drawing upon the experiences of other rights groups that had coalesced at Berkeley (as covered in 14.3), the disabled students sought to challenge their marginalisation on campus. They advised the university upon what support they would need to live independently in the community i.e. help getting up, dressed, preparing meals and many other activities of daily living.

Recognising this same injustice, concerning the barriers to accessing the community, individuals with all types of disabilities throughout the US banded together to form cross-disability advocacy groups. These groups staged protests and held rallies arguing for increased accessibility and fair treatment. This movement later became known as the Independent Living Movement. The movement was based upon the notion that disabled people should have the right to live independently within the community rather than in institutions. This example nicely exemplifies the well-known slogan of the disability movement, 'nothing about us, without us', which stipulates that no policy should be decided by any representative without the full and direct participation of members of the group affected by that policy. Disabled people are therefore best placed to advise upon any decisions made about them, their lives and their needs.

There have been many successful lobbying and campaign groups as part of the disability movement. Disabled people also used more militant tactics as part of the disability rights movement. One notable example in America are ADAPT. They were established in 1982 in Colorado and are known for being part of the militant wing of the disability rights movement due to their history of engagement in direct action in order to bring attention to the lack of civil rights held by disabled people. The group initially focused on getting wheelchair accessible lifts on buses but later expanded to consider a wider set of issues affecting disabled people across America, with the aim of ensuring that disabled people have the civil and human rights to live in freedom. Today, ADAPT is active across America with local groups scattered across the country, focusing on various issues including freeing people from institutions, fighting for access, and against discrimination. From the US, the disability movement spread globally.

The second route to reform was how the disability rights movement co-created the social model as a tactic to challenge how disability was being understood in reductionist ways – that assumes that disabled people's autonomy is limited solely due to their impairment. As we saw in Chapter 9, the Union of Physically Impaired Against Segregation (UPIAS) in the UK rejected the tragic or medical approaches to disability, which placed the problem as being with the individual, in their bodies and in their impairments. It directed attention towards the social and structural barriers that 'disable' people with impairments, rendering them 'disabled'. In other words, it was society, built and run by non-disabled people, which was preventing disabled people from actively participating in a fuller working and social life. Although this was not direct action, it provided an ideology and rallying call for the movement.

The third route to reform relates to the embodied practices of disability activists. Historically, founding members of the Disabled People's Movement and pioneers of the social model of disability (UPIAS, 1974) were reluctant to acknowledge bodily vulnerability and constraints, in fear that emphasising impairment could weaken the collective strength of the movement. Further, by turning attention back to the body, there were fears this could diminish efforts to reduce disabling barriers and risk re-introducing what Oliver (1990: 78) called the 'personal tragedy theory' of disability. Recently, Angharad Butler-Rees and Andy (Power) (2025) explore how disability activists are now advocating for the impaired and disabled body to be at the front and centre of disability protest, as it is about highlighting the body as the site which has been impacted by forms of care, treatment and exclusion. By coming together these individual bodies use their vulnerability strategically in a bid to form a larger collective body of resistance and advocate for their inclusion and right to exist alongside others. The paper seeks to challenge the implicitly ableist and masculinist accounts of protest persisting in some geographic scholarship, and questions the assumption that direct action is the opposite to vulnerability. Rather, they show how exhibiting vulnerability can be a crucial factor in demonstrating injustice (e.g. of welfare cuts and stigmatisation), along with garnering greater media attention and visibility.

14.5 Moving Beyond Parochial Accounts of Public Space and Protest

As we saw at the outset, contemporary understandings of the idea of public space in the Global North have largely been based on inherently singular accounts of ancient classical cities, particularly the Agora in Greece. Furthermore, many nations in the Global North have either directly (through colonisation) or indirectly exported the models of European public space to many other parts of the world. Viewed from the Global South or the Global East, however, provides further critique on this much-debated idea of public space (Staeheli and Mitchell, 2007), highlighting its rather parochial meaning. As we saw above, the largest concentration of mass protests have been seen in the Middle East and North Africa, and the fastest rate of growth in sub-Saharan Africa. In this section, we consider two countries, which have had distinct historical, cultural and political contexts concerning public space and protest – South Africa and China.

South Africa serves as an example of a country that has been powerfully inscribed by its postcolonial past, and the development of the Apartheid system (1948–1994) remains a crystal clear example of social and spatial injustice and exclusion. So much so that the word 'apartheid' is now applied to many current examples of socially unjust spatial separation and social marginalisation. But even before Apartheid, many of South Africa's contemporary public spaces were designed and maintained for colonial power and control. Within its legislative capital, Cape Town, the legacies of 18th century slavery and 20th century Apartheid are still evident. Apartheid was the official policy that governed the spaces of encounter between South Africa's White minority and non-White majority for much of the latter half of the 20th century, sanctioning racial segregation, and political and economic discrimination against non-Whites. Myriam Houssay-Holzschuch and Emma Thébault (2017) explore how these processes shaped a very different model of urban design than imagined within the lofty diversity ideals of public space romantics. The White regime attempted to exclude the Black majority from the *political* public space by denying it the most fundamental political and citizen rights. It also sought to deny access to *juridical* public space, where it was owned by a 'White-only' state and an African presence in the city streets was limited by law. Last, *social* public spaces were dismantled by Apartheid legislation. Although the legislation that formed the foundation of Apartheid was repealed by the early 1990s, the social, economic and *geographic* repercussions of the discriminatory policy have persisted into the 21st century. This troubled history has also manifested in criminal violence, which has been met by equally problematic responses of neglect and securitisation from a highly sensitive White population, bent on generating processes of commodification, touristification and privatisation of public spaces.

Situated within this context, South African protest movements have favoured specific spatialities, often linked to the persisting coloniality of their cities' spatial layout. As explored by Houssay-Holzschuch and Thébault (2017) for Cape Town, these have included road blockades in the form of 'service delivery' protests. Global movements like Occupy have also adopted a particularly localised flavour. Located at the heart of Cape Town's Southern Suburbs, Rondebosch Common is one of the city's more popular urban green spaces for recreation among wealthy White Capetonians. It had historically been used by colonial Dutch camps and later by British colonial troops during the two World Wars. The Common is therefore deeply imbued with a sense of control and privilege. This was where the Occupy Movement in Cape Town sought to *take* space with specific demands that related to the above context. Specifically, Houssay-Holzschuch and Thébault's case study reveals how the chosen location first did not conform to global Occupy practices of occupying municipal, central areas of large cities like New York and London. Second, it had symbolic relevance locally and the demands were specific. The global Occupy Movement's financial focus was explicitly rejected. Activists instead were calling for affordable housing in the city. Their demands to redistribute public land for the development of affordable housing for the poor directly challenged the existing, inherited and racially-skewed property regime. The point of the gathering was not to resolve ethnic or class differences, but rather to play them out, to make them visible using the theatrical quality of public space.

This fuelled a fear among nearby residents that their White areas would be invaded by poor Black people – hence, the earlier point about the specificity of the site as relevant. Unsurprisingly, the municipality declared the event illegal. The authorities and Apartheid-era armoured vehicles stopped different marches, which were planned to converge on the Common, while Casspirs (four-wheel drive vehicles used for transporting troops) met the remaining protesters on the site. The local environmentalists also expressed fears that the Occupiers would damage the *fynbos* – the local flora – but conveniently overlooked the impact of the Casspirs' heavy tires.

Although it was a short-lived event concerning a few hundred people, its local enactment of public space is illustrative. A second illustrative example is the People's Republic of China. Here, protest is firmly governed through a tightly-controlled state regime, yet it also offers a peculiarly local blend of political engagement. The Chinese Constitution, which has been amended over time, contains several statements that appear to protect citizens' right to protest. Wright (2018) outlines how the Chinese Constitution claims to protect civil liberties, echoing similar language to liberal democracies, for example, Article 35 states that 'citizens of the People's Republic of China enjoy freedom of speech, of the press, of assembly, of association, or procession and of demonstration'. Article 41 states that China's citizens have 'the right to criticise and make suggestions to any state organ or functionary'. However, the same legislation includes other statements that restrict that same right. Article 1 of China's constitution outlines that 'the state maintains public order and suppresses treasonable and other counter-revolutionary activities'. So, while the right may exist in principle, this only extends to activities that are not perceived by ruling authorities to 'sabotage' China's 'socialist system' or to 'infringe on the 'interest of the state'. Otherwise, this could be deemed 'treasonous' to disturb 'public order'. This obscurity of state laws and the deployment of punitive consequences for its citizens can stymie protest, but there is a perception that people never could engage in public protests in China that are directed at those in power.

The Tiananmen Uprising of 1989 was the largest and most dramatic of these incidents, but it marked neither the beginning nor the end of widespread unrest in the reform period. In the first decade of reform (1978 to 1988, marked by a period of economic restructuring and opening up to the world), China experienced a steady stream of collective protests, culminating in the massive demonstrations in Tiananmen Square (and many other Chinese cities) in the spring of 1989. Despite the brutal suppression of the Tiananmen Uprising, the frequency of protests has escalated in the years since. Public disturbances in China increased tenfold during the period from 1993 to 2005, from 8,700 to 87,000. Most observers believe that the actual figures are considerably higher than these official statistics – which the Chinese government ceased making public after 2005. These often take the form of manifestoes, petitions and slogans put forth by protesters in both city and countryside, initially targeted at local officials. If a local resolution is not forthcoming, protesters often take their petitions to higher levels – including all the way to Beijing – involving vocal demonstrations surrounding government offices, marching through city streets, staging sit-ins in public places and blocking traffic on busy highways and railways.

The Chinese specialist Elizabeth Perry (2009) indicates a 'rising rights consciousness' and a newfound appreciation of citizenship in these protests, leading to a gradual permeation of ideas of being a *citizen* (versus *comrade*) among the population. However, Perry warns us against drawing parallels to Western political concepts of 'rights consciousness' and 'citizenship' – unless very carefully distinguished from their European and American counterparts. Once again, we are reminded that the archetypal 'mirage' of the Agora as a model for political action does not serve us well, as it creates an unwarranted expectation of a fundamental transformation in Chinese state-society relations. When placed in historical perspective, the evidence that Perry presents from an examination of the past three decades of popular protest does not offer grounds for a prognosis that Chinese state power is eroding under the pressures of a newly awakened and assertive society animated by alien conceptions of universal human rights. Participants of these visible and vocal (and sometimes violent) protests usually go to great lengths to demonstrate their loyalty to central policies and leaders. This offers a glimpse into the lifeworld of activists, who often must tread carefully in, which we return to at the end of the chapter. Nonetheless, these protests have wrung some significant concessions from the state; for example, the abolition of an agricultural tax in 2006, a property rights law the following year, and the move to privatise collective land ownership.

14.6 Going Too Far? The Case of Social Unrest

The earlier sections considered the geographies of protest (and occupation) in the Global North, South and East. Moving beyond these two forms of activism, this section examines the phenomenon of social unrest, otherwise known as civil disobedience or popular uprisings. Social unrest is a more antagonistic, unpredictable and confused form of activism, involving both overt resistance but also looting, rioting and vandalism. Social unrest is not something that occurs as an entirely separate form of activism beyond protest and occupation. Often, it occurs at the fringes, by a small(er) minority to disrupt and some cases discredit these other forms of activism.

Social unrest is an especially intense mode of activism and resistance, creating a very hard line between order and disorder, the state and the people, the status quo and change. Unrest can be defined as 'involv[ing] at least one group publicly, and with little or no attempt at concealment, illegally assaulting at least one other group or illegally attacking or invading property' – and that this must reach a certain threshold of intensity (Halle and Rafter, 2003). There have been many instances of urban unrest just during the writing of this textbook, to say nothing of what had come before, beginning with the George Floyd unrest in late May and early June 2020 across nearly all American cities over 500,000 people, and more recently: Cardiff 2023; Paris 2023; Dublin 2023; Leeds, 2024; Nairobi, 2024; Dhaka, 2024; and multiple British cities 2024 (although the latter decidedly right-wing in nature, and covered in more detail in Chapter 7 on race in the UK and directly below).

As the most prominent manifestation of direct action, the notion of social unrest in cities needs to be unpacked, since it is not always straightforwardly a matter of social justice. Social cohesion seems critical in avoiding urban unrest and rioting; if groups feel unfairly treated by, for example, the police, or if they feel that those at the top are indifferent or exploitative, then there is greater likelihood of trouble. Of course, the more economically unequal a country is, the less socially cohesive it is, since those at the top and bottom are so far apart and rarely see eye-to-eye. A French sociologist named Loïc Wacquant argued in 2008 that the anatomy of urban unrest lies in the combined logics of 'ethnoracial injustice', 'economic deprivation and widening social inequalities'. In other words, feelings of being treated unfairly, usually because of race, ethnicity and class, combined with feelings of despair and hopelessness. People are predisposed to unrest when they have a sense of being treated illegitimately and of the uselessness of making polite complaints or conventional protests.

Margit Mayer's et al., 2016 edited collection *Urban Uprisings* makes a case for taking 'riots' and uprisings seriously as forms of protest. A point of contention they raise is that social movements should not solely be considered as movement organisations, with a leader, a constitution, a campaign or even a banner before they are recognised as such. Discounting uprisings they argue diverts attention from the more rapid and disorganised forms of political unrest and consigns them to the more shadowy realms of social resistance. Given the scale of the urban uprisings across the globe and the rapid nature that they can spread across scale, it is worth acknowledging the difficulty of making a neat, conventional distinction between (disorganised) 'riots' and (organised) movement protest. It is also worth emphasising that in repressive, autocratic regimes, led by dictators who commit systemic human rights violations and war crimes, such as Bashar Al-Assad in Syria, people may not have a choice of resistance but to invoke a popular uprising as in November 2024, for example.

However, social unrest is also the calling card of far-right, often White groups. This can also include older people manipulating younger people, who feel threatened by more progressive political change (e.g. 'war on woke') or some demonised other population (e.g. migrants or refugees or minoritised ethnic or religious groups). As we saw in Chapter 11, fundamental urban issues like a lack of housing or jobs can sometimes fuel these feelings of resentment. However, these acts are frequently inflamed by disinformation on social media (Chapters 7 and 10 on the multiple right-wing, anti-immigrant unrest in British cities in August 2024 and belonging, respectively), where false claims are reposted often millions of times in a very short timeframe. Here, social media algorithms can make these reactions more aggressive. This seeding of disinformation can amplify false narratives about a perceived threat. In the process, this can inflame hatred and caffeinate social unrest through inciting violent outcomes.

Based on our own interpretations of multiple instances of social unrest across the globe since 1992 Los Angeles, but excluding right-wing unrest in the UK in 2024, we deduce at least four consistent ways to think about unrest. The first interpretations revolves around the perpetual animosity between the police and certain male youths, usually triggered by a dramatic event: a beating, a shooting, an arrest. The second interpretation sees social unrest as vandalism and hooliganism, and focuses on the looting and rioting that frequently follow the trigger event, especially in the absence of policing. Of course, deeming all those who participate in social unrest as vandals and hooligans enables the state to wash its hands of any responsibility. The third interpretation evolves around inequality and injustice. Social unrest raises a range of political grievances, with a pervasive sense of injustice at the heart of the matter. For some it could be economic injustice, particularly the face of marginalisation: lack of money, jobs or opportunity. For others it could be more about cultural imperialism and violence: lack of respect from society, the powers-that-be. The fourth and final interpretation follows on from the third. It focuses on social divisions and being on the wrong side of that divide, producing people who feel that they have no stake in society, no solidarity except through grabbing hands.

Filtering social unrest via the social justice strands of distribution, recognition and procedure can yield some interesting results. Social unrest does little to redistribute resources that do not favour those who are already better off at the beginning, except through looting. It does, however, raise issues of unequal distributions across society in a very direct and compressed fashion. In terms of recognition, social unrest brings to the fore variously ignored populations and the injustices they perceive (especially by class and race, and both combined as marginalisation). These effects can be fleeting, however, and do not usually shift which groups and how groups are oppressed – witness the same issues in the 2023 French unrest, and only 18 years after the 2005 unrest. Finally, in terms of procedure, social unrest seems to dispense with any kind of political procedure, instead pushing for immediate action from the streets. In this respect, it is very much the opposite of a patient, quiet activism! Overall, social unrest seems unsuited at promoting social justice on a long timescale – although its immediate shock value could promote long-term change. For instance, in the aftermath of the 1992 unrest in Los Angeles, new spaces of encounter emerged between groups that had rarely interacted previously, such as African-Americans and Korean immigrants, or African-Americans and the police. While a promising first step, these interactions cannot necessarily delete the deep inequalities that pervade urban America.

14.7 Spaces *for* Activists and Hope

The final section of this chapter seeks to move beyond the boundaries between activists and wider publics, media and government, towards exploring the spaces of encounter *for* activists, harkening back to the safe spaces and counter publics from Chapter 13. By this we mean the spaces that people use, occupy and encounter in the cultivation of and engagement with their own activist practices, as well as with the wider networks in which they are embedded.

Christine Milligan et al. (2011) provide a useful inside view into the lifeworld of activists and how various places have been important in cultivating and sustaining their activism. They cite spaces of childhood as being particularly formative, e.g. being born in extreme poverty, that not only influences later career choices but which also instils an early desire to 'make a difference'. Beyond these initial experiences, they reveal the extent to which activists shift between public, private and voluntary sectors over the course of their activist 'careers'. Many start their activism in political spaces that clearly sit outside the system, yet most had over time developed a toolbox of strategies for influencing change that they variously used both within newly emerging spaces of governance and from outside – depending on the issue.

Specifically, the voluntary sector can act as a 'breeding ground' for activists. Here we briefly return to consider the place of foodbanks (which we explore in Chapter 5) as a space in which people can find more emergent forms of subjectivity and relationality that suggest a more hopeful and transformative everyday politics, while not necessarily involving explicit 'resistance' to political systems (Cloke et al, 2016). Thus, foodbanks can be regarded as being ill-serviced by binaries of incorporation into the political mainstream and resistance. More widely, the resilience that voluntary sector spaces can create in turn can serve as a platform for resistance.

Individuals who engage in activism, particularly protest, are often thought to have 'durable dispositions', given the radical practices they undertake. However, geographers have highlighted that those engaging in activism often choose not to self-identify as an activist or prefer to wear this term loosely. Protest can take its toll, given the embodied nature of *taking* space. This is particularly the case for disability activists, as noted earlier, who may have physical experiences of tiredness, pain and discomfort. Protest itself can also make people additionally vulnerable; individuals open themselves up to violence, physical pain, hostility and abuse from police and other members of the public (Butler-Rees and Power, 2025). Self-care for activists is therefore often needed, and thereby spaces of sustenance and sanctuary where activists can encounter others, either at the fringes of protests or counter publics online or in-person, to provide physical and emotional support was essential. These spaces of sanctuary are often found in grassroots organisations which can sustain their ongoing engagement through finding sources of hope through their group solidarity. Let us not forget our first strand of the social, that alongside inequality and social (in)justice, people need social encounters, to build and sustain their engagement.

14.8 Conclusions

Ultimately this chapter serves to empower you as a student with a feeling that hope is possible, through various examples of progressive change that have been borne through various forms of resistance. From the discussion, resistance is 'messy' and unpredictable. The chapter is instructive of how space is operationalised within the practices of resistance. We consider a range of spaces of encounter, some foreclosed and others more fruitful. Many societies remain ripe for protest and social unrest, given stubbornly high levels of inequality, mental health issues, limited opportunities and the rise in the cost of living in the post-pandemic world. Despite greater efforts to control public space and disincentivise protests, more and more groups are *taking* place, to challenge the status quo, and in the process creating more avenues for change. Yet, as we saw, other groups who believe they benefit from a previous or current status quo, can use these same tactics to stifle change. This reveals how messy our world is.

It also underscores the contested nature of space itself – not just physical space, but symbolic and digital arenas where ideas are expressed, identities formed, and power negotiated. In this sense, resistance is not only about confrontation but also about creativity: reclaiming the right to speak, to assemble, and to imagine alternatives. The chapter invites you to reflect on how agency is exercised not just in grand movements, but in small, everyday acts that rework and resist dominant structures. Ultimately, the messiness of resistance is not a weakness, but a reflection of its democratic potential – a reminder that change is not linear or clean, but possible, and often born in the margins of struggle.

Further Reading

Back, L. (2021) Hope's work. *Antipode*, 53, 3–20. https://doi.org/10.1111/anti.12644

Editors of Encyclopaedia Britannica (2024) Protestantism, Britannica. www.britannica.com/topic/Protestantism

Garretson, J.J. (2018) *The Path to Gay Rights: How Activism and Coming out Changed Public Opinion*. New York: New York University Press.

Ortiz, I., Burke, S., Berrada, M. and Saenz Cortés, H. (2022) *World Protests: A Study of Key Protest Issues in the 21st Century*. London: Springer (Palgrave Macmillan).

Wasserstrom, J.N. (2018) *Popular Protest and Political Culture in Modern China.* 2nd edition.

References

Barnett, C. and Low, S. editors. (2004). *Spaces of Democracy: Geographical Perspectives on Citizenship, Participation and Representation*. New York: Sage Publications.

Bey, H. (1991) *The Temporary Autonomous Zone, Ontological Anarchy, Poetic Terrorism*. New York: Autonomedia.

Butler-Rees, A. and Power, A. (2025) Vulnerability in action: The role of the disabled body in shaping geographies of protest. *Social & Cultural Geography*, 1–21.

Cloke, P., May, J. and Williams, A. (2016). The geographies of food banks in the meantime. *Progress in Human Geography*, 41(6), 703–726.

Fox, S. and Bell, A. (2016) Urban geography and protest mobilization in Africa. *Political Geography*, 53, 54–64.

Garrett, B. (2015) The privatisation of cities' public spaces is escalating. It is time to take a stand. *The Guardian* 4 August.

Graham, S. (2010) *Cities Under Siege: The New Military Urbanism*. London: Verso.

Halle, D. and Rafter, K. (2003) Riots in New York and Los Angeles, 1935–2001 In Halle, D. (ed.) *New York and Los Angeles: Politics, Society, and Culture--A Comparative View*. Chicago: University of Chicago Press.

Harvey, D. (2003) The right to the city. *International Journal of Uran and Regional Research*, 27(4), 939–941.

Harvey, D. (2012) *Rebel Cities*. London: Verso Books.

Houssay-Holzschuch, M. and Thébault, E. (2017) Dis-locating public space: Occupy Rondebosch Common, Cape Town. *Environment and Planning A: Economy and Space*, 49(3), 555–571. https://doi.org/10.1177/0308518X15603985

Isin, E. and Neilsen, G. (2008) *Acts of Citizenship*. New York: Zed Books.

Iveson, K. (2007) *Publics and the City*. London: Wiley.

Katz, C. (2004) *Growing up Global: Economic Restructuring and Children's Everyday Lives*. Minneapolis: University of Minnesota Press.

Mayer, M., Thorn, C. and Thorn, H. (2016) algrave Studies in European Political Sociology. *Urban Uprisings: Challenging Neoliberal Urbanism in Europe*. London: Springer.

McAdam, D. and Sewell, W.H. (2001) It's about time: Temporality in the study of social movements and revolutions. In R.R. Aminzade, J.A. Goldstone, D. McAdam, E.J. Perry, W.H. Sewell, S. Tarrow, and C. Tilly (Eds.), *Silence and Voice in the Study of Contentious Politics* (pp. 89–125). Cambridge: Cambridge University Press.

Milligan, C., Kearns, R. and Kyle, R.G. (2011) Unpacking stored and storied knowledge: Elicited biographies of activism in mental health. *Health Place,* 17(1), 7–16.

Mitchell, D. (1995) The End of Public Space? People's Park, Definitions of the Public, and Democracy. *Annals of the Association of American Geographers* 85(1): 108–133.

Mitchell, D. (2003) *The Right to the City: Social Justice and the Fight for Public Space*. London: Routledge.

Mitchell, D. (2013) Labour's geography and geography's labour: California as an (anti-) revolutionary landscape. *Geografiska Annaler, Series B, Human Geography*, 95(2013), 219–233.

Mitchell, D. (2023). Beautiful impossibility: A fifty-year retrospective on *Social Justice and the City* and David Harvey's – and geography's – journey into Marxism. *Scottish Geographical Journal*, 139(3–4), 363–372.

Monbiot, G. (2013) EDF's vengeful £5m No Dash for Gas lawsuit is corporate and PR suicide I EDF Energy I The Guardian 28 February.

Oliver, M. (1990) *The Politics of Disablement: Critical Texts in Social Work and the Welfare State*. Basingstoke: Macmillan.

Ortiz, I., Burke, S., Berrada, M. and Cortes, H. (2022) *World Protests: A Study of Key Protest Issues in the 21st Century*. London: Springer Nature.

Perry, E. (2009) China since Tiananmen: A new rights consciousness? *Journal of Democracy*, 20(3), n.p.

Preston (2011) Wall Street Protest Spurs Online Conversation, *New York Times* 8 October.

Smith, D.M. (1994) *Geography and Social Justice*. Oxford: Blackwell.

Staeheli, L.A. and Mitchell, D. (2007) *The People's Property? Power, Politics and the Public*. New York, NY: Routledge.

Tarrow, S. (1998) *Power in Movement: Social Movements and Contentious Politics*. Cambridge: Cambridge University Press.

Union of the Physically Impaired Against Segregation (1974) *Fundamental Principles of Disability*. London: UPIAS.

Vigneswaran, D., Iveson, K. and Low, S. (2017) Problems, publicity and public space: A resurgent debate. *Environment and Planning A: Economy and Space*, 49(3), 496–502. https://doi.org/10.1177/0308518X17690953

Wright, T. (2018) *Popular Protest in China.* Cambridge UK: Polity Press.

Younge, G. (2019) Streets on fire: How a decade of protest shaped the world. *The Guardian* 23 November.

15 CONCLUSIONS

In this final chapter, we return to our three key themes of the social, inequality and social justice which framed the book, and from there outline a series of future considerations for social geography within our current world (as of 2025). The theme of the social was explored through a relational lens, where our lives are shaped by the spaces of encounter that we occupy, move through, and shape through our embodied and intersectional presence. One takeaway is understanding that the social is framed by at least geographic scales – macro, meso and micro – which are neither neat nor separate categories but porous and simultaneously intertwined in everyday life, sometimes symbiotically linked. This means that wider macro-scale processes, including the geographies of production and welfare, can have meso-scale effects in the way urban issues are understood and dealt with within cities. At the same time, these can have immediate lived and felt realities at the micro-scale for people on the ground, the so-called spaces of encounter. Over the course of the book, we have been attuned to how spaces of encounter are often inexplicably linked, where we must remain conscious of how each side of the perceived line affects the other. Here, relations are manifest, 'made' and come into being through the shifting places that people (may be forced to) use, occupy, abandon and shape, for example via social, affective and material relations at home, in the community and at a cultural/political level. Such emplaced and embodied experiences are encountered through both individual embodied changes/difference and through multi-scalar relations with others. This understanding of the experience of the social allows us to better appreciate how we are socio-economically situated or perhaps marginalised through those relations in ways that shape or constrain experiences of hope.

The crosscutting thread of (multiple) relationalities and encounters have framed this focus on the theme of the social – and operating in a way that extends to inequality and social justice, to which we return below. We underscore how the spaces of encounter that some people experience can be often difficult, contested, confrontational and harmful. Yet for the 1%, privilege is now so entrenched that entitlement is virtually guaranteed both socially (Chapter 1) and spatially (Chapter 11). The rendering of a space of encounter usually requires some implicit controlling of power and resources

DOI: 10.4324/9781003414889-18

between groups. Sometimes this can be overt, in the form of a hate crime or subtle in the form of an uncomfortable fleeting encounter due to inexperience of those perceived as different. An encounter can involve a degree of mistrust or misunderstanding. When spaces of encounter are drawn by the state, they are typically policed by some bureaucratic reasoning, for example, through a rationing of eligibility (a procedural form of (in) justice). Hillary Arksey (2002) for instance explores how social care services can be rationed at three different scales: national government, local authority, and front line practitioner, echoing the three interconnected scales mentioned above.

Examples of rationing and scarcity that Arksey identifies are evident throughout this book. These include rationing by denial where services are denied to specific individuals, or client groups based on characteristics such as age or income, or rationing by selection where individuals thought to be the most deserving or have the most to gain from an intervention (or are least likely to cause problems) are 'selected in' (evident in disability services and homeless shelters). Rationing can also be via more subtle mechanisms, including deterrence that was raised in Chapter 4 where access to support is made difficult via street-level bureaucracy. Services are unhelpful, branch offices are closed, telephone messages are not passed on or answered, and information is not freely available. This negative version of street-level bureaucracy can be extended by dilution, where demand is managed by reducing the quantity and quality of services provided: no one is excluded, but instead everyone receives less. We also saw examples of rationing by delay (where waiting lists are in operation for health and social care services for people with mental ill-health) or simply through rationing by financial charges for accessing services. These provide a detailed insight into how encounters that we saw through the book are managed and controlled in a procedural sense.

An encounter can also appear as a material imprint of societal attitudes, for example as physical barriers to disabled people within the built environment. How these spaces unfold can thus hinge on different levels of power, resources, historical or contemporary norms associated with difference, as well as contrasting motivations behind such encounters. Sometimes each side has views that are at odds with the other. For some, relationships are obligatory – for instance, between the welfare state and the person who depends on its assistance. But sometimes relationships are not so clearcut, when both sides are either jointly debilitated from wider contexts such as care staff and disabled people (Macpherson et al., 2023; see also Chinn et al., 2024) or where people share common experiences and goals (for example, volunteers, youth workers, social care staff). As we saw in Chapters 3, 4 and 5, the outer reaches of the state become blurred, including notions of the shadow state, and incomplete neoliberalism.

Yet, these spaces of encounter can also be convivial, mutualistic and empowering where people share common ties, solidarities, motivations, characteristics and/or are merely welcoming of differences. These spaces of encounter can engender a sense of hope where groups build solidaristic ties and find allies. As we saw in Chapter 13 and Chapter 14, change is always possible, even under the most unpromising conditions. Wicked systems usually implode at some point, given enough time and patience – think of the fall of Apartheid in 1994. Bad times never last forever, as much as good times never do either. This open approach to change is absolutely essential in our post-pandemic world, where the stakes have only become higher around identity, life chances and shaping the future social and physical environment.

Our second key theme of social geography was about inequality. Throughout the book, we have considered inequalities in income, wealth, housing, racial differences, gender differences, age differences and so forth. Inequalities can also manifest as spatial expressions of difference, such as mortality rates, life chances and life expectancy and educational attainment. All of this further relates to issues of who gets what, and where. Relationalities were unearthed around who decides on the future, which necessarily involves the social identifier of age, via the concept of generations. In most democratic nations, it is well recognised that older people are more likely to vote. Moreover, some evidence (Geys et al., 2022) shows that shifting towards the left is more likely among the young (under 40 years) whereas shifting towards the right occurs at an older age (over 55 years). This can be exploited by right-leaning governments, who try to ensure that their policies favour the older demographic, such as prioritising pensions or landlord rights, or implementing anti-migrant laws and policies, or even turning back the clock on the rights of women and LGBTQ+ people. In effect, many of the rightward shifts in governments in North America and Europe are the product of stubborn inequality, especially between well-off parts and neglected parts of nations. Interestingly, right-leaning populist governments rarely seek to minimise these regional inequalities, and rarely deploy the welfare state to do so. Rather, they blame immigrants for all that ails the country but propose remarkably little in the way of long-term investment into people and places that most need it.

Our third framing of social geography was around the key theme of social justice. This abiding theme crosscuts all the chapters, showcasing the three strands (distribution, recognition, procedure) in various ways. For some chapters, distribution was key – via the welfare state, and for the increasingly precious resource of housing. For other chapters, the strands of recognition and procedure took precedence, especially as they related to social justice campaigns based on key social identifiers such as race, gender, age, disability and class, as well as their various intersections. Issues of recognition and procedure were especially important in terms of resisting (or living with) various oppressions (Young, 1990), from the economic (marginalization, exploitation, and powerlessness) to beyond (cultural imperialism, violence). As we underlined in Chapter 1, social justice is always the following: contested; spatial; practical and utopian; linked to social injustice; and especially, though not exclusively, urban. Fraser's (1998) earlier provocation, introduced in Chapter 2 that the politics of recognition, in the form of seeking a more favoured political identity, should not be the endgame for political activism. (Re)distribution should hold equal weight as it ultimately affects fairer access to (adequately) paid work, property ownership, education, health and social care and leisure time, but also more starkly in reducing exposure to environmental toxicity (Fraser, 1998: 166). Fraser's critique is a useful reminder that while reducing stigma is fundamentally important, it may not automatically create tangible outcomes for people beyond social recognition. Let us also not forget that recognition and distribution are both mutually dependent on more just procedures – our third strand of social justice – in the form of the state infrastructure and state services which underpin and enable more equitable distribution.

Taken together, we appreciate how the prospect of achieving a more socially just world is an ever-evolving and momentous task. To sustain this effort, we move beyond describing the different forms of social justice to consider how we frame this journey, particularly in terms of *hope*. Drawing on our insights from Part III of the book, we seek to move towards a more hopeful framing to end our book.

Towards a More Hopeful Framing?

We understand hope as an important guiding undercurrent for change. It can be understood as a broader ethic, a sense of imagining and/or helping create better futures. Hope has a prior quality; an immediacy that envelops people's journeys through life and adversity. It can manifest itself in other affective qualities such as care and love. Hope is generated and felt relationally through various encounters, some of which are experienced affectively. It can emerge in new possibility and potential, and in the affective feeling of bodies acting, and moving forwards (Andrews, 2017).

Moments of hope can materialise on both sides of (and be shared across) spaces of encounter. Such moments may be felt by people who are disempowered and feel at the margins. It may also emerge as a feeling among community leaders, support providers, volunteers and carers – as both a motivation to sustain their support and in the belief that they can make a difference. Hope may emerge even when people may feel, in the words of Andrews (2017: 793):

> acutely aware of the often unsustainable and seemingly insurmountable challenges they face in supporting vulnerable people with increasingly limited resources, yet consistently practice 'hope' in their persistence and innovation.

Hope is threaded through both chapters in Part III. Yet, as we saw across the various chapters in Part II, hope must always be tempered, as people are often aware that progressive change or full recovery may not be feasible. Nonetheless, when people merely persist or seek to negotiate life challenges, their positions can bolster and sustain ways of adapting to adversity. In a co-authored paper (Power et al., 2019), Andy and colleagues proposed the concept of *hopeful adaptation* to understand these myriad, often non-linear ways that people negotiate their routines and strategies to live with, reverse or counter adversity, and to (re)discover a sense of well-being – or health-in-adversity – in their everyday lives. In using the term adapting, they do not advocate more of the damaging neoliberal agenda that has placed responsibility to absorb structural shocks with the individual. Rather, we highlight how these negotiations can be profoundly transformative – at individual, community or even larger scales.

We also feel more broadly that any textbook with the (social) geographies of social justice needs to offer a sense of hope to you, the reader. Hope is often hidden, in the sense that it is not always discussed directly in geographic scholarship; hence it is not easily located via traditional access points. Yet, as Andrews (2017) argues, scholars hope for the betterment of their studies and for the good health and empowerment of the individuals they encounter and communities they interact with through their research. Research can also be an important instrument of hope in that our work might address inequalities and indignities that are lived by particular groups in particular places or contribute to interventions that drive change. Nonetheless, Andrews reminds us that there is a converse side. Sometimes we as researchers offer more than we can deliver and the notion of a promise not fulfilled can also be seen as a dashed hope or a means of sustaining a 'cruel optimism' (Berlant, 2011).

As an emerging scholar yourself, or someone keen to work towards social justice in a different field of occupation (or on an everyday basis), it is important to consider your own position and where you stand with hope. While hope arises in all aspects and areas of life, it is important to think critically about our hopes and where they may lead us. Sometimes, other motivations may propel action or sustain living, for example, desperation, urgency – even pessimism and 'weariness' (Wilkinson and Ortega-Alcázar, 2019) with the status quo, as explored further below. Sometimes people merely seek to escape adversity, through activities that may create an alternative reality (e.g. cosplay) or may be considered health-compromising. As Andy argues with colleagues in Power et al. (2019), these offer an important reminder of the tensions within people's coping strategies that may support an immediate sense of well-being (or at least release or escapism) while compromising one's physical health in the longer-term.

Despite these different orientations, what they have in common is that they can sustain a particular way of life, and because life always possesses the possibility *for* something else (i.e. for a different future), and the possibility *of* that something else (i.e. what that future will affectively feel like if it arrives) – they can serve towards an 'immanent utopianism' (Anderson, 2006). And so, strongly allied to the sense of hope is the creation of utopias – future worlds that are, well, better than the one we have now. As signposted in Chapter 13, this utopian thinking is based in the idea that in order to improve the world, one must first think up a perfect world and proceed backwards from there. This is what is also called the normative – what ought to be in an ideal world, rather than 'what is' already (and which is always subject to change). Over history, utopian worlds have been articulated through a variety of positive and negative ways – from Communism to Fascism, cults, 'back to the land' movements to most organised religions. The more progressive utopian schemes always share the same striving for a better distribution of resources, stronger recognition of misunderstood and ignored social groupings, and fairer procedures to ensure better distribution and recognition.

Cities have frequently been the sites of such utopian thinking, acting as spaces of tolerance and diversity, what Iris Marion Young (1990: 212) called 'unassimilated otherness' and an 'ideal of city life as a vision of social relations affirming group difference'. For LGBTQ+ communities, feminists, racial minorities and so forth, cities can provide alternative means ways of being without having to bow to conformity. Urban spaces offer relative anonymity, heterogeneity, openness and change, in which otherness can become unfixed from any totalising sense of community or self-identity, can help base utopian thinking. To give a quick example: for many African-Americans in the American South, New Orleans has long represented a place of freedom from the narrow-mindedness of rural life and its deep reminders of slavery, oppression and exploitation. To go to New Orleans, or eventually other cities such as Atlanta, Chicago, Los Angeles and New York, was not just to economically better oneself, but to also enjoy certain freedoms of expression and association. As we shall see, this thinking and striving can be quiet (Chapter 13) or loud (Chapter 14), but always messy and entangled.

As a reader, you may be young, hopeful, and looking toward a prosperous future. Yet, we also hope that you maintain an interest in social justice issues that may (or may not) directly affect you. Change is possible but it can take a long time, even a lifetime or over entire generations. Let us remind you of some social justice gains that have been made in many parts of the world – the rights of young people, women, LGBTQ+ groups, disabled people, Black, Asian and minority ethnic people. Overall, this arc of change does not neatly map onto a linear line of progress and is frequently stymied, disrupted or overturned, and even reversed. Injustices are also commonly repeated. Yet, injustice is rarely absolute. Martin Luther King, Jr. argued that 'the arc of the moral universe is long, but it bends toward justice'. While we can take some comfort or hope in this sentiment (as former President Obama did during his US presidency when he frequently quoted this), this does not mean that we should disengage from political activism, but it does caution against believing that history is naturally or inevitably moving toward a perfect progressive or liberal future. The various insights from Part II of the book show how messy our world is when compared to our initial premises regarding social justice and ideas of fairness. The key framing in Chapter 1 remains useful but this must be accompanied with the admission of the complexity of encounters and issues of injustice.

Equally, people's varied responses to injustice from our previous chapters shows how activism can appear in the most unlikely places. The hard work in securing justice is not easy nor is it pointless. Achieving social justice requires patience and resilience. Indeed, Power et al. (2019) discuss how hope can persevere and co-exist alongside adversity. Their concept of *hopeful adaptation* briefly introduced at the start of Part III underscores the idea that hope can coalesce with feelings of adversity. It emphasises how people employ adaptive strategies in often non-linear ways. For instance, individuals can develop and negotiate new routines and strategies to reverse or counter adversity, and (re)discover a sense of well-being – or health-in-adversity – in their everyday lives. Hope can emerge in new possibilities and potential, and in the affective feeling of bodies acting and moving forwards (Andrews, 2017). This recognises people's capacity or ability to act, to create something new and forge alternative life trajectories rather than simply withstanding or tolerating change. People's varied responses to adversity can include coping strategies that allow them to function, confront issues, share in collective storytelling, galvanise resources or support an immediate sense of well-being (or at least release or escapism).

This can work the other way though. Sometimes people's coping mechanisms may compromise their physical health in the longer-term. Examples include smoking, excessive alcohol consumption and obesogenic diets to manage stress. Protest itself can also be physically or psychologically damaging, such as being exposed to tear gas and police violence. Trauma can stay a long time in a person, and can be passed on from one generation to the next. One commonality in this journeying though is that these ways of coping can be transformative, even in small ways and over varying times. With this focus, we may come to understand how and why situations characterised by adversity become otherwise, how they are reconfigured to generate moments and experiences of hope – again at individual, community or even larger scales.

We strove in this book to offer up a buffet approach to various (and sometimes controversial) social issues and social problems. In so doing, we remind you that the social world is not immaculate and will increasingly not be in this post-pandemic world. We provided conceptual tools to make sense of the social world, but not to explain it 100%. There are no easy answers, and there are no easy answers in this final chapter. We would advocate for you to take a toolbox approach in utilising the concepts and ideas from this book and wider field of geography. This emphasises that you can and should remain selective about what concepts (e.g. exclusion, stigma, vulnerability) are most appropriate to arm yourselves with, in setting guard rails against injustice, in helping you articulate injustices to people from diverse audiences to encourage them to think about an issue in a novel way and how change may be addressed. With this in mind, we leave you with some problem examples for future action, which represent some contemporary challenges that face societies around the world. In each case, we ask that you think about what concepts might help describe the experiences being felt by those affected by the social injustice and what forms of (quiet or direct) activism that might enable the people affected to both *cope with* and *confront* each challenge. You may also consider other examples beyond those in the exercise below.

Box 15.1 Engagement Questions: Scenarios for Future Action

In this exercise, we ask you to contemplate future scenarios that might prove socially unjust. Using the three strands of social justice (distribution, recognition and procedure), we ask you to analyse the potential injustices, but also (and based on the Chapters 13 and 14 in particular) extend your analysis to include potential avenues of obtaining justice. These avenues need to consider (1) the spatial (where, for which places), (2) the temporal (long vs. short-term), (3) strategies (active, passive, etc.) and (4) outcomes (transformation, reworking within current systems, resilience to current systems).

- Housing evictions mount in the private rental market due to a cost-of-living crisis
- Young people are blamed for causing a crisis in mental health services due to rising demand
- Disability benefits are reduced in line with government rhetoric about them being 'work shy'
- Homeless people are removed in a newly gentrified neighbourhood
- Social support is rationed for family carers deemed the most in need
- Unionised care workers refusing to implement the deinstitutionalisation of people with mental health issues
- New tax cuts for high-income earners are implemented
- Inheritance taxes are cut to zero
- Undocumented immigrants are deported
- Voter rights for young people are curtailed
- Legal recognition of trans people is rescinded
- Women's rights are rolled back

Moving beyond the exercise above, this toolbox approach can help move you and your fellow students from being merely passive observers of society to occupying an interested position in political struggles which shape society. This book should thus serve as one step on your journey from passive learner to active campaigner or activist citizen through your engagement and journey through this text. On this journey, it is important though to remain sensitive towards others and their views, especially when we find those people to be different or their views objectionable. This is not moral complicity but a necessary tool for understanding other people's positions (that we may not agree with) and bringing people on board.

References

Anderson, B. (2006) 'Transcending without transcendence': Utopianism and an Ethos of Hope. *Antipode*, 38, 691–710

Andrews, G.J. (2017) Health geographies I: The presence of hope. *Progress in Human Geography*, 42(5), 789–798. doi: 10.1177/0309132517731220 (Original work published 2018).

Arksey, H. (2002) Rationed care: Assessing the support needs of informal carers in English social services authorities. *Journal of Social Policy*, 31(1), 81–101.

Berlant, L. (2011) *Cruel Optimism.* Durham NC: Duke University Press.

Chinn, D. Power, A., Brickley, K. and Ali, S. (Accepted/In press 02 July 2024) What does 'feeling at home' mean for adults with intellectual disabilities living in group homes in England? *Journal of Applied Research on Intellectual Disability (JARID)*, 37(5), e13274. doi: 10.1111/jar.13274.

Fraser, N. (1998) *Social Justice in the Age of Identity Politics: Redistribution, Recognition, Participation.* London: Sage.

Geys, B., Heggedal, T-R. and Sørensen, R.J. (2022) Age and vote choice: Is there a conservative shift among older voters? *Electoral Studies*, 78, 102485.

Macpherson, H., Hall, E., Kaley, A. and Power, A. (2023) Debilitating landscapes of care and support: Envisaging alternative futures. *Social & Cultural Geography*, 24(1), 140–156.

Power, A., Bell, S.L., Kyle, R.G. and Andrews, G.J. (2019) 'Hopeful adaptation' in health geographies: Seeking health and wellbeing in times of adversity. *Social Science & Medicine*, 231, 1–5. doi: 10.1016/j.socscimed.2018.09.021.

Wilkinson E., Ortega-Alcázar I. (2019) The right to be weary? Endurance and exhaustion in austere times. *Transactions of the Institute of British Geographers,* 44, 155–167.

Young, I.M. (1990) *Justice and the Politics of Difference*. Princeton: Princeton University Press.

INDEX

Note: Page locators in **bold** refers to Tables.

For Product Safety Concerns and Information please contact our
EU representative GPSR@taylorandfrancis.com Taylor & Francis
Verlag GmbH, Kaufingerstraße 24, 80331 München, Germany